T0180576

Lecture Notes
in Business Information Processing

502

LNBIP reports state-of-the-art results in areas related to business information systems and industrial application software development – timely, at a high level, and in both printed and electronic form.

The type of material published includes

- Proceedings (published in time for the respective event)
- Postproceedings (consisting of thoroughly revised and/or extended final papers)
- Other edited monographs (such as, for example, project reports or invited volumes)
- Tutorials (coherently integrated collections of lectures given at advanced courses, seminars, schools, etc.)
- Award-winning or exceptional theses

LNBIP is abstracted/indexed in DBLP, EI and Scopus. LNBIP volumes are also submitted for the inclusion in ISI Proceedings.

Maria Papadaki · Marinos Themistocleous ·
Khalid Al Marri · Marwan Al Zarouni
Editors

Information Systems

20th European, Mediterranean,
and Middle Eastern Conference, EMCIS 2023
Dubai, United Arab Emirates, December 11–12, 2023
Proceedings, Part II

 Springer

Editors
Maria Papadaki
British University in Dubai
Dubai, United Arab Emirates

Marinos Themistocleous 🆔
University of Nicosia
Nicosia, Cyprus

Khalid Al Marri
British University in Dubai
Dubai, United Arab Emirates

Marwan Al Zarouni
Dubai Blockchain Center
Dubai, United Arab Emirates

ISSN 1865-1348   ISSN 1865-1356 (electronic)
Lecture Notes in Business Information Processing
ISBN 978-3-031-56480-2   ISBN 978-3-031-56481-9 (eBook)
https://doi.org/10.1007/978-3-031-56481-9

This Springer imprint is published by the registered company Springer Nature Switzerland AG
The registered company address is: Gewerbestrasse 11, 6330 Cham, Switzerland

Paper in this product is recyclable.

Preface

The European, Mediterranean, and Middle Eastern Conference on Information Systems (EMCIS) is an annual research event that explores the field of Information Systems (IS) from practical, theoretical, regional and global perspectives. Renowned for its success in bringing together researchers from across the globe, EMCIS fosters a friendly atmosphere that encourages the exchange of innovative ideas. Recognized as one of the premier conferences in Europe and the Middle Eastern region for IS academics and professionals, EMCIS comprehensively addresses technical, organizational, business, and social issues in the application of information technology. The conference is committed to defining and establishing IS as a discipline of high impact for professionals and practitioners in the field. EMCIS places a strong emphasis on facilitating the identification of innovative research that holds significant relevance to the IS discipline.

EMCIS 2023 marked the celebration of two significant milestones: (a) the conference's 20th edition and (b) the return to an in-person format. Held in December and hosted by the British University in Dubai, this year's conference was particularly special. Out of the submissions received, 43 papers were accepted, representing an acceptance rate of 35%. These papers spanned various tracks, including:

- Artificial Intelligence
- Big Data and Analytics and Machine Learning
- Blockchain Technology and Applications
- Climate Change and Emerging Technologies
- Cloud Computing
- Digital Services and Social Media
- Digital Governance
- Emerging Computing Technologies and Trends for Business Process Management
- Enterprise Systems
- Information Systems Security and Information Privacy Protection
- Healthcare Information Systems
- Managing Information Systems
- Innovative Research Projects
- Metaverse
- Smart Cities

The submitted papers underwent a double-blind review process with a minimum of two reviewers. Additionally, submissions from track chairs were assessed by a member of the EMCIS Executive Committee and a member of the international committee. Finally, contributions from conference chairs underwent review by two senior external reviewers.

EMCIS has once again demonstrated its truly international nature, with authors originating from 22 different countries and participants representing 29 countries. The diversity of contributions is highlighted in the summary below, organized by the country of origin of authors / participants:

- Austria
- Brazil
- Bulgaria
- Cameroon
- Canada
- Croatia
- Cyprus
- Estonia
- Finland
- France
- Germany
- Greece
- The Netherlands
- India
- Italy
- Norway
- Pakistan
- Poland
- Romania
- Russia
- Saudi Arabia
- Slovakia
- South Africa
- Sweden
- Switzerland
- Tunisia
- United Arab Emirates
- UK
- USA

The papers were accepted for their theoretical and practical excellence and promising results. We hope the readers will find them interesting and consider joining us at the next edition of the conference.

December 2023

Maria Papadaki
Marinos Themistocleous
Khalid Al Marri
Marwan Al Zarouni

Organization

Conference Chairs

Khalid Almarri   British University in Dubai, UAE
Maria Papadaki   British University in Dubai, UAE

Conference Executive Committee

Richard Kirkham   University of Manchester, UK
  (Publications Chair)
Marinos Themistocleous   University of Nicosia, Cyprus
  (Program Chair)
Nikolay Mehandjiev   University of Manchester, UK
  (Public Relations Chair)

International Committee

Marwan Al Zarouni	Dubai Blockchain Center, UAE
Charalampos Alexopoulos	University of the Aegean, Greece
Nikolaos Bakas	GRNET, Greece
Paulo Henrique de Souza Bermejo	Universidade de Brasília, Brazil
Lasse Berntzen	University of South-Eastern Norway, Norway
Yannis Charalabidis	University of the Aegean, Greece
Savvas Chatzichristofis	Neapolis University Pafos, Cyprus
Klitos Christodoulou	University of Nicosia, Cyprus
Paulo Rupino Cunha	University of Coimbra, Portugal
Vasiliki Diamantopoulou	University of the Aegean, Greece
Irenee Dondjio	Hague University, The Netherlands
Catarina Ferreira da Silva	University Institute of Lisbon, Portugal
Besart Hajrizi	University of Mitrovica, Kosovo
Elias Iosif	University of Nicosia, Cyprus
Muhammad Kamal	Coventry University, UK
Angeliki Kokkinaki	University of Nicosia, Cyprus
Przemysław Lech	University of Gdańsk, Poland
Euripidis N. Loukis	University of the Aegean, Greece
Nikolay Mehandjiev	University of Manchester, UK

Paulo Melo
Vincenzo Morabito
Andriana Prentza
Maribel Yasmina Santos
Alan Serrano
António Trigo
Horst Treiblmaier
Aggeliki Tsohou
Piotr Soja
Gianluigi Viscusi

University of Coimbra, Portugal
Bocconi University, Italy
University of Piraeus, Greece
University of Minho, Portugal
Brunel University London, UK
Coimbra Business School, Portugal
Modul University Vienna, Austria
Ionian University, Greece
Cracow University of Economics, Poland
Linköping University, Sweden

Contents – Part II

Managing Information Systems

Contents – Part I

Digital Governance

Healthcare Information Systems

Artificial Intelligence

Big Data and Analytics

Data Analytics and Data Science: Unlocking the Open Data Potential of Smart Cities

Larissa Galdino de Magalhães Santos[(✉)] [iD] and Catarina Madaleno

Operating Unit on Policy-Driven Electronic Governance, United Nations University,
4810225 Guimarães, Portugal
larissamagalhaes@unu.edu

Abstract. This article explores the integration of innovative data-driven technologies into digital governance at the local level, with a focus on open data, smart cities, and public sector data analytics and processing. Governments globally strive for digital transformation with emerging technologies, and data play a crucial role in improving service delivery and decision-making processes at the local level. However, there needs to be a proper debate about data analytics and data science in the public sector and the crucial aspects of smart cities' interoperability and open data governance. This research aims to fill this gap, proposing a systematic approach to connect and link these innovations, promoting interoperability and data governance at the local level. Based on a multi-method literature analysis, including Portugal's remarkable digitalization journey, this study sheds light on the importance of comprehensive data analytics in the public sector. The findings indicate that the existing debate on data analytics in the public sector needs more depth and synergy from the point of view of data analytics techniques. By presenting propositions on the challenges for interoperability and open data governance of smart cities, this article provides valuable information for policy-makers, decision-makers, and implementers looking for solutions to governance challenges at the local level.

Keywords: Open Data · Data Analytics · Data Science · Smart Cities · Public Sector

1 Introduction

Governments worldwide recognize the power of emerging technologies to transform the public sector and its capabilities to operate and deliver services. Therefore, governments use data in all aspects of their operation [1] and they also use data analytics to capture the value of data and improve its performance [2].

Public sector data analytics is a digital innovation source for dealing with economic, social, political, and ecological pressures. The continued tension in public organizations generated different innovations, which were gradually incorporated by those under the umbrella of digital government [3]. The pressure for the modernization of public administration was reflected in expanding access to information, data processing, and

M. Papadaki et al. (Eds.): EMCIS 2023, LNBIP 502, pp. 3–15, 2024.
https://doi.org/10.1007/978-3-031-56481-9_1

organization of open government databases. Pressures for more effective public services have driven the use of data collection, extraction, interpretation, and extraction technologies as input to make intelligent decisions and inform policy. The pressure to engage citizens in the execution of government programs and the provision of personalized services has enabled the use of contextualized technologies to engage the local community through smart cities.

Thus, the use and analytics of public sector data has been driven by the need for governments to respond to these pressures effectively and efficiently, mainly through evidence [2]. However, there is evidence that the availability of data - open, closed, or large sets, and the improvement of data processing techniques, interpretation, and visualization of data is facilitating government innovations [1]; there is an inaccuracy in the literature from the point of view of defining the approach of applied techniques. Technological, organizational, and environmental variables have been poorly developed in the literature on data-driven innovation [4].

Studies indicate that the public sector data analytics debate still needs more depth and scope [2]. The literature also discusses data analytics techniques in the public sector [2] and data science applied to the public sector [1] as distinct innovations, while both approaches suggest collecting, processing, analyzing, and presenting data to inform decision-making and government operations. There are few studies on the design of data publications, decision-making processes [5], organizational processes, or the continuous cycle of open data utility [6]. Therefore, the debate on data analytics as part of a smart city open data process or cycle is addressed in isolation.

Is an integrated approach to public sector data analytics and data science in government possible? How do we integrate data analytics or data science in the public sector? To overcome these issues, this article analyzes and informs this gap between data analytics in the public sector and data sciences in government through multi-method. The article also offers final propositions to support the attempt to link government innovations systematically, addressing interoperability and governance for government digital transformation at the local level.

These propositions come from the "INOV.EGOV-Digital Governance Innovation for Inclusive, Resilient and Sustainable Societies" project, which includes the production of an open knowledge repository that brings together diverse documents, categorized and organized to support the community interested in opening quality data and purpose, data analytics for the public sector, and making cities smarter.

This article is organized as follows. After introducing the inaccuracy of the literature on data analytics and data science, the following section discusses the theoretical assumptions in depth. In the theoretical debate session, we explained the debate on open data related to smart cities and data analytics in the public sector. Next, we present the methodological design of this research associated with the production of the INOV.EGOV knowledge repository. In the fourth session, we discuss the results; we offer propositions based on the literature and case studies for the identified challenges. The conclusion discusses the limitations of the study and future research.

2   Theoretical Assumptions

2.1   Open Data Background

Open data is digitized and structured data based on non-proprietary and user-friendly formats. Open data is popularly known as open government data made available to the public and private sectors through government portals or platforms [7].

Open data initiatives were launched to increase transparency, citizen participation, and access to information and create an ambitious, diffuse offer of data-based services [8]. In the current context, the objectives associated with open data, collected and processed by government organizations, are facilitating and encouraging the development of tools, generating of creative applications, making cities smarter in different environments and scales [9], creating new business standards and public-private cooperation [7], innovating services, promoting analytical capabilities for resource optimization, and finally creating value [10].

Open data has become the most significant and decisive element for smart cities to face the urban challenges of sustainable development, which require fast or real-time decisions from different frameworks, from the environment to technology and building infrastructure [10].

Data applicability must be considered when implementing open data policies since governments are the main disseminators of these resources [7]. Furthermore, public sector organizations are working towards many different goals, which may include combining public-private ecosystems of government data platforms and external technology providers [11], open data services industry [7], capturing data value to improve government performance [2].

A list of open data commitments must be undertaken to harness the potential of government open data in community engagement and smart city development: data accessibility, open licensing, bulk downloading, API endpoints, and data visualization tools [12]. In any case, for governments to carry out the profound digital transformation, data must be of high quality, easy to access and share [11, 13], complete, accompanied by records of metadata, published in standardized schemas and semantics [14].

Faced with the global trend of data-driven decision-making, the variety and volume of data produced by the city are essential for gaining insights and knowledge [13] as an open data ecosystem, followed by analytical processing, which generates valuable information, provides added value, optimizes resources and promises benefits for sustainable urban development [7, 10, 12].

Open data is digitized and structured data based on non-proprietary and user-friendly formats. Open data is popularly known as open government data made available to the public and private sectors through government portals or platforms [7].

Open data is an urban innovation because it can be used freely and for all kinds of ideas [13]. Companies also combine government and other open data to develop innovative data-driven applications that facilitate access to public services or provide public services [4]. Many organizations, including public organizations, focus on data processing and analysis, particularly open data [15], as they can create sophisticated modeling capabilities [12]. Open data is expected to bring advanced forms of political decision-making through data [6].

While recent attention has been paid to how to incorporate artificial intelligence based techniques and technologies into data analytics, there are misalignments between open data and smart city initiatives that directly impact the implicit and embedded information in the data, potentially useful and valuable that scientific, mathematical or statistical methods of analytics can discover.

2.2 Smart Cities Background

A smart city is recognized for using technological solutions to improve the management and efficiency of the urban environment [16]. There are several definitions and models to conceptualize smart cities. The most commonly used definition is related to the smart, sustainable city [16], which uses information and communication technologies to boost the quality of life, the efficiency of urban services, and competitiveness following the needs of current and future generations. This definition is related to the potential of technologies for collecting, analyzing, and integrating data from different domains and systems of the city to obtain improved and predictive information for decision-making and to develop urban intelligence [13].

Categories, levels, skills, and dimensions to understand smart cities have expanded over the years, usually related to information technology and aspects of the urban context [17]. The literature has also developed approaches on how the architecture of smart cities, that is, the use of technologies and the vast amount of data and information, are associated with the functions of governance, services, resilience, and others, extending the urban transformation to meet the needs of the need of institutions, citizens and companies [13].

Overall, recent literature agrees that data has become a key feature for smart cities due to the development of cities connected by government digital sensor installations, dissemination of mobile technologies, popularization of social media and web applications, acceleration of distributed computing techniques and data storage [18]. Therefore, the smart city is recognized as a plethora of data collected through the urban data ecosystem, with incorporated sensors, machines, and devices, which are the object of analysis, information, and forecasts [13].

Benefits were linked to cities via open data strategies: data transparency that would allow citizens to access more information and encourage decision-making; access would also contribute to encouraging more active and informed participation; data would provide service improvements and efficiency gains; and would enable companies and developers to create products, services, and applications [19].

However, in practice, the objectives of government open data strategies prove to be generic, whose results only sometimes result in the expected benefits [19]. In addition to the limitations of access and reuse of urban data [18], there is also a lack of a consistent, systematic, and structured debate to assess the impacts of open data in smart city contexts [13].

Faced with the growing volume and variety of data produced in the urban ecosystem, crucial for obtaining insights [12], new, more sophisticated, dynamic, and large-scale analytical methods are being developed to understand urban issues [11]. For example, extracting hidden insights and correlations from big data has become a growing trend to provide better services to citizens and support decision-making processes [16].

There are several limitations regarding open data and operations of government organizations under the umbrella of data science, such as data quality, costs of maintaining the infrastructure and analysis tool, ethics regarding data-driven algorithms, lack of skilled workforce to handle the related projects, lack of organizational skills, instability of regulatory and policy frameworks, barriers to use, and lack of methods and techniques to manage the complex projects [5] that need to be explored by the literature.

3   Methodology

This article is associated with producing the open knowledge repository "Open data and data science technologies for digital governance at the local level". The Design Science Research approach, Fig. 1, is used to generate the repository designed with focus on organizing, categorizing, and disseminating public sector data analytics knowledge at the local level. Concerning the following Methodological scheme, we established the preliminary design and process to the repository during the design cycle. After design validation, the requirements were defined to guide the analytical framework at rigor cycle. Moving on to the relevance cycle, the literature review and documental analysis work simultaneously as a research contribution through the repository and this article.

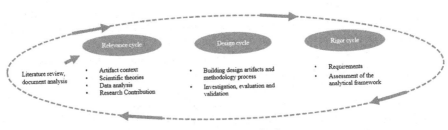

Fig. 1. Methodological scheme

The research design cycle was based on previous literature and good practices to subsidize a descriptive mental map of categories and functionalities related to the back office and front office of the repository. The protocol for collecting, retrieving, structuring, and categorizing documents was defined in the rigor cycle. Data was collected between May and August 2023 in the Scopus database and Portugal's Open Access Scientific Repositories. The search string was based on Boolean operators and the keywords open data, smart cities, data analytics, data science, public sector, and government. Complementary methods were used, such as cross-referencing and snowballing based on citations for academic articles, theses, and dissertations; mapping of local initiatives in Portugal[1]. The material collected generated a matrix (in a spreadsheet) of 273 documents in English or Portuguese.

---

[1] The cases in Portugal were mapped as the platform is a product of the project in partnership with the Regional Coordination and Development Commission for the North of Portugal "INOV. EGOV-Innovation in Digital Governance for Inclusive, Resilient and Sustainable Societies".

The categories were consolidated and associated with Q&A to guide the repository user to complete the relevance cycle. Thus, as the projected solution was being developed, the methodological framework for reviewing and debating with the literature was systematized by categorization, content, and document analysis. Specific taxonomies have been developed to relate open data, smart cities, and data analytics, such as government level, content type, project phases, open data type, data analytics type, smart city type, country, city or region, and associated SDGs.

With the matrix of categorized documents, it was possible to observe which themes could be better explored; for example, about the phases of open data or smart cities projects, fewer documents refer to "explore data analytics," and most of them are related to initiatives at the local level. Initial evidence resulted in research questions for this article; documental and content analysis occurred as the database categorization was reviewed. Therefore, in addition to enabling the organization of content for the repository, the matrix became a compass for future research.

4 Results

The literature on data analytics related to smart cities' open data initiatives is recent and has been increasing since 2016. However, in 2019, there was a significant leap, possibly associated with the context of COVID-19, followed by government digital transformation strategies. Regarding the topics covered, 24 academic articles address open data, smart cities, and data analytics simultaneously, but only four are dedicated exclusively to data analytics. Open data and data analytics are a much more recurrent theme than smart cities and data analytics. Most periodicals discuss government issues relating to operations or general services. Mobility, public transport, geography, and the environment are among the most relevant topics of the debate.

Regarding the potential contribution of the articles, the categorization indicates that most design analyzes improvements in open data, a significant part of the debate on how to make the city smart and explores data analytics. The most discussed topics among open data are data governance, data sharing, and access. There is a diversity of topics concerning data analytics, from edge computing to principal component analysis. About smart cities, smart mobility, smart tourism, smart urban planning, GIS, and territory development are among the most debated topics. Interestingly, there needs to be a more in-depth analysis of open health and education data, possibly due to data governance and personal data protection issues.

4.1 Related and Rising Concepts

In general, the articles and case studies are dedicated to something other than the epistemological discussion of the concept of data analytics. Roughly speaking, both discuss the process of extracting, recovering, analyzing, interpreting and publishing the results [1, 2]. Also, there is no explicit integration in the literature between the definition of public sector data analytics and data science in government.

From a data point of view, the data analytics domain is evolving under various names associated with big data, IoT, data visualization, and data mining, which are driven by the speed, volume, and need to incorporate unstructured data or semi-structured secondaries. Consequently, more sophisticated analytical methods were introduced under the scope of data science [21].

For example, traffic management has become more efficient due to the application of deep learning and machine learning techniques that complement other analytical and statistical techniques [10]. The heterogeneous sources of urban traffic data, generated by mobile devices, inductive loop counters, and integrated AFC systems, provide important complementary insights into traffic dynamics [22] and, therefore, require a multimodal analysis of big data. Furthermore, government-generated and developed data often do not capture characteristics of urban form and functions based on public perception, so urban data includes social networks, voluntary data, and others [18]. Data from traffic cameras can offer enormous value but require the application of data science, statistics, and data engineering techniques that can produce robust results [23]. Thus, it is known as data science in its broad sense, based on modern methods characterized by dealing with structured and unstructured data, i.e., big data [24, 25].

So far, the most integrated data science and data analytics approach is defined as a "view of the conceptualization, thinking, modeling, and processing of smart city data science". Thus, data analytics is the method of analytical reason and logic applied to extract information from data, i.e., to find meaning. Data science is a scientific and analytical study of data that uses different methodologies, such as machine learning. Thus, in data science, interpreting the raw data collected for a given problem domain is data analytics. Creating a model, defining a method, and analyzing the technique depend on the data format, quantity, volume, access, metrics, and the problem to be solved [26].

However, there needs to be more understanding of the entire process or cycle of analytics of these data and the techniques employed. The field of GovTechs and CivicTechs remains unexplored; the private sector and other stakeholders are seen as crucial partners in data aggregation, analysis, and creation of data services [16, 27]. It is also important to emphasize that articles on data analytics discuss numerous challenges of using open data, such as quality, access, updating, standardization, and metadata. While the articles that use the use of data science deepen the technical discussion on statistical, mathematical, or computational methods, they refer to the use of open data as data that can be complemented by unstructured data, such as big data, without connecting with the elaboration of policies, organizational resources, legislation, and public values [1], or risks of stigmatization [2].

The findings indicate that concerns about personal data protection, algorithm ethics, organizational processes, culture, skills, quality, and access to open data remain open. For example, data science applications already use convolutional neural networks to recover missing data or errors from public domain data [28]. Although computational power is relevant, artificial intelligence can benefit from the breadth and depth of open data if there is minimal access and quality. Otherwise, AI systems trained on incomplete data may make biased predictions and be unreliable [29].

4.2 Gaps and Challenges of Smart Cities Open Data Approach to Data Analytics

There is a global and systematic trend towards the production and use of data. Smart cities permeated by devices and digital public services produce increasingly varied, fast, and extensive data. The COVID-19 crisis has highlighted how open data has become a critical factor for decision-making, transforming services, improving citizen satisfaction, and artificial intelligence developments. Data has become an essential infrastructure for public policies [30]. Meanwhile the potential of open data analysis from smart cities depends on data quality, management of the entire process or data cycle analytics, interoperability, and governance data.

Open data has accelerated the debate on the use of public data and the openness of high-value datasets, particularly in Europe, as part of digital strategies and data policy. Data openness relates to data availability without restrictions or obstacles by public authorities, aiming at better and transparent decision-making processes and, mainly, interoperable services. In addition, using open standards avoids blocking proprietary technologies while promoting an ecosystem of digital solutions, enabling data sharing with non-public actors and vice versa.

However, substantial challenges limit the effective implementation and application of open data to support smart cities [12, 13] and, consequently, impact the value of the techniques applied in the analytics of actionable data. Concerning the benefits provided by government open data strategies, there is a mismatch, as the objectives are generic, probably copies of other initiatives, and need more context with the social problems to be solved with data [20].

In general, the data are of low quality [12, 13], usually below the minimum threshold compatible with the creation of value, that is, outdated, incomplete, and fragmented [8]. Lack of infrastructure capacity, security, and privacy are blocking the development of government open data initiatives [14]. In addition, accessibility gaps for any user and the lack of interoperability for any data processing and analysis system to work well affect the credibility and reliability of the data [31]. Finally, critical issues are related to urban open data dispersion, heterogeneity, and provenance [18].

Many governments open data initiatives still need to be fully implemented, as they are prone to releasing low-value data for fear of exposing administrative activities or limiting data provision to the number and type of data.

Proposition 1. Maintaining the quality of open data is a fundamental step for responsible data. However, there needs to be more understanding of the entire process or cycle of analytics of these data. Although governments are attracted by big data and emerging technologies based on data science, the minimum quality to ensure the benefits of open data needs to be achieved.

The conversion of data into social value is not done automatically. It requires a series of actions between actors of different functions to activate, connect, and use them. Slow implementation is due to government administrative and technical capacity gaps [32]. Therefore, data portals have an informal approach to data management, making publication insufficient through the recourse to disclose certain data sets and subject to errors, as they are not machine-readable [8]. Therefore, more design principles [5] regarding the data life cycle and organizational processes must be considered. First, a data life cycle must include acquisition, access, analytics, and application, operationalized through data capture, indexing, storage, sharing, archiving, analysis, manipulation, and support tools [9]. Design principles must be contextualized with the organization and the city; after all, creating a successful smart city strategy depends on existing data and correct benchmarking [17].

Proposition 2. Analytics of public sector data wanting organizational vision contextualized in opening, maintaining, analyzing, sharing, using, and acting capable of guiding stakeholders through a smart city strategy.

The use of urban data is complex and always involves interactions across disciplines. There are challenges in integrating data from heterogeneous sources and interoperating diverse sources with stakeholders. Particularly in cities, data is continually changing in terms of attributes, spatial and temporal information [28], and real-time data such as "smarter" government open data [27].

Therefore, extraction, modeling, interpretation, and presentation of data become crucial for smart cities [1, 26], in addition to other initiatives that have become relevant, such as data virtualization in smart cities [16]; missing data recovery methods based on machine learning approaches [28]; urban data mining algorithms to accurately classify land transport modes [33]; 3D virtual models to represent cities and urban areas for storage and exchange of geometric data between buildings [34]; predictive, descriptive and prescriptive analysis of traffic data as multimodal analysis of big data [22].

Added to this scenario is the growing development of innovative industries and markets based on the Internet of Things, cyber-physical systems, artificial intelligence, big data, cloud computing, and establishing connections with city components. All these processes produce data that is entirely, partially, or not processed by Fourth Industrial Revolution technologies [12] through smart city data science [26] or government data science [1]. "Openness" has become synonymous with intelligence aligned with the recent smart government movement supported by GovTechs and CivicTechs [27]. Thus, a prosperous smart city must establish the objectives and conditions at the strategic, operational, and tactical levels [24].

Proposition 3. The open data analytics of the smart city is a decision-making facilitator, which must be associated with the strategic level for the standard and long-term goals of the cities; at the tactical level on the implementation of decisions, workflow, financial resources; and the operational level of implementation.

Interoperability depends on adopting and maintaining common principles and standards, capable of facilitating business processes based on data exchange and information and communication technology systems [9, 16]. Furthermore, Organizations must be empowered and involved in sharing information and knowledge. Interoperability [35] also focuses on the local context facilitating the adoption and compatibility of developed and implemented technologies between city sectors and services. As the digitiza tion of local public administration expands, cities hold or advance in the ability to use technologies and data to make the local context favorable to sustainable development, such as a smart city.

Proposition 4 Interoperability management is fundamental to obtaining the value of the data because then, the standardized data can be processed appropriately and encapsulated so that the analytical tools - techniques, and technologies can extract insights.

In addition to interoperability, open data depends on governance; data can exist but with low quality or inappropriate format. Therefore, governance is vital to define the policy and infrastructure for classifying public sector data and data available in the public domain. Governance must enable internal approval and ensure qualified data is. available or shared [8]. Local data governance offices ensure risk mitigation regarding data privacy and government strategies.

A reflection related to data governance in cities is that cities are spaces where most data businesses can be created and developed; in addition, political and legal initiatives can contribute to the development of capacities at the citizen and government levels. The lack of resources and skills is often mentioned at the government level as a barrier to the openness and use of data [16]. Although this is a purely organizational attribution (Proposition 1), governance based on an emerging data, services, and innovation eco-system can facilitate detecting, capturing, and transforming dynamic capabilities in the public sector. Therefore, a governance structure is a knowledge and alliance structure. The data governance framework should support citizens' data literacy and stimulate debate at the community level [8] on public sector data analytics, data protection and privacy, and information security issues. The knowledge and insights extracted from datasets are critical components for decision-making processes, solution development, and technological advances to address complex challenges such as climate change, social issues, urban mobility, energy efficiency, and health.

Proposition 5. Open data governance can support the activation of data analytics projects, including sophisticated data science techniques, in a qualified, safe, and auditable way.

5 Conclusion, Limitations, and Future Research

Data openness is a systematic and global trend accompanied by the voluminous data production at the city level. Emerging technologies and public emergencies such as COVID-19 have shown that the public sector must move beyond the use of data for public operations and services. In addition, cities must serve the purpose of fostering the innovation ecosystem.

Reviewing the literature and building the open knowledge repository, the current investigation offers a framework for the public sector's open data analytics debate and data science techniques. The results contribute to the literature by clarifying the evolution of data analytics techniques parallel to the growing volume and diversity of data.

The study contributes to the field of knowledge from the propositions related to the challenges of open data analytics of smart cities. First, the quality of open data, standardization, must be the primary condition guaranteed to reach the proper data analytics process, and consequently the responsible decision-making. Second, we make explicit the importance of establishing an organizational process that links the flow of data between smart city devices and sensors and public sector databases. Organizational processes must rely on data governance to acquire transformation skills and capabilities and rely on interoperability requirements for sharing and reusing data. Next, the data governance structure ensures citizen scrutiny by promoting transparency as a rule, establishing ethical legal and administrative foundations, and fostering an innovation eco- system. Finally, in practice, by making this separation, we endorse the creation of connected digital environments in favor of vital interoperability for digital government. Consequently, policymakers, decision-makers, executors, and researchers find potential in exploring the value of data associated with processing, analysis, and intelligent solutions at the local level. Therefore, this scheme can ensure effective communication between data, devices, and networks at the local level and data analytics through mathematical, statistical, and computational methods.

This article prioritized the analysis of these dissertations and use cases related to Portugal. Although this clipping has not limited the articles collected in the Scopus database, therefore, the analysis extends to general literature on public sector data analytics.

Acknowledgments. This study was funded by project "INOV.EGOV-Digital Governance Innovation for Inclusive, Resilient and Sustainable Societies NORTE-01–0145-FEDER-000087", supported by Norte Portugal Regional Operational Program (NORTE 2020), under the PORTUGAL 2020 Partnership Agreement, through the European Regional Development Fund (EFDR).

References

1. Matheus, R., Janssen, M., Maheshwari, D.: Data science empowering the public: Data- driven dashboards for transparent and accountable decision-making in smart cities, Government Information Quarterly, Volume **37**, Issue 3, (2020), 101284, ISSN 0740–624X, https://doi.org/10.1016/j.giq.2018.01.006
2. van Veenstra, A.F., Grommé, F., Djafari, S.: The use of public sector data analytics in the Netherlands, Transforming Government: People, Process and Policy, Vol. **15** No. 4, (2021), pp. 396–419. https://doi.org/10.1108/TG-09-2019-0095
3. Janowski. T.: Digital government evolution: From transformation to contextualization. Government Information Quarterly, Volume **32**, Issue 3, (2015), Pages 221–236, https://doi.org/10.1016/j.giq.2015.07.001
4. Hossain, M., Quaddus, M., Hossain, M., Gopakumar, G.: Data-driven innovation development: an empirical analysis of the antecedents using PLS-SEM and fsQCA. Ann. Oper. Res. (2022). https://doi.org/10.1007/s10479-022-04873-3
5. Matheus, R., Janssen, M., Janowski, T.: Design principles for creating digital transparency in government, Government Information Quarterly, Volume **38**, Issue 1, (2021), 101550, ISSN 0740–624X, https://doi.org/10.1016/j.giq.2020.101550
6. Park, S., Gil-Garcia, J. R.: Open data innovation: Visualizations and process redesign as a way to bridge the transparency-accountability gap, Government Information Quarterly, Volume **39**, Issue 1, (2022), 101456, ISSN 0740–624X, https://doi.org/10.1016/j.giq.2020.101456
7. Chang, Y.-T., Chen, M.-K., Kung, Y.-C.: Evaluating a Business Ecosystem of Open Data Services Using the Fuzzy DEMATEL-AHP Approach. Sustainability **14**(13), 7610 (2020). https://doi.org/10.3390/su14137610
8. Concilio, G., Molinari, F.: The Unexploitable Smartness of Open Data. Sustainability **13**(15), 8239 (2021). https://doi.org/10.3390/su13158239
9. Gessa, A., Sancha, P.: Environmental Open Data in Urban Platforms: An Approach to the Big Data Life Cycle. J. Urban Technol. **27**(1), 27–45 (2020). https://doi.org/10.1080/10630732.2019.1656934
10. Peyman, M., Copado, P.J., Tordecilla, R.D., Martins, L.D., Xhafa, F., Juan, A.A.: Edge Computing and IoT Analytics for Agile Optimization in Intelligent Transportation Systems. Energies **14**(19), 6309 (2021). https://doi.org/10.3390/en14196309

11. Conde, J., Munoz-Arcentales, A., Choque, J., Huecas, G., Alonso, Á.: Overcoming the Barriers of Using Linked Open Data in Smart City Applications, in Computer, vol. **55**, no. 12, pp. 109–118, Dec. (2022), https://doi.org/10.1109/MC.2022.3206144

12. Lnenicka, M., Nikiforova, A., Luterek, M., Azeroual, O., Ukpabi, D., Valtenbergs, V., Machova, R.: Transparency of open data ecosystems in smart cities: Definition and assessment of the maturity of transparency in 22 smart cities, Sustainable Cities and Society, Volume**82**, (2022), 103906, ISSN2210–6707, https://doi.org/10.1016/j.scs.2022.103906

13. Neves, F. T., de Castro Neto, M., Aparicio, M.: The impacts of open data initiatives on smart cities: A framework for evaluation and monitoring. Cities, **106**, 1–15. (2020) [102860]. https://doi.org/10.1016/j.cities.2020.102860

14. Roa, H. N., Loza-Aguirre, E. Flores, P.: A Survey on the Problems Affecting the Development of Open Government Data Initiatives, 2019 Sixth International Conference on eDemocracy & eGovernment (ICEDEG), Quito, Ecuador, (2019), pp. 157–163, https://doi.org/10.1109/ICEDEG.2019.8734452

15. Escobar, P., Candela, G., Trujillo, J., Marco-Such, M., Peral, J.: Adding value to Linked Open Data using a multidimensional model approach based on the RDF Data Cube vocabulary, Computer Standards & Interfaces, Volume **68**, (2020), 103378, ISSN 0920-5489, https://doi.org/10.1016/j.csi.2019.103378.

16. Huang, Y., Peng, H., Wen, L., Xing, T.: Using digital technologies to plan and manage the pipelines network in city. IET Smart Cities (2023). https://doi.org/10.1049/smc2.12054,5,2,(95-110)

17. Correia, D., Marques, J.L., Teixeira, L.: The State-of-the-Art of Smart Cities in the European Union. Smart Cities **5**, 1776–1810 (2022). https://doi.org/10.3390/smartcities5040089

18. Zhou, Y., Long, Y.: SinoGrids: a practice for open urban data in China, Cartography and GeographicInformationScience,43:5,379–392,(2016). https://doi.org/10.1080/15230406.2015.1129914

19. Pagaime, R. G. T.: Data quality management in an open data context: Lisbon case study. Master's Dissertation in Information Management, specialization in Knowledge Management and Business Intelligence. NOVA Information Management School (NIMS). (2019). http://hdl.handle.net/10362/64939

20. Zuiderwijk, A., Shinde, R., Janssen, M.: Investigating the attainment of open government data objectives: is there a mismatch between objectives and results? International Review of AdministrativeSciences,85(4),645–672.(2019). https://doi.org/10.1177/0020852317739115 Author, F.: Article title. Journal **2**(5), 99–110 (2016)

21. Gudivada, V., N.: Data Analytics: Fundamentals. Charter 2. In: Data Analytics for Intelligent Transportation Systems. Editor(s): Mashrur Chowdhury, Amy Apon, Kakan Dey, Else vier, (2017), Pages 31–67. https://doi.org/10.1016/B978-0-12-809715-1.00002-X

22. Lemonde, C., Arsenio, E., Henriques, R.: Integrative analysis of multimodal traffic data: addressing open challenges using big data analytics in the city of Lisbon. Eur. Transp. Res. Rev. **13**, 64 (2021). https://doi.org/10.1186/s12544-021-00520-3

23. Chen, L., et al.: Estimating Vehicle and Pedestrian Activity from Town and City Traffic Cameras. Sensors **21**(13), 4564 (2021). https://doi.org/10.3390/s21134564

24. Shahat Osman, A.M., Elragal, A.: Smart Cities and Big Data Analytics: A Data-Driven Decision-Making Use Case. Smart Cities **4**, 286–313 (2021). https://doi.org/10.3390/smartcities4010018

25. Watson, R.B., Ryan, P.J.: Big Data Analytics in Australian Local Government. Smart Cities **3**, 657–675 (2020). https://doi.org/10.3390/smartcities3030034

26. Sarker, Iqbal H. Smart City Data Science: Towards data-driven smart cities with open research issues, Internet of Things, Volume 19, (2022), 100528, ISSN 2542-6605, https://doi.org/10.1016/j.iot.2022.100528.

27. Nikiforova A. Smarter Open Government Data for Society 5.0: Are Your Open Data Smart Enough? Sensors. (2021); **21**(15):5204. https://doi.org/10.3390/s21155204

28. Ho, D., Lee, Y., Nagireddy, S., Thota, C., Never, B., Wang, Y.: OpenComm: Open community platform for data integration and privacy preserving for 311 calls. Sustainable Cities and Society. **83**. (2022). 103858. https://doi.org/10.1016/j.scs.2022.103858

29. European Commission. Open data and AI: A symbiotic relationship for progress. Unleashing the potential of this powerful duo. In: data.europa.eu - The official portal for European data. (2023). Available at: https://data.europa.eu/en/publications/datastories/open-data-and-ai-symbiotic-relationship-pro-gress#:~:text=Open%20data%20and%20AI%20have,return ing%20accu-rate%20and%20useful%20predictions Accessed 20 Aug 2023

30. Santiso, C. Moving toward startup States? Govtech startups and the transformation of public action. Article n°7. IGDPE Editions publications. (2020). Available at: https://www.econo-mie.gouv.fr/igpde-editions-publications/thearticle_n7 Accessed 20 Aug 2023

31. Fernández-Ardèvol, M., Rosales, A. Quality Assessment and Biases in Reused Data. American Behavioral Scientist, 0(0). (2022). https://doi.org/10.1177/00027642221144855

32. Fusi, F., Zhang, F., Liang, J.: Unveiling environmental justice through open government data: Work in progress for most US states. Public Administration. (2022). https://doi.org/10.1111/padm.12847

33. Balbin, Paul Patrick F. et al.: Predictive analytics on open big data for supporting smart transportation services, Procedia Computer Science, Volume **176**, (2020), Pages 3009–3018, ISSN 1877–0509, https://doi.org/10.1016/j.procs.2020.09.202

34. Malhotra, A., Raming, S., Frisch, J., Van Treeck, C.: Open-Source Tool for Transforming CityGML Levels of Detail. Energies **14**(24), 8250 (2021). https://doi.org/10.3390/en1424 8250

35. European Commission, Directorate-General for Informatics, New European interoperability framework: promoting seamless services and data flows for European public administrations, Publications Office, (2017), https://data.europa.eu/doi/https://doi.org/10.2799/78681

Data Mesh Adoption: A Multi-case and Multi-method Readiness Approach

Isabel Ramos[1]([✉]) [iD], Maribel Yasmina Santos[1] [iD], Divya Joshi[2] [iD], and Sheetal Pratik[3] [iD]

[1] ALGORITMI Research Centre, University of Minho, Guimarães, Portugal
{iramos,maribel}@dsi.uminho.pt
[2] Thoughtworks, Chicago, USA
divya.joshi@thoughtworks.com
[3] Adidas, Haryana, India
sheetal.pratik@adidas.com

Abstract. Data Warehousing systems have been used to support Business Intelligence applications by ingesting operational data and providing analytical data. As data volume, variety, and velocity increased in Big Data contexts, this data architecture needed to be modernised, and Big Data Warehouses emerged as scalable, high-performance, and highly flexible processing systems capable of handling ever-increasing volumes of data. These monolithic techniques, however, create major challenges to data engineering teams in terms of design, development, management, and evolution. Data Mesh emerged as a novel and disruptive concept aimed at data-driven businesses. The research detailed in this paper seeks to characterise Data Mesh readiness by examining the elements that influence the adoption choice using the technology-organization- environment (TOE) paradigm. A survey and a set of interviews were used in a multi-case and multi-method approach. Researchers and data triangulation were implemented to ensure rigour and arrive at a comprehensive understanding of Data Mesh adoption. The obtained results demonstrate the successful adoption of Data Mesh once its benefits are well understood, with increased teams' creativity, data accuracy, data security, data governance and interoperability.

Keywords: Data Architectures · Data Mesh · Technology Adoption

1 Introduction

With increased data volume, variety and velocity there was a demand to modernize data architectures to address issues associated with traditional mechanisms of data warehousing, data lake, data governance, data quality, and other data management scenarios which were primarily centrally managed and were posing challenges to scalability and speed [11]. Data Mesh is a disruptive and innovative concept aiming at data-oriented organizations [7], requiring data governance and standardization for data interoperability among the Mesh nodes. In a Data Mesh implementation [1], the core principle of 'Data as a Product' uses the DAT- SIS (Discoverable, Addressable, Trustworthy, Self-describing, Interoperable, and Secure) principles to enhance data quality and efficiency.

© The Author(s), under exclusive license to Springer Nature Switzerland AG 2024
M. Papadaki et al. (Eds.): EMCIS 2023, LNBIP 502, pp. 16–29, 2024.
https://doi.org/10.1007/978-3-031-56481-9_2

The main goal of this research is the characterization of the factors influencing the adoption of the Data Mesh (architecture, infrastructure, and principles). Information was gathered using a multi-case and multi-method approach in which a survey and interviews were used to inquire how business and technology decision-makers in organizations perceive the implementation process and its impact on the organization. The research question *"What are the factors affecting the decision of Data Mesh adoption?"* guides this study, trying to provide insights to different dimensions, such as: a) understanding what is prompting organizations, both business and tech side, to consider Data Mesh; b) what organizations want to achieve out of it or what specific problems they want to solve; c) what main obstacles do tech and business stakeholders see in the path of such adoption; d) what tech challenges they foresee for the implementation; and, e) what organizational challenges - people and processes - need to be addressed.

The questionnaire was developed based on the Technology-Organization- Environment (TOE) framework [16], which guides the elicitation and analysis of the factors affecting IT adoption by organizations. Since the TOE framework is a higher-level conceptual framework, we draw on relevant studies to identify predictors of attention to be used within the TOE framework. These predictors were used to define the questions of the questionnaire. After a first characterization of the Data Mesh adoption, the survey was complemented with six interviews undertaken in two organizations, targeting three types of users: business users, tech decision-makers, and tech implementors. Previously predictors of Data Mesh adoption were expanded during the interviews to include new adoption factors and impacts mentioned by interviewees. The state of Data Mesh readiness was analyzed considering the three dimensions of the framework and looking for insights about the benefits, expectations and threats associated with the adoption process.

This paper is organized as follows. Section 2 summarises the related work. Section 3 describes the adopted methodological approach. Section 4 presents the survey results. Section 5 details the analysis of the interviews. Section 6 concludes with some remarks and guidelines for future work.

2   Related Work

In a Data Mesh, data are organized into domains and different data teams manage those domains. For this paradigm shift, organizations need to change the way teams are organized and decentralized in domains [7]. In 2019, Zhamak Dehghani proposed the concept and the core principles [6]. The first one addresses the domains and the decentralized nature of the concept, *domain-oriented de- centralized data ownership and architecture*, changing the way data are foreseen in organizations, *data as a product*. The third concept demands a *self-serve data infrastructure* to support data teams in making available data products. To manage all the available resources, the *federated computational governance* layer avoids chaos in the Mesh. As data governance concerns are of utmost relevance in a Data Mesh, the work of [9] addresses data cataloguing, data quality, and data ownership to provide transparency, foster trust, and manage access and control of the data.

From the seminal core concepts and the DATSIS principles, [10] proposed a domain model and a conceptual architecture towards the design and development of this decentralized data architecture. The domain model includes components for the Mesh operation (such as cataloguing and communication), and the domains, data products, data product owners, and data product teams available from the Mesh nodes. The conceptual architecture supports the domain model with four key components, *Security Mechanism*, *Nodes and Catalog*, *Self-serve Data Platform*, and *Infrastructure*. In [1], a technological architecture for the implementation of a Data Mesh is proposed and evaluated with a proof-of-concept, making available a wide range of technologies suitable to implement the components of the conceptual architecture [10]. The work of [4] also proposes a Data Mesh architecture with a domain layer, a governance layer, a consumption layer, and an instantiation on the cloud using Amazon Web Services.

Regarding Data Mesh implementations, Zalando's work [14] evolved from a Data Lake to the Data Mesh due to the lack of ownership over the data, the poor quality of the data, and the organizational scalability. The changes were experienced in the evolution towards decentralized data ownership, the prioritization of data domains (over pipelines), data seen as a product and not as by-product, and teams organized by domains [14]. For Netflix, the company experienced problems in data pipelines and unnecessary overload in the maintenance of those pipelines, lower latency, and data quality [5]. With the Data Mesh approach, data pipelines can be developed in an infrastructure that abstracts configuration complexity, for instance. In [15], Saxo Bank implementation with data as a product follows seven main principles, data ownership by domain teams, discoverable high-value data assets, data curation, analysable data in the enterprise data warehouse, reliability in the quality of the data, authoritative data sources, and data as an immutable resource once it is created.

Given the potential impact of a Data Mesh implementation, [2] conducted 15 semi-structured interviews with industry experts from various business areas (Sportswear, E-commerce, Consulting, Information Technologies, Automotive, Software, Food delivery, and Healthcare) and with different roles (Solution Architect, Senior Consultant, Director of Engineering, Chief Data Architect, among others). In this study, the authors highlight that industry experts experienced difficulties with the transition to a federated governance approach, the shift of responsibility for the development, provision, and maintenance of data products, and also with the data products model. In [8], the authors collected, analysed, and summarised 114 industrial gray literature articles. The results provide insights into practitioners' perspectives associated with the four fundamental principles of Data Mesh and present open research issues proposed by the practitioners, Standardizing Data Mesh, Methodologies and Tools for Data Mesh Development and Operation, Data Product Life Cycle, Self-serve Platform Services, Data Mesh Governance, and Organizational Change Management.

In this work, we started by conducting a survey with 21 questions. Only after this first characterization of the adoption of the Data Mesh concept, a set of six interviews was conducted to analyse the Data Mesh. These six interviews considered six individuals, from two different organizations, with three different roles: Business Users, Technical Decision Makers, and Technical Implementers.

3 Methodological Approach

3.1 The TOE Framework

The TOE framework is a comprehensive and adaptable model encompassing the key constructs of technology, organizational factors, and environmental elements that influence the adoption of innovation [3] (Fig. 1). Initially introduced by Tornatzky and Fleischer [16], the TOE framework provides a solid theoretical foundation and proves useful for analysing various factors influencing technology adoption. The framework's value can be further enhanced when the determinants within each context are tailored to the specific technology under examination [13]. The framework's constructs have been measured using Likert scales and extended to incorporate new perspectives and contexts [12].

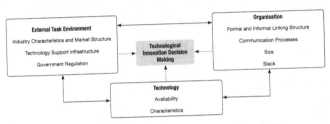

Fig. 1. The TOE Framework. Adapted from [16].

The TOE framework's technology construct evaluates the characteristics of the technology to be adopted. The perception of a new technology's favourable attributes, when compared to the benefits offered by its predecessor, plays a crucial role in driving its successful adoption [16]. This perception of relative advantage reflects how an innovation is perceived as superior to the previous approach [16]. The organizational construct considers the organizational factors that may have contingent effects on the adoption. Perceiving the advantages associated with an innovation provides economic and political legitimacy for its adoption [13, 16]. The environmental construct examines the macro-environmental elements that shape the organization's context. Top management support, previous experience with similar technologies or processes, existing technological capabilities within the organization, and adoption costs are commonly identified as critical factors for successful technology adoption [16]. Moreover, larger organizations tend to have a higher inclination towards adopting innovations due to their greater adaptability and a higher tolerance for increased risk [13]. In the environmental context, to gain a competitive advantage in the face of intense competition and dynamic market changes, organizations recognize the need to strategically align their business practices with IT [13]. Embracing innovation is crucial to mitigate the risk of falling behind competitors and underperforming in the market [16]. These contexts are relevant due to the fact that Data Mesh is a socio-technical construct [7], entailing both technological and social aspects. Its adoption does not depend only on the quality of the technological solution or its alignment with organizational processes; the success of this adoption also depends on the perception that organizational decision-makers have of its business benefits, support from top management, and market pressure.

Our research question is focused on understanding how organizations adopt Data Mesh and on identifying factors that attract or hinder its adoption. In our exploration of the research question, we relied on the TOE framework as a guiding framework. It served as the cornerstone of our thought process throughout the entire research journey. From the development of the initial survey questionnaire to the formulation of the interview guide and the subsequent analysis of the recorded interviews, the TOE framework provided us with a consistent and reliable reference point. Due to its extensive use in research, it was possible to identify the pattern of questions typically used in studies, adapting this pattern to the questions formulated in our study, focusing on a specific technological approach, Data Mesh. Step 1 involved creating a survey based on the TOE framework, which consisted of objective and closed questions (evaluated on a seven-point Likert scale). The survey was distributed to a wide pool of participants, reaching over 2000 individuals, and we received 33 responses (the questions and the results are summarised in Sect. 4). To ensure a comprehensive perspective of Data Mesh adoption, step 2 focused on gathering qualitative data through recorded interviews. This allowed us to delve deeper into the subject and gain valuable insights from firsthand experiences.

3.2 The Research Design: A Multi-case Study

The survey, the quantitative part of our study, allowed us to identify common success factors among companies that have adopted or are in the process of adopting Data Mesh. To gain more insights into these factors, we also conducted a qualitative study with two companies that are already using Data Mesh solutions. This work was conducted by four researchers that established an industry and academic-oriented framework of questions, focused on structuring the tiers of stakeholders to gather diverse perspectives and create a comprehensive picture. Three key roles to interview were identified: a strategic tech decision maker responsible for implementing Data Mesh, a hands-on tech implementer experienced in executing the data mesh framework, and a business user who has experienced or is currently experiencing the benefits of the Data Mesh approach. We conducted a targeted search for organizations that possessed the desired complexities, including a workforce of over 2,000 employees, a sizable tech team (more than 50), and a global presence, and implemented a Data Mesh approach. These criteria were derived from the TOE framework, which emphasizes the propensity of large organizations to adopt innovative solutions. Ultimately, we selected Saxo[1] Bank and Delivery Hero[2] because they aligned with the specified criteria and had successfully implemented Data Mesh practices.

For gathering and analyzing qualitative data on Data Mesh adoption, the following steps were followed:

- Developing the interview guide[3], aligned with the TOE framework. This guide served as the foundation for our interviews. In some cases, we included additional questions to gather specific information, particularly if it was relevant to participants' roles or

---

[1] https://www.home.saxo

[2] https://www.deliveryhero.com

[3] The interview guide is available here.

circumstances. We aimed to ensure comprehensive coverage and capture insights that aligned with the overarching framework.

- Interviewing the participants and presenting the standardized set of questions to all participants. In certain cases, based on the participants' roles, we incorporated follow-up questions to delve deeper into specific areas of interest. To ensure accuracy and maintain transparency, all interviews were recorded with explicit consent from the participants.
- Gathering information from the interviews, where each researcher independently analyzed the transcripts using the predetermined analysis framework. This individual analysis was conducted without consulting one another, ensuring impartiality and minimizing bias. Subsequently, we engaged in collective reviews and discussions, addressing gaps and resolving disagreements. Through multiple rounds of discussion, we reached a consensus and developed a collective viewpoint. This process is called researchers triangulation.
- Analysing the recorded interviews by organizing content into various themes that emerged from our analysis. To aid in visualization, these themes were represented as mind maps. The main categories of our mind maps were drawn from the TOE framework and complemented with new success factors/barriers mentioned during the interviews. The descriptions of each factor or barrier provided by the interviewees were added at the end of the visualization, thereby facilitating the emergence of patterns and key findings. In this way, it was possible to compare the perceptions of the various interviewees, compare perceptions across organizational roles, and compare perceptions between the two companies that participated in our study. Equally important was the ability to compare the information collected from the survey with that obtained from the interviews, connecting them and identifying any potential inconsistencies. This process is called data triangulation.

4   Step 1: Data Mesh Survey

The first step of this study applied a questionnaire with 21 questions grouped into 3 categories, Technological, Organizational, and Environmental, to characterize the state of adoption of Data Mesh. As already mentioned, all questions use a seven-point Likert-scale, ranging from 1 (strongly disagree) to 7 (strongly agree). Figure 2 shows the obtained results with the average value by question.

Analysing Fig. 2, the organizational context highlights as the one without any question with an average value equal to or higher than 5 and, also, as the one with the lower value (3,09) associated with Q14, "The employees of my organization have a high level of Data Mesh-related knowledge". The technological and environmental contexts are the ones in which the participants were able to point out the advantages of adopting a Data Mesh (speed up decisions or provide high value to customers) but also the main difficulties in doing so, such as the complexity to implement or the integration in the daily activity of the organization. From the technological perspective, the major challenges are associated with the information infrastructure (Q9) and with the current practices (Q10). In the environmental context, several answers point to competitive pressure to introduce Data Mesh (Q16), while globally perceiving no connection between Data Mesh adoption and customers loyalty (Q17). Also, it is worth mentioning that there is a strong focus

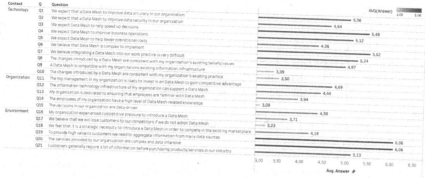

Fig. 2. Overview of the Survey Results by Question.

on data-driven decision-making (Q15 and Q20), with the need to integrate data from several sources (Q19).

5   Step 2: Deep Dive Interviews for Data Mesh Adoption

As mentioned previously in the research design, the multi-case study tries to gather different perspectives and create a comprehensive picture of the adoption of Data Mesh. Six interviews were conducted in two organizations, targeting three different roles (business user, tech decision maker, and tech implementor) to collect the needed information. The several themes that were identified in the analysis were represented in six mind maps. To have a comprehensive overview of all the perspectives and, at the same time, be able to compare them, the information available in the mind maps was integrated into a visual table that helped in the construction of the knowledge that is next shared.

Starting with the technological construct, Fig. 3 highlights the themes that emerged during the analysis of the several dimensions of this construct. As already mentioned, the main dimensions of each construct were drawn from the TOE framework and complemented with new success factors/barriers mentioned during the interviews. The adopted visual representation goes from the dimensions of analysis of each construct to the common identified themes.

In this construct (fig. 3), it is recognized that Data Mesh presents Direct Benefits, such as:

- Improve Data Accuracy by eliminating data redundancies or other data-related problems (Data Quality) and enhance access to trustworthy data (Data Ownership). Both are possible due to the adoption of Data Products that comply with Standardization practices and that are supported by the Central Platform.
- Improve Data Security by adopting Data Governance mechanisms specific to Data Mesh and with the support of the Central Platform and the Standardization practices.
- Improve Operation Efficiency, with a better alignment between business and data, and with data analytics processes that are easier to implement. This is motivated by the availability of decentralized data (Data Products) that makes available curated (Data Curation) and well-documented data (Data Lineage), complying with Standardization

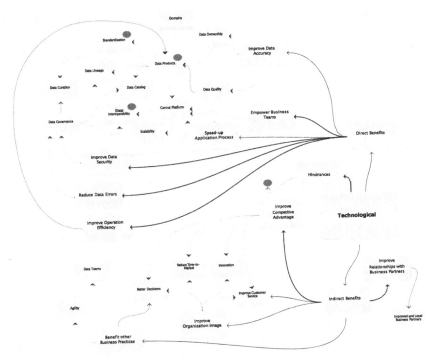

Fig. 3. Technological Construct.

practices and with a scalable and interoperable environment made available by the Central Platform.

- Speed-up Application Process, with collaborative development that is supported by the Central Platform that complies with Data Governance requirements, where Data Products are catalogued and available for different purposes, allowing fast development by the local teams.
- Reduce Data Errors, as data are easy to find due to the Data Governance mechanisms and errors in data are easy to find and report due to Data Products and their ownership.
- Empower Business Teams, as the Central Platform, and all the predefined practices and available resources, made local teams/brands more autonomous in making available and using Data Products.

In the Indirect Benefits, the analysis highlights:

- Improve Organization Image and Improve Competitive Advantage, as the organization is now more responsive, with Reduced Time-to-Market, Better Decisions that seem to be influenced by improved data quality for the clients, and Innovation that has been unlocked.
- Benefit other Business Practices with more Agility for the Data Teams, as these teams now have autonomy and work with data products that have cleaned data, provide an alignment between the business and the data, and are trustable, supporting Better Decisions.

- Improve Customer Services, a construct that is part of the derived theory, but also a relevant concept identified in the Improve Organization Image.
- Improve Relationships with Business Partners having as result Improved and Loyal Business Partners that benefit from the advantages already mentioned previously in the Direct Benefits.

In terms of Hindrances, these are mainly associated with dimensions not yet fully realized, such as Interoperability, Standardization, and Data Products, either by the domains to which they are associated or to the full access to data that may be critical when access to event data is needed. For the sake of clarity, these relations in the concept map are represented by a red circle.

After the systematization of these findings, it is important to compare them with the results reported previously in Sect. 4, related to the survey. Globally, the results seem to be aligned, with improved data accuracy (Q1), improved data security (Q2), speed-up decisions (Q3), and improve business operations (Q4). For Q5 to Q10, costs may be reduced in the future (Q5), but there are Set-up, Running and Training costs (presented next in the organizational construct). The analysis of the collected data expresses the difficulty in adopting and implementing Data Mesh (Q6), due to the resistance to change (Q7), the existing mindset (Q8), the changes in existing infrastructures (Q9), and the existing practices (Q10).

For the organizational construct, Fig. 4 highlights two main dimensions, the Perceived Costs and the Perceived Technical Competence. Starting with the costs, these consider Set-up, Running, and Training costs. For the Set-up costs, the analysis of the interviews allowed us to verify that both organizations high- lighted the:

- Migration Costs, as there is the need to adopt new technologies (available in the Central Platform), define and develop Data Products, implement new Data Pipelines, and consider new Data Governance procedures.
- Mindset Change, which includes looking into data in a different way (now with Data Products), the definition of new roles (e.g. central data team and data governance team), and the need to Nurturing Motivation, as the benefits of adopting and implementing Data Mesh are not immediately seen, requiring negotiation of interests and perspectives, adoption of new working practices, and restructuring some existing working practices.
- Standardization is fundamental to establishing the principles that ground Data Sharing and Data Ownership.

For the Running costs, there must exist continuous efforts in data governance, data cataloguing, data interoperability, data products maintenance, and in ensuring the data teams for these several roles. Besides the roles already mentioned, there is the need to make the central technological infrastructure (and its continuous evolution) widely available, with a central team that also needs to be available to support the several local teams. These local teams usually offer some resistance in this migration, as they were used to having autonomy in the technologies and procedures to follow. For the running costs, there is also the need to consider maintaining some legacy systems that may still be needed. In the Training costs, hands-on learning is required to support the local teams and maintain them motivated (reducing resistance), as these teams are new to adopting new working practices, using new technologies, developing new pipelines with those technologies, etc. Also, training includes a very interesting and important concept that emerged in the interviews, which is associated with Data Literacy. It is meant to disseminate and recall the importance of data and all the dimensions that need to be considered along the several maturity stages of data, such as quality, security, governance, and ownership, among others.

In the Perceived Technical Competence, the constructs are associated with the Performance in Providing IT Support, mainly with the central data teams making available the supporting infrastructure and predefined working procedures; Experience in Supporting Data Mesh, with previous knowledge about data architectures and distributed ownership in helping Data Mesh adoption; Expertise in Supporting Data Mesh, with technology cloning and data engineering as a service as being important factors for supporting the domain teams; and, Data Mesh Champions, with CTO, CIO or similar roles as key for the Data Mesh adoption, as well as existing data governance directors and data engineers.

In the analysis of technical competence (experience and expertise), it is important to look back at the results previously obtained in the survey (Sect. 4). In the organizational questions, the survey highlights the enrollment of the top management, the available technology as being able to support the Data Mesh approach, but also the not-so-high level of Data Mesh-related knowledge. Q14, addressing this question, was the one with the lowest average value of all the. survey questions. In the analysis of the environmental construct, Fig. 5 highlights that the Data Mesh adoption was Recommended by Important Business Partners (as there was a need to support the autonomy of local brands, to have the support of a central governance strategy and better data quality, or the will of business partners to adopt the concept) or resulted from Industry Adoption Approaches to Data Mesh (with no rigid approach or technologies to this adoption), Perceived Government Pressure (with GDPR being a major concern in both organizations), and Competitors using or soon using Data Mesh.

Comparing these findings with the ones obtained in the survey, it was possible to verify the presence of some competitive pressure towards the Data Mesh adoption (Q16–Q17), to realize the strategic need for this adoption (Q18) or the expressive importance of providing high value to customers (Q19). Q20 and Q21 reinforce enhanced competitive advantage.

Besides the analysis of the three contexts, this study addressed the identification of the Perceived Success in the perspective of the six interviewees. The perceived success considered four dimensions, Achievements, Expectations, Threats, and What Went Wrong. Nothing was mentioned for the What Went Wrong dimension, which may be interpreted as a successful adoption of the Data Mesh concept, although there are difficulties in the process as already mentioned in the findings of the three contexts and also highlighted next in the Expectations and Threats. For presenting the results, Fig. 6 adopts a performance indicators-based approach, with the Achievements highlighted in green, the Expectations in yellow reinforcing that more work/engagement is needed in them, and the Threats in red to point out that these are aspects that may need immediate attention to overcome or minimize these difficulties. The first achievement is associated with the adoption of Data Mesh itself, with its acceptance once the benefits are well

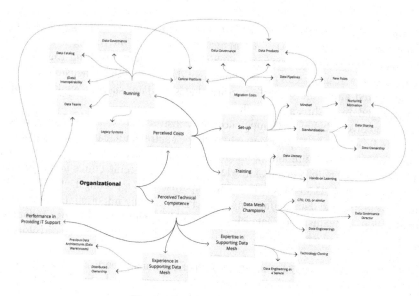

Fig. 4. Organizational Construct.

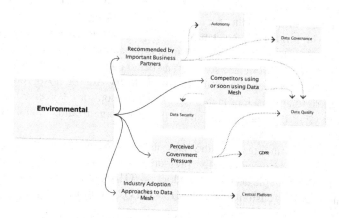

Fig. 5. Environmental Construct.

understood and the local teams are able to see the balance between lower flexibility and increased data quality. Moreover, it increased teams' creativity. Data Governance and (Data) Interoperability are now the result of the successful implementation of the federated model. The Domains and the Central Platform provide the information needed for handling data-related activities, replacing tribal knowledge dependence. The Central Platform also provides scalability. Data Products increased efficiency, with data closer to those that need them.

For the Expectations, Domains still need improvements to ensure collaboration between teams focusing on similar data products. Improvements are also needed in

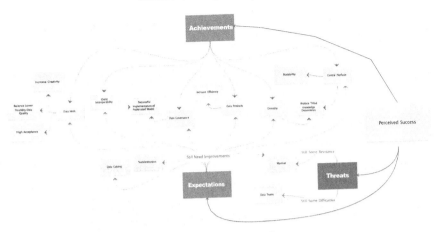

Fig. 6. Perceived Benefits.

Data Products, Standardization, Data Catalog, Data Governance and (Data) Interoperability as, for instance, automated data governance is seen as an important benefit not yet realized. The Central Platform needs a better alignment between business and technology. The Data Mesh needs to increase the maturity of the adoption process enhancing the organization's efficiency and a better perception of the Data Mesh value. Work still needs to be done in the existing Mindset, as a full behavior change was not yet achieved, which is also seen as a Threat, as the interests of all Data Teams were not all addressed yet.

6 Conclusions

This paper characterized Data Mesh readiness with a multi-case and multi- method study grounded by the Technology-Organization-Environment frame- work. This study included a survey and a set of interviews, in which the triangulation of researchers and data were key to rigorously uncover qualitative insights. Through theory construction, it was possible to uncover the successful adoption of Data Mesh and its contributing factors. Table 1 summarizes the main contribution of our study, the systematization of factors that influence the adoption of Data Mesh. As result of Data Mesh adoption, increased team creativity, data accuracy, data security, and data governance and interoperability were verified. As some expectations and threats are still in place, future work includes follow- up surveys and more detailed interviews to monitor the adoption evolution, as there is the need to increase the maturity of the adoption process for enhancing the organization's efficiency and the better perception of the Data Mesh value.

Table 1. Enablers and Barriers of Data Mesh Adoption.

Construct	Enabler	Barrier
Technology	Improved data accuracy	Loss of autonomy
	Improved data security	Resistance to new ways (data and technology)
	Faster decisions	Impact on existing infrastructures
	Improved business operations	Prior users mindset
Organization	Top management championing	Lack of Data Mesh knowledge
	Availability of required technology	Tech users find in difficult to explain the concept to business users
	Availability of domain teams support	Lack of understanding of the concept and implementation methods
	Well-defined data governance strategy	
Environment	Business partners adopting Data Mesh	Lack of a standard adoption approach
	Adoption approaches shared by the industry	
	Motivation to provide high value to customers	

Acknowledgements. This work has been supported by *FCT – Fundação para a Ciência e Tecnologia* within the R&D Units Project Scope: UIDB/00319/2020. We acknowledge the valuable contributions of Delivery Hero, Saxo Bank, Thoughtworks, Brian Leonard, Francisco Sanchez, Kristian Frederiksen, Paul Makkar, Pavel Rabaev, Rasmus Aagaard, and Sean Gustafson.

References

1. Arau´jo Machado, I., Costa, C., Santos, M.Y.: Advancing Data Architectures with Data Mesh Implementations. In: De Weerdt, J., Polyvyanyy, A. (eds.) Intelligent Information Systems, vol. 452, pp. 10–18. Springer International Publishing (2022). https://doi.org/10.1007/978-3-031-07481-3 2
2. Bode, J., Ku¨hl, N., Kreuzberger, D., Hirschl, S., Holtmann, C.: Data Mesh: Motivational Factors, Challenges, and Best Practices (Apr 2023), http://arxiv.org/ abs/2302.01713, arXiv: 2302.01713 [cs]
3. Bryan, J.D., Zuva, T.: A Review on TAM and TOE framework progression and how these models integrate data. Adv. Sci. Technol. Eng. Systems J. 6(3), 137–145 (2021)
4. Butte, V.K., Butte, S.: Enterprise Data Strategy: A Decentralized Data Mesh Approach. In: 2022 International Conference on Data Analytics for Business and Industry (ICDABI). pp. 62–66. IEEE (Oct 2022)
5. Cunningham, J.: Netflix Data Mesh: Composable Data Processing (2020), https://www.youtube.com/watch?v=TO_IiN06jJ4

6. Dehghani, Z.: How to Move Beyond a Monolithic Data Lake to a Distributed Data Mesh (2019). https://martinfowler.com/articles/data-monolith-to-mesh.html
7. Dehghani, Z.: Data Mesh: Delivering Data-Driven Value at Scale. O'Reilly (2022)
8. Goedegebuure, A., Kumara, I., Driessen, S., Di Nucci, D., Monsieur, G., Heuvel, W.j.v.d., Tamburri, D.A.: Data Mesh: a Systematic Gray Literature Review (Apr 2023), http://arxiv.org/abs/2304.01062, arXiv:2304.01062 [cs]
9. Joshi, D., Pratik, S., Rao, M.P.: Data Governance in Data Mesh Infrastructures: The Saxo Bank Case Study. In: 21st International Conference on Electronic Business. pp. 599–604. Nanjing, China (2021)
10. Machado, I.A., Costa, C., Santos, M.Y.: Data-Driven Information Systems: The Data Mesh Paradigm Shift. In: 29th. International Conference of Information Sys- tem Development (ISD'2021). p. 6 (2021)
11. Madera, C., Laurent, A.: The Next Information Architecture Evolution: The Data Lake Wave. In: In 8th. international conference on management of digital ecosystems (MEDES 2016). pp. 174–180. France (2016)
12. Malik, S., Chadhar, M., Vatanasakdakul, S., Chetty, M.: Factors affecting the organizational adoption of Blockchain technology: extending the Technology–Organization–Environment (TOE) framework in the Australian context. Sustainability 13(16) (2021)
13. Oliveira, T., Martins, M.F.: Literature review of information technology adoption models at firm level. Electron. J. Inf. Syst. Eval. 14(1) (2011)
14. Schultze, M., Wider, A.: Data Mesh in Practice: How Europe's Leading Online Platform for Fashion Goes Beyond the Data Lake (2020), https://www.youtube.com/watch?v=eiUhV56uVUc
15. Sheetal, P.: Saxo Bank: Data mesh (2021), https://blog.datahubproject.io/enabling-data-discovery-in-a-data-mesh-the-saxo-journey-451b06969c8f
16. Tornatzky, L.G., Fleischer, M.: The Processes of Technological Innovation. Issues in organization and management series, MA: Lexington Books, Lexington (1990)

Distributed Representational Analysis in Support of Multi-perspective Decision-Making

Olga Menukhin[(⊠)] [iD] and Nikolay Mehandjiev [iD]

Alliance Manchester Business School, The University of Manchester, Booth Street West, Manchester M13 0PB, UK
{olga.menukhin,n.mehandjiev}@manchester.ac.uk

Abstract. This paper describes research that explores principles of designing visual representations of data in sup-port of strategic decisions from the perspective of distributed decision makers. It uses the concept of distributed cognition (DC) to understand requirements of different stakeholders for data and information needed for making decision, and the concept of boundary objects to understand the shared utility of the visual representations in cross-functional decision-making. The study examines a product development setting in which decision-making is based on data resulting from digital modeling. Visualizations are widely used to explore data and make sense. However, research lacks studies connecting data visualization to decision-making. This research aims to propose visualization principles that are based on investigating information and visualization requirements of distributed decision makers. This empirical study uses the human-centered distributed infor-mation design (HCDID) methodology to analyze stake-holder requirements in a case study company and contributes to the research of decision-making with big data and analytics, particularly to corporate multi-perspective decision-making.

Keywords: Data visualization · Multi-perspective requirements · Distributed cognition · Decision-making

1 Introduction

Strategic decision-making research acknowledges that the increasing use of data and business analytics affects decision-making. Visualizations have become a norm for data exploration and understanding [21, 28], however creating actionable insights from data happen within existing organizational structures, decision-making routines [46] and managers' frames of reference [29]. This has prompted visualization researchers to explore human cognition and decision-making [9, 27, 36, 40, 41], but visualization is not explicitly linked to decision-making in the existing studies [11]. The information systems (IS) and strategic decision-making literature contains limited studies on the usage of visualizations in support of decision-making. This research aims to bridge this gap by examining a crucial aspect of decision-making aided by visualizations and focusing on the development of a feasible approach to designing visual representations that can facilitate multiple viewpoints. Multi-perspective decision-making occurs in a shared decision space where cognitive, physical, or digital artifacts are used to provide information that is relevant to stakeholders with varying managerial expertise and accountability. Our research explores a case study of new product development, and by investigating a field problem as an opportunity to create knowledge, we pose the following question:

How can a distributed cognition perspective aid the development of a data visualization approach in support of multi-perspective decision-making?

To answer this question we use the distributed information analysis methodology to investigate information requirements for visualizing multi-dimensional data for research and development (R&D) teams and senior decision makers. In addition to understanding various needs of stakeholders, another objective of this study is to derive a set of principles from their diverse requirements that could be converted into a feasible visualization approach reflecting distributed decision-making.

2 Related Work

Our search for literature on the work related to identifying stakeholder requirements for data visualization design provided a limited number of articles relevant to the topic. When existing literature regarding a new subject is scarce, its examination is usually shorter and researchers are allowed an opportunity to make a contribution to a developing area [46]. We reviewed key academic journals in the field of information systems, information visualization, as well as selected leading journals in management and decision-making, see Table 1. Searchers have been done using the Web of Science database. Our selection criteria did not impose any limitations based on the year of publication. We combined the keywords "decision-making" OR "decision*", "visualization", and "requirement*" to conduct our search. We also run the search in combination with the term "-centered" to identify articles that referenced either user-centered or human-centered approaches. This search produced 19 articles that mentioned the design of visualizations for decision-making based on user requirements. In addition, we explored the proceedings of the most prestigious IS conferences, specifically those found in the AIS eLibrary [1] and Springer EMCIS (EMCIS Homepage), complementing the database search with 25 articles. A number of articles by publication is provided next to the title in Table 1.

van Wijk [48] suggested that the essence of visualization lies in its efficacy and efficiency, as well as in its alignment with the intended context. Being influenced by our belief in the significance of the real context for decision makers, as well as the notion that cognitive processes related to a decision situation may be distributed among a group of stakeholders [16], in our examination of the literature we focused on identifying approaches or methodologies used for understanding the data and information needs in collaborative decision-making. In this research, stakeholders are individuals who are potential or intended consumers of data visualizations meant to facilitate their decisions. After examining the titles, abstracts and keywords of the collected articles, 29 that did not meet these considerations were excluded.

Table 1. Publications titles used in the literature search.

Information Systems (IS)	Information Visualization	Management Decision-Making
Information Systems Research (0)	IEEE Trans.on Visualization and Computer Graphics (8)	Academy of Management Journal (0)
MIS Quarterly (0)	Information Visualization (4)	Academy of Management Review (0)
Journal of Information Technology (0)	Journal of Visualization (5)	British Journal of Management (0)
Journal of Management IS (0)	Information and Organization (0)	Journal of Behavioral Decision Making (0)
European Journal of IS (0)	AIS eLibrary Conference Proceedings (2)	Judgement and Decision Making (0)
Information Systems Journal (0)	Springer EMCIS Proceedings (23)	Journal of Management (0)
Journal of Strategic IS (0)		
Journal of the AIS (0)		
Decision Support Systems (2)		
TOTAL 2	TOTAL 42	TOTAL 0

The emergence of various data analysis and visualization tools and techniques to facilitate decision-making has made the search outcome quite unexpected showing that the **information systems** and **management/decision-making** scholarly communities have not given adequate attention to the ways of understanding decision stakeholder needs.

The **information visualization** research community deals with numerous research interests, such as the management of complex and large datasets, the handling of uncertainty, the process of validation, and the seamless integration with the user task. This holds even more true in the realm of visual analytics, where the focal point shifted towards sensemaking, as highlighted by van Wijk [48]. The literature suggests that mainly qualitative approaches are used by visualization designers to characterize the target domain of their users. To describe and understand their problems, tasks and data visualization designers interview and observe their target audience [33, 34], conduct ethnographic studies and surveys to learn how the users achieve their professional objectives [47], analyze current practices [4], examine documentation and business processes [30] to identify domain situations for problem-driven visualization projects [13, 32]. Describing technical challenges of designing and implementing visualizations as a key tool for collaborative interactions with data, scholars emphasized the necessity for a broader understanding of user needs and universally applicable overview of collaborative visualization requirements and best practices [8, 18, 57]. Researchers derive design requirements from literature and close collaboration with expert scholars, practitioners and managers [37, 38, 50, 51], observe sequences of user actions to find regularities in the user behavior and based on findings create models to predict information displayed by a decision support system [20].

The reviewed literature, however, indicates that many visual analytics systems or visualization tools are aimed at individual users. Even though such systems are intended for multiple users, the use happens individually. The challenges in designing visualizations for collaborative environments have been documented, however, research into structured holistic approaches to understanding the requirements of multiple stakeholders with varying experiences and perspectives has been insufficient. The need for applications with more human-centered approaches for complex cognitive activities is growing [39]. This need originates from the demands of modern knowledge-driven environments because a decision-making process involves numerous interactions, and the organization and administration of these interactions necessitate the appropriate technological assistance [19]. Indeed, recent studies indicate the growing use of algorithmic aids to augment decision-making [7]. At the same time, difficulties to convince business stakeholders and managers to use requirements-driven big data methodologies remain as their main business focus continue to be on increasing return on investment [31].

Limited examples show that researchers tried different approaches when designing visual analytics systems for multi-perspective decision-making. One research attempted to classify participants into distinct groups in order to gain insights into the different patterns exhibited by the individuals [35]. Requirements of business value chain modeling were used to identify business opportunities for stakeholders using big data visualizations [5]. One study proposed that visualizing the results of neural networks models' predictions of future disruptive innovations based on historic big data and an organization's investment plans could support business decision-makers in predicting technology-based product success before proceeding with investments [44]. As indicated above, multiple interactions occur in knowledge intensive multi-perspective decision-making environments involving people, groups of people and technology artifacts. Therefore, the current paper proposes to examine stakeholder requirements through the lens of distributed cognition.

3   Theoretical Perspective

To examine the information requirements for decision-making that takes into account diverse perspectives, we take on the theory of *distributed cognition* (DC) that views cognition as extending beyond an individual and into the environment [15]. Distributed cognition emerged as a theoretical and methodological approach to analyze complex organizational activities that necessitate the utilization of material and technological tools. Building on the earlier research, Hutchins (1995) further developed this approach to encompass cognitive systems that exist within an individual's mind, between individuals and their use of tools, within a group of individuals engaged in interaction, or between a group and their use of tools. It primarily concerns how the information is represented, transformed and disseminated throughout the cognitive system [15]. According to the theory, the distributed cognitive system involves the relationship between the cognitive processes occurring within the human mind(s) and the social and material context of cognitive activities [16]. This approach has gained significance in the *information systems, human-computer interaction,* and *strategic management* research communities.

In the IS field, it contributed to the development of design and implementation principles. Scholars have conducted research on the supportive role of information technology in distributed cognition, which allows individuals to create comprehensive representations of their understanding, reflect on these representations, engage in dialogues with others, and utilize them to guide their actions [6]. Additionally, the DC approach was used to analyze collaborative work within organizations that introduced new technologies where the coordination of distributed work activities was essential [43], and to explore intuitive decision-making of managers in order to develop executive support systems [25]. It has been positioned as a fundamental framework for exploring the interactions between humans and technologies as it provides a focused approach to the design of human-computer interaction [15, 42].

Distributed cognition has also been applied to examine decision-making in the context of big data decision support systems (DSS). Researchers have shifted the focal point of strategic decision-making from the cognitive processes of individuals making decisions to the interconnected network of artifacts and individuals involved in a decision task. Investigating decision-making with big data DSS through the lens of distributed cognition could aid strategic IS researchers in gaining a better understanding of the specific circumstances of decision situations, as well as the ways in which the relationships between human (e.g., decision makers) and non-human (e.g., models, screens, software) entities influence decisions made with DSS and big data [3]. When distributed cognition occurs among groups of minds, externalizations are important for collaborative activity as they create a record of mental efforts and enable critique and negotiation. The challenge, however, is to integrate various perspectives from stakeholders by supporting reflection within a shared context, which enhances the creation of shared understanding and the quality of the designed artifact, providing future stakeholders with rich contextualized information [2].

Building upon Hutchins's research on the processing of information, Zhang and Norman [54] formulated a theoretical framework of distributed representations whereby both internal and external representations are involved in any given distributed cognitive task. Each representation forms an internal and external representational space. When combined, they create a distributed representational space that serves as the depiction of an abstract space for a task. The effectiveness of this framework was further examined and validated on its ability to describe and explain the concept of affordance as an abstract task space [16].

4 Method

4.1 Single Case Study

A consumer goods manufacturing company is used as a case study to explore information requirements and principles of creating visualizations for collaborative decision-making. Defined as 'an in-depth inquiry into a specific and complex phenomenon set within its real-world context' [53], a single case study allows close examination of facts, evidence and information within a specific setting [52] p. 40). This research is an exploratory case study set in a product development environment.

For this research, we consider augmented decision-making (i.e. decision-making using big data analytics, algorithmic decision aids and data visualizations) a distributed cognitive task that is spread out across both internal and external structures [16]. To analyze the decision-making requirements, we use the concept of distributed representations [54]. The external structures, which can be observed, take the form of data, predictive models, a decision-aiding tool, computer-generated recommendations that contain crucial product and business performance information visualized for cross-functional decision-making. The internal structures, which are not directly observable, are represented by the expertise of different stakeholders, all of whom comprise a strategic decision-making team, their level of confidence, information requirements and perspectives. All associated cognitive processes are dispersed throughout the business and innovation realities that exist within the organization, which in themselves constitute an external environmental structure. In this context, the distributed representational space [54] is the representation of the collaborative decision-making space.

4.2 Data Collection

Being the most suitable for an in-depth qualitative study [14], semi-structured interviews were conducted online [24] to collect research data. The purpose of these interviews was twofold: 1) to explore a decision-making scenario and 2) to gather the information requirements of cross-functional stakeholders involved in the decision-making task. By interviewing relevant employees and senior managers within our case study company, we engaged with key practitioners and users from various departments and teams. Eleven interviews were conducted with data scientists, informaticians, modelers responsible for building predictive model capabilities; project team leaders, R&D managers involved in product development; and R&D directors who make global decisions regarding strategic direction and implementation. Since defined topics allow for the alignment of questions with the flow of the interview [45] p. 391, an interview guide was developed based on themes derived from existing literature and informed by the distributed information design methodology [54–56]. All interviews began with questions about the roles and functional responsibilities of the interviewees, which contributed to the analysis of user characteristics. Subsequent questions focused on the decision-making process, data, predictive models, digital decision aid, the most important information for decision-making within the participants' scope, the alignment between information needed for decision-making and information received, and the role of visualization in understanding digital predictions. These questions were designed based on the research problem definition, exploratory literature, and research assumptions.

4.3 Data Analysis

Collected data was analyzed using the Human-centred Distributed Information Design (HCDID) methodology [55], which is based on the theory of distributed cognition [15, 54–56], and which offers an approach requiring multiple levels of analysis. According to the HCDID, the unit of analysis is a system composed of human and artificial agents, and their relations distributed across time and space [54–56]. The distributed information analysis was used to determine which aspects of digital product recommendations were relevant to cross-functional stakeholders, and what information and data they required to make the decision. This analysis provides a framework for (a) studying diverse perspectives occurring in any business decision-making situation, and (b) understanding cognitive aspects of decision-making in a distributed collaborative business environment, such as new product development. Key components of the analysis that included four levels were adapted from the HCDID methodology. We drew on the practice of template analysis [22, 23], and our template shown in Table 2 incorporated the HCDID structure as a priori themes:

Distributed User Analysis. Helps to identify the characteristics of existing and potential users, such as expertise, skills and knowledge base. While user analysis can be very deep going into education background, cognitive capacities and limitations, perceptual variations, age related skills, cultural background, personality, time available for learning and training, frequency of system [54], for the purpose of this research we only focused on roles, work responsibilities, spheres of decision making, overlap of knowledge and skills, pattern of communication, work relationships, information and data visualization requirements.

Distributed Functional Analysis. Helps to understand a domain, identify top level interrelations and constraints of decision makers and digital tools (artifacts), information flow. Functional analysis is carried out based on interactions with domain experts and practitioners who contributed to the product development decision making process.

Distributed Task Analysis. Is conducted to identify functions of the system and functions of users that need be performed, actions that need to be taken to achieve task goals, information that needs to be processed, input and output formats, and any potential constraints. Also, it is important to take into consideration the space and time distributions of activities, functions and actions, information flow across the decision making process, decision makers and digital tools.

Distributed Representational Analysis. Is done to identify the best information display and flow for a decision-making tasks, distribution of information across decision makers and digital tools under different implementations, and which implementation is most efficient for which decision-making task.

Table 2. Distributed information analysis template.

Main theme	Defining scope of theme
User	• Role and work responsibilities • Areas of decision making • Knowledge, expertise • Patterns of communication and work relationships • Information and data visualization requirements • Use of digital tools and data in decision-making
Distributed function	• Interactions with other domain experts and practitioners who contributed to the decision-making process
Distributed task	• Actions to be taken to achieve decision task goals • Information that needs to be processed • Task input and output • Potential constraints • Information and data sharing needs
Distributed representation	• Distribution of information and data across stakeholders, overlap • Visual formats convenient to understand data • Visual representations most efficient for which decision task

5 Findings

The distributed information and visualization requirements analysis focused on four key template components in Table 2. The analysis results are mapped across the three identified decision-making spaces – Digital/Analytics, User, and Cross-functional decision-making, see Table 3.

5.1 User

Three categories of individuals have been identified. In the first category, Digital/Analytics Space, there exists a team of digital experts, including data scientists, statisticians, informaticians, modelers, data analysts. The second category, User Decision-making Space, consists of direct users of the digital predictions, analytics output, data visualizations who possess various levels of technical expertise in their respective domains. Although these users are expected to understand the meaning of predictive output and data, they lack knowledge regarding the inner workings of digital modeling algorithms that generate recommendations. The third group, Cross-functional Decision-making Space, encompasses non-technical domain experts, such as business managers, who are unlikely to directly interact with specialized digital tools and workflows. Nevertheless, they form a cross-functional decision-making team that evaluates and makes judgments about the digital predictions and output. Thus, it is crucial to design visual representations of the relevant data and modeling output in a manner that caters to the needs of the different stakeholder types participating in a decision situation.

Table 3. Distributed information analysis for visualization design.

Distributed information analysis	Digital/Analytics Space	User Space (organizational function level)		Cross-functional Space (senior level)		
	Digital, Analytics teams	Domain Expert teams	Domain Expert management	Organization function 1	Organization function 2	Organization function 3
Case study	*e.g. Digital team*	*e.g. R&D teams*	*e.g. R&D managers*	*e.g. Procurement*	*e.g. Marketing*	*e.g. Manufacturing*
User	Data scientists, statisticians, modelers, data analysts	Technical domain experts (scientists). Some are unfamiliar with digital tools	Managers-domain experts. Decision making using predictive models and analytics is new	Technical, non-technical domain and business experts. Some stakeholders may be unfamiliar with and/or not used to making decisions based on predictions and models		
Distributed Functional Analysis	Building predictive model capabilities	Developing and testing new products	Developing product solutions, investment accountability	Controlling raw material supply purchasing and storage	Monitoring changes in consumer attributes, demand, sales	Managing production capacity
Distributed Decision Task	Evaluate and validate data accuracy and performance of predictive models	Create product options, assess product parameters, consider alternatives	Ensure products keep to strategic business objectives, cost, tech. Validity	Determine supply volumes and storage capacity	Perform consumer validations, propose marketing campaigns	Evaluate any impact on production, manufacturability and scalability
Distributed Representational Analysis	out of scope	Product characteristics, performance, cost. Parity/difference to the benchmark. E.g. Which predicted alternative is the cheapest/most sustainable?	Overall R&D, production, business metrics. E.g. Is a product alternative stable, sustainable and meets actual standards?	Raw materials variations, impact on purchasing costs and supply chain E.g. What do subtle variations mean?	Product attributes meet consumer demand. Product costs and evidence of achieving performance	Production needs, any limitations. Evidence that recommended alternative can be manufactured without problems

5.2 Distributed Functional Analysis

The primary function of the Digital/Analytics teams is to build predictive modeling and analytics capabilities in order to support technical domain experts with various data related aspects in product development projects. Domain experts on the functional level use relevant digital tools and need to interpret and validate recommendations generated by the digital workflow or algorithmic aid. Their functional responsibility is to test and select suitable product candidate(s) that can be recommended to the cross-functional stakeholders as potential products for manufacturing and market launch. This final recommendation must provide information relevant to all senior level stakeholders with different functional expertise within the business. Once a successful candidate is confirmed, a product proposal is passed on to senior functional managers (e.g. Function 1, Function 2, etc., see in Table 3). Such proposal with the recommended option must provide information relevant to all stakeholders with different functional expertise, for example, R&D teams expertise is in developing and testing new products; R&D management in developing solutions, capital, investment and budget accountability; Procurement in controlling all aspects of supply chain operations; Marketing in monitoring changes in consumer attributes, market demand, sales; and Manufacturing in managing production capacity.

5.3 Distributed Decision Task

The task of decision-making about a predicted product recommendation is a shared decision-making space for stakeholders with different functional expertise within the business. An exchange of relevant contextual information by R&D and other functional departments allows interpretation of common goals, potential market responses, supply and manufacturing aspects. However, our participants in the Cross-functional Decision Space each need to make decisions within their functional area before a collaborative decision is made. This may include clarifications and risk evaluation. New product development strategic decisions, which are the decisions about approval of a new or modified product for manufacturing, sustainability and cost, result from the process that propagates the representational state of product information (e.g. a product proposal describing facts related to business considerations and constraints concerning a product innovation proposed by R&D in all respective domains) through the cognitive system of human actors and the digital tools. Therefore, presented information needs to be relevant to decision-making tasks of cross-functional stakeholders. For example in our case study analysis in Table 3 decision tasks differ by function: for Procurement decision need to be made regarding supply/purchasing volumes, storage capacity; for Marketing – decisions regarding performing consumer tests and designing marketing claims; for Manufacturing – evaluating any impact on production, manufacturability and scalability of the proposed product.

5.4 Distributed Representational Analysis

The analysis of representation for the Digital/Analytics Space team falls outside the scope of this study as participants in this category provide support to the users and other stakeholders with all aspects of using data. For the participants from the technical User Space domain, it was essential that visual representations of an object of decision-making showcase physical characteristics and attributes of the product, any similarities or differences from the benchmark, as well as a variety of business metrics. This particular group of stakeholders should be able to assess alternatives with a high degree of data granularity on a technical level. For instance, in our cases study, domain experts were interested to see product properties, performance characteristics, cost, and comparison to the existing products. Examples of questions, see in Table 3, that needed to be answered to make a decision at this level had a more technical focus i.e. 'Is a recommended product alternative stable and sustainable?', 'Does it meet actual performance standard?', and 'Which predicted product alternative is the cheapest?'.

In contrast, in the Cross-functional Space, data visualizations for the senior stakeholders should address their primary concerns at a business, operational or managerial level, rather than delve into the technical specifics of each alternative. For instance, for Procurement decisions, data visualizations need to show variations in supply volumes to answer potential questions such as 'What do subtle variations mean?', 'What is the impact on purchasing volumes and costs?'; for Marketing 'Are the proposed product attributes meet consumer demand?'; and for Manufacturing 'What are the production needs?' and 'What are the limitations?'. Finally, there existed a set of aspects in which all stakeholders showed interest, such as product and business costs, evidence of achieving performance, and the evidence that the recommended product alternative can be manufactured without problems.

6 Conclusion

The result of our study are the following three principles for a data visualization approach based on the distributed information analysis. Firstly, an essential element of the approach to designing visualizations for multi-perspective decision-making is the understanding of diverse significant viewpoints that emphasize distinct needs of different stakeholders involved in the decision situation and process. It is also useful to understand their shared interests and concerns. This can be done on four levels – users, their functional analysis, their decision tasks and representations of relevant data to reflect the distributed nature of stakeholder interactions with regards to the object of collaborative decision-making. Secondly, the approach must be able to provide necessary details, and show perspectives of all relevant stakeholders to accommodate multiple viewpoints. Lastly, the approach must provide functionalities that will allow the stakeholders read visualizations in a 'language' they are familiar with in their daily responsibilities to confidently select decision alternatives.

The existing research is short of studies exploring decision makers' perceptions of the utility of visual representations for decision-making so that they are connected to and support relevant business decisions. Our study contributes to the big data, analytics and data visualization research area by exploring multi-perspective stakeholder requirements for visual representation designs in support of collaborative decision-making. The proposed design principles are based on the distributed cognition approach that takes into account 1) the functional relationships among multiple elements that contribute to the decision-making process and 2) the integration of a person's internal cognitive processes with external objects. Our research shows that the information and visualization requirements of distributed decision makers can be investigated using the HCDID methodology. Thus the connection of visual representations to strategic decisions is rooted in the decision situation needs and questions for which data visualizations need to provide answers. The proposed approach is going to be tested in a subsequent visualization prototype development and validation study.

References

1. AIS Research: Senior Scholars' List of Premier Journals webpage: https://aisnet.org/page/SeniorScholarListofPremierJournals. Accessed 29 July 2023
2. Arias, E., Eden, H., Fischer, G., Gorman, A., Scharff, E.: Transcending the individual human mind—creating shared understanding through collaborative design. ACM Trans. Comput.-Hum. Interact. 7(1), 84–113 (2000)
3. Aversa, P., Cabantous, L., Haefliger, S.: When decision support systems fail: insights for strategic information systems from Formula 1. J. Strat. Inf. Syst. 27, 221–236 (2018)
4. Bergner, S., Sedlmair, M., Moller, T., Abdolyousefi, S.N., Saad, A.: ParaGlide: interactive parameter space partitioning for computer simulations. IEEE Trans. Visual. Comput. Graphics 19(9), 1499–1512 (2013). https://doi.org/10.1109/TVCG.2013.61
5. Berntzen, L., Krumova, M.: Big data from a business perspective. In: Themistocleous, M., Morabito, V. (eds.) Proceedings 14th European, Mediterranean, and Middle Eastern Conference, Information Systems. EMCIS 2017, LNBIP, vol. 299, pp. 119–127. Springer, Cham. (2017)
6. Boland, R., Tenkasi, R., Te'eni, D.: Designing information technology to support distributed cognition. Organ. Sci. 5(3), 456–475 (1994)
7. Burton, J.W., Stein, M.K., Jensen, T.B.: A systematic review of algorithm aversion in augmented decision making. J. Behav. Decis. Mak. 33(2), 220–239 (2020)
8. Chinchora, N., Pike, W.: The science of analytic reporting. Inf. Vis. 8(4), 286–293 (2009)
9. Dimara, E., Bezerianos, A., Dragicevic, P.: Conceptual and methodological issues in evaluating multidimensional visualizations for decision support. IEEE Trans. Visual Comput. Graphics 24(1), 749–758 (2018)
10. Dimara, E., Zhang, H., Tory, M., Franconeri, S.: The unmet data visualization needs of decision makers within organizations. IEEE Trans. Visual. Comput. Graphics 28(12), 4101–4112 (2022). https://doi.org/10.1109/TVCG.2021.3074023
11. Dimara, E., Stasko, J.: A critical reflection on visualization research: where do decision making tasks hide? IEEE Trans. Visual Comput. Graphics 28(1), 1128–1138 (2022)
12. Dwyer, T., Marriott, K., Isenberg, T., Klein, K., Riche, N., Schreiber, F., Stuerzlinger, W., Thomas, B.H.: Immersive analytics: an introduction. In: Marriott, K., et al. (eds.), Immersive Analytics, pp. 221–257. Springer Nature, Switzerland AG (2018)

13. Hakone, A., et al.: PROACT: iterative design of a patient-centered visualization for effective prostate cancer health risk communication. IEEE Trans. Visual Comput. Graphics **23**(1), 601–610 (2017)
14. Healey, M., Rawlinson, M.: Interviewing business owners and managers: a review of methods and techniques. Geoforum **24**(3), 339–355 (1993)
15. Hollan, J., Hutchins, E., Kirsh, D.: Distributed cognition: toward a new foundation for human-computer interaction research. ACM Trans. Comput.-Human Interact. **7**(2), 174–196 (2000)
16. Hutchins, E.: How a cockpit remembers its speeds. Cogn. Sci. **19**, 265–288 (1995)
17. Hutchins, E.: Distributed cognition. In: Smelser, N., Baltes, P. (eds.) International Encyclopedia of the Social & Behavioral Sciences, pp. 2068–2072. Elsevier (2001) https://doi.org/10.1016/B0-08-043076-7/01636-3
18. Isenberg, P., Elmqvist, N., Scholtz, J., Cernea, D., Ma, K., Hagen, H.: Collaborative visualization: definition, challenges and research agenda. Inf. Vis. **10**(4), 310–326 (2011)
19. Karacapilidis, N., Tsakalidis, D., Domalis, G.: An AI-enhanced solution for large-scale deliberation mapping and explainable reasoning. In: Papadaki, M., Rupino da Cunha, P., Themistocleous, M., Christodoulou, K. (eds.) Information Systems. EMCIS 2022. LNBIP, vol. 464. Springer, Cham (2023). https://doi.org/10.1007/978-3-031-30694-5_23
20. Kamsu-Foguem, B., et al.: User-centered visual analysis using a hybrid reasoning architecture for intensive care units. Decis. Support Syst. **54**(1), 496–509 (2012)
21. Keim, D., Andrienko, G., Fekete, J. D., Görg, C., Kohlhammer, J., Melançon, G.: Visual analytics: definition, process, and challenges. In: Kerren, A., Stasko, J., Fekete, J.D., North, C. (eds.) Human-Centered Issues and Perspectives, pp. 154–175. Springer-Verlag, Berlin Heidelberg (2008)
22. King, N.: Using templates in the qualitative analysis of text. In: Cassell, C., Symon, G. (eds), Essential Guide to Qualitative Methods in Organizational Research, pp. 256–270. Sage, London (2004)
23. King, N., Brooks, J.M.: Template Analysis for Business and Management Students. SAGE Publications Ltd, 1 Oliver's Yard, 55 City Road London EC1Y 1SP (2017). https://doi.org/10.4135/9781473983304
24. King, N., Horrocks, C.: Interviews in Qualitative Research. Sage, London (2010)
25. Kuo, F.: Managerial intuition and the development of executive support systems. Decis. Support Syst. **24**, 89–103 (1998)
26. EMCIS Homepage. https://link.springer.com/conference/emcis. Last accessed 10 Jan 2023
27. Luo, W.H.: User choice of interactive data visualization format: the effects of cognitive style and spatial ability. Decis. Support Syst. **122**, 1–11 (2019)
28. Lurie, N., Mason, C.: Visual representation: implications for decision making. J. Mark. **71**(1), 160–177 (2007)
29. Lycett, M.: 'Datafication': making sense of (Big) data in a complex world. Eur. J. Inf. Syst. **22**(4), 381–386 (2013)
30. Mayer, J.H., Meinecke, M., Quick, R., Kusterer, F., Kessler, P.: Applying predictive analytics algorithms to support sales volume forecasting. In: Papadaki, M., da Cunha, P.R., Themistocleous, M., Christodoulou, K. (eds.) Information Systems: 19th European, Mediterranean, and Middle Eastern Conference, EMCIS 2022, Virtual Event, December 21–22, 2022, Proceedings, pp. 63–76. Springer Nature Switzerland, Cham (2023). https://doi.org/10.1007/978-3-031-30694-5_6
31. Ben Aissa, M.M., Sfaxi, L., Robbana, R.: DECIDE: a new decisional big data methodology for a better data governance. In: Themistocleous, M., Papadaki, M., Kamal, M.M. (eds.) EMCIS 2020. LNBIP, vol. 402, pp. 63–78. Springer, Cham (2020). https://doi.org/10.1007/978-3-030-63396-7_5

32. Meyer, M., Sedlmair, M., Quinan, P.S., Munzner, T.: The nested blocks and guidelines model. Inf. Vis. **14**(3), 234–249 (2015)
33. Munzner, T.: A nested model for visualization design and validation. IEEE Trans. Visual Comput. Graphics **15**(6), 921–928 (2009)
34. Musleh, M., Chatzimparmpas, A., Jusufi, I.: Visual analysis of blow molding machine multivariate time series data. J. Visualization **25**, 1329–1342 (2022)
35. Onoue, Y., Kukimoto, N., Sakamoto, N., Koyamada, K.: E-Grid: a visual analytics system for evaluation structures. J. Visualization **19**, 753–768 (2016)
36. Padilla, L., Creem-Regehr, S., Hegarty, M., Stefanucci, J.: Decision making with visualizations: a cognitive framework across disciplines. Cogn. Res.: Principles Implications **3**(1), 1–25 (2018)
37. Pajer, S., Streit, M., Torsney-Weir, T., Spechtenhauser, F., Muller, T., Piringer, H.: Weightlifter: visual weight space exploration for multi-criteria decision making. IEEE Trans. Visual. Comput. Graphics **23**(1), 611–620 (2017). https://doi.org/10.1109/TVCG.2016.2598589
38. Park, H., Bellamy, M.A., Basolea, R.C.: Visual analytics for supply network management: system design and evaluation. Decis. Support Syst. **91**, 89–102 (2016)
39. Parsons, P., Sedig, K.: Adjustable properties of visual representations: improving the quality of human-information interaction. J. Am. Soc. Inf. Sci. **65**(3), 455–482 (2014)
40. Patterson, R., et al.: A human cognition framework for information visualization. Comput. Graph. **42**(1), 42–58 (2014)
41. Perdana, A., Robb, A., Rohde, F.: Interactive data and information visualization: unpacking its characteristics and influencing aspects on decision-making. Pac. Asia J. Assoc. Inform. Syst. **11**(4), 75–104 (2019)
42. Rinkus, S., et al.: Human-centered design of a distributed knowledge management system. J. Biomed. Inform. **38**, 4–17 (2005)
43. Rogers, Y., Ellis, J.: Distributed cognition: an alternative framework for analysing and explaining collaborative working. J. Inform. Technol. **9**(2), 119–128 (1994)
44. Saade, M., Jneid, M., Saleh, I.: Enhancing decision-making in new product development: forecasting technologies revenues using a multidimensional neural network. In: Themistocleous, M., Papadaki, M., Mustafa Kamal, M. (eds.) EMCIS 2020. LNBIP, vol. 402, pp. 715–729. Springer, Cham (2020). https://doi.org/10.1007/978-3-030-63396-7_48
45. Saunders, M., Lewis, P., Thornhill, A.: Research Methods for Business Students, 7th edn. Pearson Education Limited, Essex (2015)
46. Sharma, R., Mithas, S., Kankanhalli, A.: Transforming decision-making processes: a research agenda for understanding the impact of business analytics on organisations. Eur. J. Inf. Syst. **23**(4), 433–441 (2014)
47. Shneiderman, B., Plaisant, C.: Strategies for evaluating information visualization tools: multidimensional in-depth long-term case studies. In: Proceedings of the AVI Workshop on Beyond time and errors: novel evaluation methods for Information Visualization (BELIV). ACM, Article 6 (2006)
48. van Wijk, J.: Views on visualization. IEEE Trans. Visual Comput. Graphics **12**(4), 421–432 (2006)
49. Webster, J., Watson, R.T.: Analyzing the past to prepare for the future: writing a literature review. MIS Q. **2**(26), 13–23 (2002)

50. Weng, D., Chen, R., Deng, Z., Wu, F., Chen, Z., Wu, Y.: SRVis: towards better spatial integration in ranking visualization. IEEE Trans. Visual Comput. Graphics **25**(1), 459–469 (2019)
51. Xu, P., Mei, H., Ren, L., Chen, W.: ViDX: visual diagnostics of assembly line performance in smart factories. IEEE Trans. Visual Comput. Graphics **23**(1), 291–300 (2017)
52. Yin, R.K.: Case study Research: Design and Methods, 2nd edn. Sage, Newbury Park, CA (1994)
53. Yin, R.K.: Validity and generalization in future case study evaluations. Evaluation **19**(3), 321–332 (2013)
54. Zhang, J., Norman, D.: Representations in distributed cognitive tasks. Cogn. Sci. **18**, 87–122 (1994)
55. Zhang, J., Patel, V., Johnson, K., Smith, J.: Designing human-centered distributed information systems. IEEE Intell. Syst. **17**, 42–47 (2002)
56. Zhang, J., Patel, V.: Distributed cognition, representation, and affordance. Pragmat. Cogn. **14**(2), 333–341 (2006)
57. Zhuang, M., Concannon, D., Manley, E.: A framework for evaluating dashboards in healthcare. IEEE Trans. Visual Comput. Graphics **28**(4), 1715–1731 (2022)

Digital Services and Social Media

Mobile Application Diffusion: An Exploration of Trust and Privacy Amongst Rural Enterprises in South Africa

Wellington Chakuzira(✉) and Marcia Mkansi

University of South Africa, Pretoria, South Africa
{chakuw,mkansm}@unisa.ac.za

Abstract. This review paper presents academic literature exploring why there is a slow adoption of mobile technologies amongst rural enterprises in South Africa. The researchers conducted a thorough literature review of studies related to trust and privacy on mobile applications published in journals between January 2020 and April 2023. The study reviewed two main external variables: trust and privacy of technology adoption. The systematic review shows that common trust concerns for rural enterprises include untrusted service providers, weak security, and security attacks. Additionally, the review revealed common privacy issues and concerns, including integrity, user awareness, unobservability, and deniability as obstacles to the fast uptake of mobile applications in rural marketplaces in South Africa. Considering these factors, the paper concludes by offering practical and theoretical suggestions that can assist rural entrepreneurs in enhancing the diffusion of mobile technology.

Keywords: mobile applications · technology diffusion · trust issues · privacy concerns

1 Introduction

Marketers in many developing countries are still trying to gain a better understanding of the motivations and methods of consumer engagement with modern technologies such as mobile applications or "apps". It is interesting to note that the study of the nature of mobile applications and, consequently, the reasons for and ways in which users utilize them, has dominated the literature on this developing technology [1–3]. Understandably, the proliferation of mobile applications, akin to various digital entrepreneurial pursuits and the ubiquity of internet infrastructure, has ushered in fresh prospects and entrepreneurial models in both rural and urban marketplaces [4, 5], although the introduction of novel technological applications often encounters more skepticism within rural areas compared to urban markets [6, 7, 9]. Alavion and Taghdisi [8] contend that the distinctive characteristics and capabilities of and demand for advanced technical proficiencies make mobile application diffusion challenging in rural areas. Evidently, substantial infrastructural developments focusing on connectivity upgrades are imperative for mobile applications to thrive in rural regions [10].

M. Papadaki et al. (Eds.): EMCIS 2023, LNBIP 502, pp. 47–64, 2024.
https://doi.org/10.1007/978-3-031-56481-9_4

Undoubtedly, mobile-application use is on the rise in rural areas, serving a multitude of purposes. These include providing access to information about government initiatives, agricultural techniques, market prices, and weather forecasts, as well as facilitating money transfers and banking [3, 11, 12]. Despite the apparent opportunities generated by mobile applications, the complexities associated with technology trust and the security issues associated with mobile-application implementation have resulted in a relatively slow adoption rate among rural South African entrepreneurs [13]. While technology significantly impacts product utility [14], Rogers [15] cautions that making people use new technology for the first time can be challenging. Therefore, Sisi and Souri [13] contend that mobile applications must gain recognition among retailers and customers in these digitally challenged markets to remain competitive in rural markets. The critical question is, what are the trust and privacy peculiarities underlying rural entrepreneurs' mobile applications adoption? In subsequent sections, we review mobile applications in rural areas through different dimensions of trust and privacy concerns, for mobile applications in rural markets. Then we present methodology, followed by findings and future research implications for entrepreneurs and theory to advance diffusion on rural markets. Finally, the paper ends with concluding remarks.

2 Literature Review

Previous studies indicate that rural marketers face the most pronounced impact of the digital divide [16]. This situation often stems from factors such as limited data connectivity, power interruptions, and socio-economic disparities [17, 18]. Unless effective measures are enacted to narrow this gap, the uptake of technology in rural regions will continue to be slow. Vasileiadou, Huijben, and Raven [19] highlight the high cost of technology in rural markets as contributing to widespread technology poverty and exacerbating the digital divide. This divide is notably conspicuous among many rural residents in South Africa who lack the means to afford information and communication technologies (ICTs), perpetuating technological impoverishment across the majority of rural areas in the country. It is important to recognize that such technological poverty has the potential to increase South African communities' overall resistance to technology, which will also hinder adoption [20].

Recognizing that the adoption of technology is a multifaceted and socially intricate process, this study focuses on the need to address logical, sensitive, and contextual apprehensions [21]. This paper borrows from technology theories to address the trust and privacy concerns prevalent in rural markets. The paper acknowledges the conceptual underpinnings, practical applications, and evolutionary nature of technology adoption models and theories. These include the unified theory of acceptance and use of technology (UTAUT) [22, 23], the diffusion of innovation theory (DIT) [15], the theory of reasoned action (TRA) [24], the theory of planned behavior (TPB) [25, 26], and the technology acceptance model (TAM) [27].

These theories illuminate potential applications for the adoption of technology and provide a conceptual framework that can aid future researchers in comprehending the underlying technology models and theories shaping past, present, and future technology adoption trends. However, this paper studies the trust and privacy dimensions as potential extension variables to any technology adoption theory mentioned earlier. As postulated by Nguyen et al., [28] there is continued uptake of technology in conducting business and in most cases the electronic data is exposed to theft, falsification, or unauthorized access. A situation which leaves customers lacking user's confidence and judgement that a specific service is free from privacy and security threat [29]. Notably trust and privacy concerns may pause a threat in all forms of technology adoptions making them important variables in technology adoption theories. Considering the foregoing postulations, the current paper reviews different dimensions of trust and privacy concerns. Now, the paper presents the relevant and recent literature for each stream.

2.1 Trust Concerns Among Rural Enterprises

Lankton, McKnight, and Tripp [30] define trust in technology as the user's confidence and judgement that a specific service is free from privacy and security threats. This public confidence plays an important role in the populace's selection of mobile applications. The collective disposition towards trust significantly influences the utilization of mobile applications in rural markets. Entrepreneurs recognize the potential of mobile applications in enhancing rural business operations [29]. However, their willingness to embrace these applications is often impeded by concerns that transmitting enterprise data over the internet might result in data tampering and privacy breaches. Al-Azawei and Alowayr [31] affirm that trust in mobile internet and data quality (essentially trust in technology) impact users' inclination to use mobile applications. This assertion is reinforced by Mkansi and Nsakanda [3], who emphasize that security and privacy concerns and distrust in mobile application services pose substantial challenges for rural South African enterprises.

In addition, trust in technology stands as a notable catalyst in decreasing the perceived risks associated with technology utilization, particularly when novel technologies and transactions with uncertain statuses are involved [31]. Given that the adoption of mobile applications in rural markets of South Africa is still in its early stages, users harbor uncertainties concerning the technical abilities of their service providers and the security and privacy aspects of the services offered [3]. Consequently, this lack of assurance frequently fosters reluctance among rural entrepreneurs to embrace and use mobile applications due to the inherent risks involved [32]. As a result, this study acknowledges the main challenges associated with untrusted cloud service providers and uses them as the basis of the analysis in the current review paper.

2.2 Privacy Concerns Among Rural Enterprises

Recognizing the transformative impact of the internet on our lives [29, 33], people now view the internet as a dependable source of information on products and services. However, the use of the internet for mobile applications in rural areas in developing countries, including South Africa, has not been growing as fast as the internet's other uses [34, 35]. One possible rationale for this uneven growth could be the hesitancy of entrepreneurs to divulge personal business information on crowd-based mobile application platforms. For example, Khan, Mihovska, Prasad and Velez, [36] indicates that the most important reason why entrepreneurs do not use mobile applications is their concern about sending out their private business information. The survey findings demonstrate that merely 24.9% of entrepreneurs felt comfortable using mobile applications [36]. However, the media's focus on online fraud, hacking, and identity theft has likely made entrepreneurs more conscious of the dangers of using mobile applications.

While many entrepreneurs perceive mobile applications as either a menace or a nuisance, these applications hold substantial value as tools for both online and offline businesses [37]. The rapid advances in information technology and the surge in internet usage empower companies to gather, retain, and exchange consumer data, which can be harnessed for crafting more accurately targeted marketing strategies [3]. It is noteworthy that a typical entrepreneurial approach revolves around cultivating long-term relationships and a series of business transactions. This approach is facilitated through the accumulation of data pertaining to entrepreneurs' purchasing behaviors, preferences, and personal information on the internet.

Importantly, information databases necessitate entrepreneurs to share their personal information, whether by choice or circumstance. In most cases, the nature of mobile applications mandates entrepreneurs to reveal a certain amount of personal business details (for example, name, address, telephone number) and payment particulars (such as credit card numbers) [38, 39]. The problem lies in entrepreneurs' escalating concerns about privacy owing to the surge in questionable and unlawful activities associated with application usage, including a notable increase in failed deliveries, identity theft, and fraud [29]. As a result, researchers in various fields have been paying more attention to privacy and security issues in mobile applications. For instance, entrepreneurs' growing concern over their privacy is becoming a significant issue for the potential use of mobile applications [40]. The critical question is then, what motivates entrepreneurs to voluntarily share their personal business information on mobile application platforms? To address this query, this paper examines diverse facets of privacy and their impact on the usage of mobile applications within rural markets. Privacy is a broad concept that covers, amongst other things, freedom of thought, control over personal information, freedom from surveillance, protection of reputation, and immunity from searches and interrogations [6, 16, 38, 40]. The privacy dimensions previously proposed for evaluating the efficacy of overarching guidelines aimed at augmenting privacy include privacy categories, privacy principles, privacy concerns, and privacy enhancements. Figure 1 highlights elements of privacy dimensions and provides the conceptualized flow of the current study.

Figure 1 clearly displays the four privacy dimensions (privacy concern, privacy principles, privacy enhancement, and privacy categories). Embedded in each dimension are the key elements that rural marketers should understand when implementing mobile applications. Closely linked to the privacy dimensions are security challenges (untrusted customer service providers, weak security models, and security attacks), as shown in Fig. 1. Mitigating security challenges by rural marketers might increase trust among users and ensure the successful implementation of mobile applications. Considering the trust and privacy matters in mobile application usage, this study used a systematic review method to understand trust and security matters and suggest the best form of technology

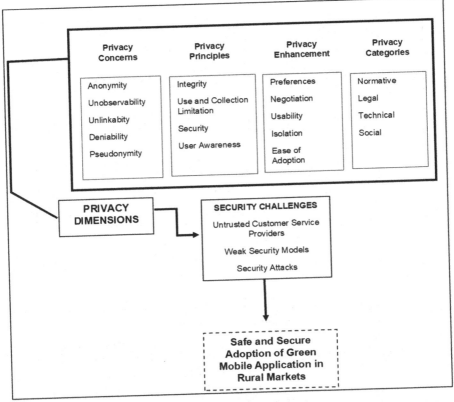

Fig. 1. Trust and privacy issues in mobile application usages

changes that is suitable for emerging rural markets. The study's methodology is detailed in the following section.

3 Methodology

This paper endeavors to provide a comprehensive and impartial synthesis of many relevant studies pertaining to trust and privacy challenges within mobile applications for rural enterprises. In line with the recommendations by Papaioannou, Sutton, and Booth [41], systematic techniques for literature review were implemented. In particular, systematic reviews are a type of literature review that involve thorough, methodical searching and are specified by predetermined eligibility criteria in accordance with guidelines [42–44]. In this study, the systematic review aims to unearth all pertinent insights relating to trust and privacy concerns surrounding mobile applications, with particular emphasis on contributions shedding light on the improvement of theory [45]. The key stages of the review process are as follows: (1) selection, (2) specification, and (3) summarizing, as shown in Fig. 2.

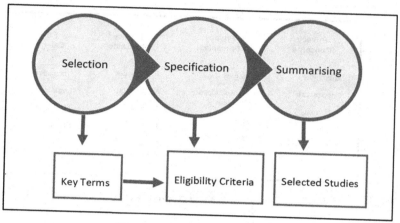

Fig. 2. Systematic review process

3.1 Selection

As recommended by Delgado-Rodríguez and Sillero-Arenas [46], the study selection criteria of a systematic review should flow directly from the review questions and be specified prior to selection. In this process, the paper considers two significant aspects throughout the selection. Firstly, the study chooses specific key terms associated with the research scope, including: "privacy concern among rural entrepreneurs;" "mobile technology in rural areas;" "trust issues among rural entrepreneurs;" "mobile applications for rural enterprises;" "user awareness;" "security attacks;" "privacy concerns;" "trust;" "anonymity;" and "integrity". Secondly, the paper employed multiple well-known digital library databases to collect resources from: Web of Science, Directory of Open Access Journals, Google Scholar, ProQuest, Research Gate, Science Direct, and Wiley.

3.2 Specification

Two straightforward standards of validation were adopted by the paper; (1) the date of publication and (2) the applicability of the study was used to govern the search results retrieved from database sources. Literature published from 2020 which addressed external variables (trust and privacy) of technology were considered. The study only used academic papers that were available online published from 2020 which included mobile application adoptions in different markets. This allowed for a thorough understanding of contemporary trust and privacy issues that marketers face regarding new technology adoptions. The search was also limited to articles released between January 2020 and April 2023. The rationale of selecting papers from 2020 was to attempt to provide critical contemporary knowledge on mobile applications, which allows for better conclusions in the modern, fast-paced technological market environments. Additionally, the selected papers contained enough information and did not stray outside the purview of crowd mobile applications among rural enterprises. Table 1 shows an analysis of the specifications of the current paper.

Table 1. Specification criteria

Source(s)	Discusses trust as an external variable	Discusses privacy as an external variable	Mobile application adoption focus area
Hajian et al. [38]		The research suggests a solution for mobile crowd-sensing-based spectrum monitoring comprised of a privacy-preserving protocol with secure rewarding capability and a trust mechanism against malicious players	
Asti, Handayani, and Azzahro [47]	This study discusses the influence of trust, perceived value, and attitude on customers repurchase intention for e-grocery		
Stocchi et al. [1]			The research presents an integrated overview of the available mobile marketing research, elaborating and clarifying what is known about how applications affect consumer experiences and value across iterative customer journeys
Gunawardana andFernando [48]	The research discusses the role of customer trust on e-service quality		
Kurniasari and Riyadi [49]	The research discusses trust issues of Indonesian e-grocery shoppers after the Covid-19 pandemic	The study discusses privacy issues of Indonesian e-grocery shoppers after the Covid-19 pandemic	

(continued)

Table 1. (*continued*)

Source(s)	Discusses trust as an external variable	Discusses privacy as an external variable	Mobile application adoption focus area
Singh, Gupta, Kumar, Sikdar, and Sinha [50]			The research identifies antecedents of customer satisfaction and patronage intentions in the context of e-grocery retailing through mobile applications
Mkansi, de Leeuw, and Amosun [4]			The research presents a mobile application supported by township and urban e-grocery distribution models that uses a software application
Nguyen, Hoang, and Vu Mai [28]			The study explores technology transfer through the perceptions of both business managers and technology specialists
Yang, Liu, Zhang and Yin [51]	The study empirically investigates the effects of social trust on technology innovation		
Sharma et al. [39]	The paper discusses the challenges and open issues related to security, privacy, and trust in mobile applications	The paper discusses the challenges and open issues related to security, privacy, and trust in mobile applications	

3.3 Summarizing

After screening the studied papers, the researchers finalized a full review of the findings, listed each paper's reference in summary tables, and discussed future research directions. Notably, the search results on the online database delivered a total of 123 original research papers. The researchers retained those that specifically discussed trust and privacy concerns among rural entrepreneurs adopting mobile applications, roughly 12% (15) of the total. The next section discusses the findings of this study. The articles selected mainly hailed from the fields of information management (40%), business management (33%, or five articles), marketing (14%), and entrepreneurship (13%) (see Fig. 3).

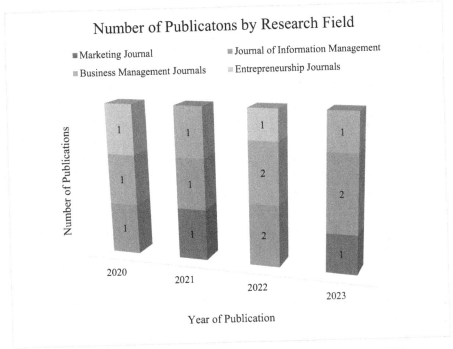

Fig. 3. Number of publications by research field from 2020 to April 2023

A pattern in publication was not evident; the number of publications were not spread evenly across the years. There were four publications in 2023, five publications in 2022 (the most in a year), and three publications in both 2021 and 2020. Most articles were published in information management and business management journals, indicating that academics have a particular interest in the interface between information management and small-business development. Figure 4 also shows that studies in these two fields were generally published every year during the period January 2020–April 2023. The next section will discuss the main findings of the research.

4 Findings

As mentioned, the findings of this paper are discussed in relation to trust challenges and privacy dimensions (see Fig. 1). Past literature on technology implementations discussed trust and privacy issues in line with innovation theories such as the TAM, UTAUT, uses and gratifications theory (UGT) [49, 50]. Indeed, these theories form a strong basis for understanding technology adoption variables, including trust and privacy issues on mobile technology implementations. However, this paper will discuss its findings under the two external variables of technology, namely security and privacy concerns.

Table 2. Mobile application-based trust and privacy analysis for entrepreneurs from 2020 to 2023

Categorization	Year	Author	Research focus	Trust and privacy dimensions													
				Untrusted service providers	Weak security	Security attacks	Integrity	User awareness	Use and collection limitation	Anonymity	Unobservability	Deniability	Preferences	Negotiation	Usability	Ease of adoption	Social and legal mitigations
Trust concerns	2023	Mkansi and Nsakanda [3]	Mobile apps, e-grocery challenges and strategies Trust and privacy-specific focus: trust and privacy challenges and strategies	✓			✓	✓		✓					✓	✓	
		Asti et al. [47]	Influence of trust, perceived value, and attitude on customers repurchase intention for e-grocery	✓	✓			✓			✓				✓	✓	
		Akinola and Asaolu [52]	A trust, privacy, and security model for e-commerce in Nigeria		✓	✓				✓	✓	✓					
	2022	Stocchi et al. [1]	Mobile apps, clarifying and expanding what is known around how apps shape customer experiences and value across iterative customer journeys Trust and privacy-specific focus	✓	✓									✓	✓		
		Khaw et al. [53]	Importance of specific determinants of trust in mobile commerce contributes to providing deep insight into the factors of customer trust in mobile commerce	✓	✓				✓	✓					✓	✓	✓
	2021	Gunawardana and Fernando [48]	The impact of e-service quality dimensions and customer satisfaction with e-groceries with the role of customer trust as a mediating variable during the epidemic			✓	✓	✓			✓		✓			✓	

(continued)

Table 2. (*continued*)

Categorization	Year	Author	Research focus	Trust and privacy dimensions														
				Untrusted service providers	Weak security	Security attacks	Integrity	User awareness	Use and collection limitation	Anonymity	Unobservability	Deniability	Preferences	Negotiation	Usability	Ease of adoption	Social and legal mitigations
		Yang, Liu, Zhang and Yin [51]	Social trust and green technology innovation: evidence from listed firms in China											✓	✓	✓	✓
	2020	Sharma et al. [39]	Challenges and open issues related to security, privacy, and trust in mobile applications	✓	✓	✓			✓	✓			✓			✓	
		Kurniasari and Riyadi [49]	Trust issues of Indonesian e-grocery shoppers after the Covid-19 pandemic	✓	✓		✓	✓		✓				✓	✓	✓	
Privacy concerns	2023	Hajian et al. [38]	Method proposed that includes a privacy-preserving protocol with secure rewarding capability as well as a trust mechanism against malicious participants for mobile crowd-sensing based spectrum monitoring	✓			✓	✓			✓	✓					
	2022	Shaw et al. [54]	A simplified model proposed for investigating mobile wallet adoption, with extensions for security, privacy, and ubiquity Tested for Canada, Germany, and the United States, and provides country-specific guidance for practitioners				✓	✓			✓			✓	✓	✓	✓
		Su et al. [55]	The role of technology acceptance models, mobile service quality factors, and personalization and privacy in promoting customers' trust and their loyalty towards mobile food delivery apps				✓		✓		✓				✓	✓	✓

(*continued*)

Table 2. (*continued*)

Categorization	Year	Author	Research focus	Trust and privacy dimensions													
				Untrusted service providers	Weak security	Security attacks	Integrity	User awareness	Use and collection limitation	Anonymity	Unobservability	Deniability	Preferences	Negotiation	Usability	Ease of adoption	Social and legal mitigations
	2021	Libaque-Sáenz et al. [56]	The effect of intervention strategies, fair information practices, and the data-collection method on privacy-related decisions							✓		✓		✓		✓	
		Tay et al. [57]	How users' risk perception can be shifted towards more privacy-aware decisions through generation fluency and framing manipulations	✓	✓			✓	✓				✓				✓
	2020	Balapour et al. [58]	The relationship between users' perceived privacy concerns and their perceptions of mobile app security		✓	✓			✓	✓		✓			✓	✓	

The study discovered three possible trust challenges for mobile application adoptions among rural South African enterprises, especially in rural markets with untrusted service providers, weak security models, and security attacks. Certainly, literature from the selected journals identifies untrusted service providers and security attacks as the main two challenges for mobile application implementations (see Table 2). The study also identifies four privacy issues that are pertinent for mobile technology usage in rural markets, namely integrity, security, user awareness and use, and collection limitation (see Table 2).

As shown in Table 2, there are two privacy concerns pertinent to rural markets that the study discovered, among others: unobservability and deniability. Other privacy dimensions, such as privacy enhancement, had limited reports in the selected papers. However, it is also important for marketers to understand enhancement issues such as preferences, negotiation, and isolation, as well as privacy categories such as legal, normative, technical, and social privacy before implementing mobile applications. Privacy is a broad notion that covers, among other things, one's right to freedom of thinking, control over one's body, privacy in one's home, control over one's personal information, freedom from surveillance, preservation of one's reputation, and immunity from searches and interrogations. Therefore, it becomes imperative to understand all privacy concerns before implementing mobile applications.

4.1 Implications for Rural Entrepreneurs

A key finding of this review paper is that, while the existing literature has indeed confirmed the significance of consumer-centric concerns in propelling the uptake of mobile application transactions within rural markets, these concerns are further amplified when entrepreneurs are assured of safeguarding their business information privacy. In addition, rural entrepreneurs should remain aware of the critical role of implementing robust security mechanisms as an integral facet of successful mobile application adoption. Lastly, when the business and financial information of rural entrepreneurs is effectively harnessed, it not only enriches their interaction with mobile applications but also cultivates a sense of trust.

Rural entrepreneurs should view data privacy and security more broadly, not only as risk-management issues but also as potential sources of competitive advantage that could greatly aid in brand building and the creation of a solid company reputation. Neglecting security issues, such as unreliable customer service providers, inadequate security models, and security breaches, can often lead to harm to brand image, business losses, and valuable time consumed by ongoing legal actions. In addition, rural entrepreneurs can ensure the security of their personal business and financial information by collaborating with mobile application service providers in conducting regular data privacy audits. While this may entail costs for rural enterprises, it is worth noting that these businesses often accumulate more data than they realize, so it is crucial to discern the necessary data, how it is stored, and the level of security protecting it.

It is also crucial for rural businesses to recognize that neglecting the importance of data collection can often ignite significant crises. For instance, overlooking privacy concerns, principles, enhancement, and categories can potentially trigger data privacy scandals. Rural entrepreneurs and mobile application service providers must make sure that they develop and enforce clear and transparent guarantees of privacy in order for rural areas to successfully utilize mobile applications. These measures should serve as signals to businesses, assuring them of the security of their confidential information. To accomplish this, rural entrepreneurs should embrace modern marketing strategies that prioritize establishing a brand and fostering robust data privacy and security practices.

Compelling findings from the analyzed articles indicate that the adoption of mobile applications for conducting business by rural consumers is still not widespread. To tackle this issue, rural entrepreneurs must understand the nuances of mobile application functionality. Moreover, the business information presented on these mobile applications should be relevant and tailored to align with the preferences and attributes of the specific consumer segments they aim to reach. It is also imperative that these mobile applications remain consistently operational and connected, providing consumers with a sense of ubiquitous accessibility.

4.2 Implications for Theory

The findings reveal that, for rural entrepreneurs, both perceived information security and privacy play significant roles as predictive factors for consumer choice behavior and the adoption patterns of mobile application innovations. This study thus contributes to the growing body of evidence, consistent with previous works such as Balapour et al. [58] and Akinola and Asaolu [52], highlighting the influential role of privacy and information security as precursory elements impacting the acceptance of mobile applications in rural markets. Moreover, the study provides evidence suggesting the applicability of the dimensions of trust and privacy in shaping entrepreneurs' decisions regarding the adoption of mobile applications. This aligns with previous research by Shaw et al. [54] and Su et al. [55]. Consequently, the present study addresses recent calls for the exploration of alternative theories, models, and insights pertaining to the innovation adoption behaviors of contemporary entrepreneurs, particularly within rural markets.

A number of notable adoption models have been identified in the literature [52, 53]. However, these models only appear to partially account for the phenomenon of choosing to adopt new marketing strategies. What appears to be lacking is the incorporation of external variables, such as trust and privacy, which could amplify the models' explanatory capacity, mitigate misinterpretations, and offer a more comprehensive understanding of the intricate aspects typically associated with research on the diffusion of mobile applications across diverse markets. In light of this, the study recommends the integration of a set of external variables into the innovation adoption models to enrich its explanatory capabilities. Particularly noteworthy is the association of trust and privacy with the success or failure of mobile application adoption, a point emphasized by researchers and practitioners alike [52–53, 55, 57]. This underscores the potential widespread applicability of trust and privacy concerns as external variables in various studies concerning the adoption of mobile applications.

5   Conclusion

This paper offers a systematic review of studies on rural enterprises' adoption strategies for mobile applications between January 2020 and April 2023. Understandably, many contributions have empirically explored trust and privacy issues on mobile application implementations in line with technology theories. This paper systematically reviewed trust and privacy issues, bringing out the trust issues as a discussion of external variables determining adoption for rural entrepreneurs. The paper also reviewed the privacy dimension (privacy concern, privacy principles, privacy enhancement, and privacy categories), which several scholars overlook in this discourse. Notably, the study still highlights a distance in the reviewed studies between their theoretical bases and analyses of trust and privacy issues on mobile application implementations in rural markets.

There are undoubtedly some shortcomings in this literature review. First, it does not cover the full spectrum of academic publications in the management, marketing, and entrepreneurship sectors because it is based on a selection of journals chosen based on specified criteria. Second, the timeline is restricted to the years 2020–2023. This constraint was addressed by the paper's attempt to correlate findings from contemporary literature with pertinent contributions that had been published prior to 2020. This paper calls for further research on related grounds. Firstly, focused research should empirically investigate the entrepreneurial trust and privacy issues on mobile application implementations in rural markets. Secondly, focused research which explains the difference between structured and unstructured crowds is also necessary.

References

1. Stocchi, L., Pourazad, N., Michaelidou, N., Tanusondjaja, A., Harrigan, P.: Marketing research on Mobile apps: past, present and future. J. Acad. Mark. Sci. **50**(2), 195–225 (2021)
2. Deepa, N., et al.: A survey on blockchain for big data: approaches, opportunities, and future directions. Future Gener. Comput. Syst. **131**, 209–226 (2022)
3. Mkansi, M., Nsakanda, A.L.: Mobile application e-grocery retail adoption challenges and coping strategies: a South African small and medium enterprises' perspective. Electron. Commer. Res. (2023). https://doi.org/10.1007/s10660-023-09698-1
4. Mkansi, M., de Leeuw, S., Amosun, O.: Mobile application supported urban-township e-grocery distribution. Int. J. Phys. Distrib. Logist. Manag. **50**, 26–53 (2020)
5. Abubakre, M., Faik, I., Mkansi, M.: Digital entrepreneurship and indigenous value systems: an Ubuntu perspective. Inf. Syst. J. **31**, 838–862 (2021)
6. Marnewick, C.: Information and communications technology adoption amongst township micro and small business: the case of Soweto. South Afr. J. Inf. Manag. **16**, 1–12 (2014)
7. Alavion, S.J., Taghdisi, A.: Analysis of rural e-marketing based on geographic model of planned behavior. Geogr. Space **20**, 57–84 (2021)
8. Alavion, S.J., Taghdisi, A.: Rural E-marketing in Iran; Modeling villagers' intention and clustering rural regions. Inf. Process. Agric. **8**, 105–133 (2021). https://doi.org/10.1016/j.inpa.2020.02.008
9. Bruwer, L.A., Madinga, N.W., Bundwini, N.: Smart shopping: the adoption of grocery shopping apps. Br. Food J. **124**, 1383–1399 (2022)
10. Saskia, S., Mareï, N., Blanquart, C.: Innovations in e-grocery and Logistics Solutions for Cities. Transp. Res. Procedia **12**, 825–835 (2016)

11. Kumar, V.: Mobile application in agriculture development in india: policy, practices and the way forward. In: Das, K., Mishra, B.S.P., Das, M. (eds.) The Digitalization Conundrum in India. ISBE, pp. 233–247. Springer, Singapore (2020). https://doi.org/10.1007/978-981-15-6907-4_13

12. Febrianda, R.: Mobile app technology adoption in indonesia's agricultural sector. an analysis of empirical view from public R&D agency. J. STI Policy Manag. **6**, 31–40 (2021)

13. Sisi, Z., Souri, A.: Blockchain technology for energy-aware mobile crowd sensing approaches in Internet of Things. Trans. Emerg. Telecommun. Technol. e4217 (2021). https://doi.org/10.1002/ett.4217

14. Shambare, R.: The adoption of WhatsApp: breaking the vicious cycle of technological poverty in South Africa. J. Econ. Behav. Stud. **6**, 542–550 (2014)

15. Rogers, E.M.: Diffusion of Innovations: modifications of a model for telecommunications. In: Stoetzer, M.W., Mahler, A. (eds.) Die Diffusion von Innovationen in der Telekommunikation, pp. 25–38. Springer Berlin Heidelberg, Berlin, Heidelberg (1995). https://doi.org/10.1007/978-3-642-79868-9_2

16. Goncalves, G., Oliveira, T., Cruz-Jesus, F.: Understanding individual-level digital divide: evidence of an African country. Comput. Hum. Behav. **87**, 276–291 (2018)

17. Meagher, P., Upadhyaya, K., Wilkinson, B.: Combating Rural Public Works Corruption: Food-for-Work Programs in Nepal. SSRN (2001). https://ssrn.com/abstract=260044. https://doi.org/10.2139/ssrn.260044

18. Faulkner, K.T., et al.: South africa's pathways of introduction and dispersal and how they have changed over time. In: van Wilgen, B., Measey, J., Richardson, D., Wilson, J., Zengeya, T. (eds.), Biological Invasions in South Africa. Invading Nature – Springer Series in Invasion Ecology, vol 14. Springer, Cham (2020). https://doi.org/10.1007/978-3-030-32394-3_12

19. Vasileiadou, E., Huijben, J.C.C.M., Raven, R.P.J.M.: Three is a crowd? Exploring the potential of crowdfunding for renewable energy in the Netherlands. J. Cleaner Production **128**, 142–155 (2016). https://doi.org/10.1016/j.jclepro.2015.06.028

20. Kumar, A., Sikdar, P., Gupta, M., Singh, P., Sinha, N.: Drivers of satisfaction and usage continuance in e-grocery retailing: a collaborative design supported perspective. J. Res. Interact. Market. **17**, 176–194 (2023)

21. Vahdat, A., Alizadeh, A., Quach, S., Hamelin, N.: Would you like to shop via mobile app technology? The technology acceptance model, social factors and purchase intention. Australasian Market. J. **29**(2), 187–197 (2021)

22. Venkatesh, V., Ramesh, V., Massey, A.P.: Understanding usability in mobile commerce. Commun. ACM **46**, 53–56 (2003)

23. Lai, P.C.: The literature review of technology adoption models and theories for the novelty technology. JISTEM-J. Inf. Syst. Technol. Manag. **14**, 21–38 (2017)

24. Ajzen, I., Fishbein, M.: A Bayesian analysis of attribution processes. Psychol. Bull. **82**, 261 (1975)

25. Manstead, A.S., Parker, D.: Evaluating and extending the theory of planned behaviour. Eur. Rev. Soc. Psychol. **6**, 69–95 (1995)

26. Ajzen, I.: The theory of planned behavior. Organ. Behav. Hum. Decis. Process. **50**, 179–211 (1991)

27. Venkatesh, V., Davis, F.D.: A model of the antecedents of perceived ease of use: development and test. Decis. Sci. **27**, 451–481 (1996)

28. Nguyen, T.T.M., Phan, T.H., Nguyen, H.L., Dang, T.K.T., Nguyen, N.D.: Antecedents of purchase intention toward organic food in an Asian emerging market: a study of urban Vietnamese consumers. Sustainability **11**, 4773 (2019)

29. Willie, M.M.: Customer Trust and Other Factors Associated with Digital Platforms & Marketing during a Pandemic. OAJBS Publishers (2022)

30. Lankton, N.K., McKnight, D.H., Tripp, J.: Technology, humanness, and trust: rethinking trust in technology. J. Assoc. Inf. Syst. **16**, 1 (2015)
31. Al-Azawei, A., Alowayr, A.: Predicting the intention to use and hedonic motivation for mobile learning: a comparative study in two Middle Eastern countries. Technol. Soc. **62**, 101325 (2020)
32. Miranda, I.T.P., Moletta, J., Pedroso, B., Pilatti, L.A., Picinin, C.T.: A review on green technology practices at BRICS countries: Brazil, Russia, India, China, and South Africa. SAGE Open **11**, 21582440211013780 (2021)
33. Tahaei, M., Frik, A., Vaniea, K.: Deciding on personalized ads: nudging developers about user privacy. In: SOUPS@ USENIX Security Symposium. pp. 573–596 (2021)
34. Nhamo, G., Mukonza, C.: Opportunities for women in the green economy and environmental sectors. Sustain. Dev. **28**, 823–832 (2020)
35. Randall, L., Ormstrup Vestergård, L., Wøien Meijer, M.: Rural perspectives on digital innovation: Experiences from small enterprises in the Nordic countries and Latvia (2020)
36. Khan, B., Mihovska, A., Prasad, R., Velez, F.J.: Overview of network slicing: Business and standards perspective for beyond 5g networks. IEEE Conf. Stand. Commun. Network. (CSCN) **2021**, 142–147 (2021)
37. Talwar, S., Kaur, P., Ahmed, U., Bilgihan, A., Dhir, A.: The dark side of convenience: how to reduce food waste induced by food delivery apps. Br. Food J. **125**, 205–225 (2023)
38. Hajian, G., Shahgholi Ghahfarokhi, B., Asadi Vasfi, M., Tork Ladani, B.: Privacy, trust, and secure rewarding in mobile crowd-sensing based spectrum monitoring. J. Ambient. Intell. Humaniz. Comput. **14**, 655–675 (2023)
39. Sharma, V., You, I., Andersson, K., Palmieri, F., Rehmani, M.H., Lim, J.: Security, privacy and trust for smart mobile-Internet of Things (M-IoT): a survey. IEEE Access **8**, 167123–167163 (2020)
40. Xu, G., et al.: Trust2Privacy: a novel fuzzy trust-to-privacy mechanism for mobile social networks. IEEE Wirel. Commun. **27**, 72–78 (2020)
41. Papaioannou, D., Sutton, A., Booth, A.: Systematic approaches to a successful literature review. Syst. Approaches Success. Lit. Rev. 1–336 (2016)
42. Howie, C.: Conducting your first systematic review. PsyPAG Q. Goes Electron. **113**, 32–35 (2019)
43. Khan, K.S., Kunz, R., Kleijnen, J., Antes, G.: Five steps to conducting a systematic review. J. R. Soc. Med. **96**, 118–121 (2003)
44. Page, M.J., et al.: The PRISMA 2020 statement: an updated guideline for reporting systematic reviews. Int. J. Surg. **88**, 105906 (2021)
45. Munn, Z., Peters, M.D., Stern, C., Tufanaru, C., McArthur, A., Aromataris, E.: Systematic review or scoping review? Guidance for authors when choosing between a systematic or scoping review approach. BMC Med. Res. Methodol. **18**, 1–7 (2018)
46. Delgado-Rodríguez, M., Sillero-Arenas, M.: Systematic review and meta-analysis. Medicina Intensiva (English Edition) **42**, 444–453 (2018)
47. Asti, W.P., Handayani, P.W., Azzahro, F.: Influence of trust, perceived value, and attitude on customers' repurchase intention for e-grocery. J. Food Prod. Mark. **27**, 157–171 (2021)
48. Gunawardana, P.K.A.T.D.R., Fernando, Imali: Does customer trust mediate the impact of e-service quality dimensions? Lessons during COVID-19 Pandemic. SSRN Electron. J. (2021). https://doi.org/10.2139/ssrn.3907878
49. Kurniasari, F., Ryadi, W.T.: Determinants of indonesian e-grocery shopping behavior after covid-19 pandemic using the technology acceptance model approach. United Int. J. Res. Technol. (UIJRT) **3**(01), 12–18 (2021)
50. Singh, P., Gupta, M., Kumar, A., Sikdar, P., Sinha, N.: E-Grocery retailing mobile application: discerning determinants of repatronage intentions in an emerging economy. Int. J. Hum.-Comput. Interact. **37**, 1783–1798 (2021)

51. Yang, Y., Liu, D., Zhang, L., Yin, Y.: Social trust and green technology innovation: evidence from listed firms in China. Sustainability **13**, 4828 (2021)

52. Akinola, O., Asaolu, O.: A trust, privacy and security model for e-commerce in Nigeria. Niger. J. Technol. **42**, 152–159 (2023)

53. Khaw, K.W., et al.: Modelling and evaluating trust in mobile commerce: a hybrid three stage Fuzzy Delphi, structural equation modeling, and neural network approach. Int. J. Human-Comput. Interact. **38**, 1529–1545 (2022)

54. Shaw, N., Eschenbrenner, B., Brand, B.M.: Towards a mobile app diffusion of innovations model: a multinational study of mobile wallet adoption. J. Retail. Consum. Serv. **64**, 102768 (2022)

55. Su, D.N., Nguyen, N.A.N., Nguyen, L.N.T., Luu, T.T., Nguyen-Phuoc, D.Q.: Modeling consumers' trust in mobile food delivery apps: perspectives of technology acceptance model, mobile service quality and personalization-privacy theory. J. Hosp. Mark. Manag. **31**, 535–569 (2022)

56. Libaque-Sáenz, C.F., Wong, S.F., Chang, Y., Bravo, E.R.: The effect of Fair information practices and data collection methods on privacy-related behaviors: a study of Mobile apps. Inf. Manage. **58**, 103284 (2021)

57. Tay, S.W., Teh, P.S., Payne, S.J.: Reasoning about privacy in mobile application install decisions: risk perception and framing. Int. J. Hum.-Comput. Stud. **145**, 102517 (2021)

58. Balapour, A., Nikkhah, H.R., Sabherwal, R.: Mobile application security: role of perceived privacy as the predictor of security perceptions. Int. J. Inf. Manag. **52**, 102063 (2020)

How Can Favorite Digital Services Enhance Users' Digital Well-Being? A Qualitative Study

Tiina Kemppainen[1](✉) [iD] and Tiina Paananen[2] [iD]

[1] School of Business and Economics, University of Jyväskylä, Jyväskylä, Finland
tiina.j.kemppainen@jyu.fi
[2] Faculty of Information Technology, University of Jyväskylä, Jyväskylä, Finland

Abstract. Digital services play a pivotal and multifaceted role in today's everyday life. Understanding digital services' impact on their users' well-being has thus become essential, and the concept of digital well-being has recently been introduced. This study aims to deepen our understanding of digital well-being by investigating how users' favorite digital services can enhance their digital well-being. The data were collected through 14 interviews. The interviewees were young Finnish adults (aged 22–31 years) who actively use digital services in their everyday lives. The findings demonstrate that favorite digital services can enhance their users' digital well-being at psychological, social, and cognitive levels. Digital services can contribute to psychological well-being by providing retreat, serenity, and enhancement to daily tasks. Digital services also contribute to social well-being by fostering connectedness and sense of unity. Finally, digital services enhance cognitive well-being by promoting knowledge and understanding and by offering inspiration.

Keywords: Digital well-being · Digital services · User perspective

1 Introduction

Digital services play a pivotal and multifaceted role in today's everyday life, permeating nearly every aspect of our routines and interactions. These services, delivered through digital channels like the internet and mobile devices, are now an inseparable part of work, leisure and interactions. Thus, digital services are increasingly crucial for affecting the well-being of their users—each engagement between a user and a digital service yields outcomes of varying nature. These outcomes might lean towards the favorable, as a digital service can facilitate communication or assist in managing timelines diligently. Conversely, there's also a possibility of unfavorable repercussions, such as a digital service promoting late-night messaging and lack of sleep. All in all, understanding digital services' impact on user wellbeing has become essential in today's technologically driven world.

The exploration of how services, digital services included, can impact the well-being of various stakeholders has gained momentum in diverse research areas over the past few years. There has been a demand to delve into the ways services can improve well-being

[1], particularly due to concerns that digital services negatively affect user well-being by causing addiction [2, 3], for instance. To tackle this issue, the concept of *digital well-being* [4, 5] has been lately introduced in the context of digital services. According to Burr et al. [4] digital well-being refers to an individual's well-being in an information society, encompassing the impact of digital technologies on an individual's well-being.

Given the prevalence of individuals increasingly using multiple devices and digital services, it's imperative to enhance our comprehension of digital well-being [5]. However, we currently have only a limited theoretical understanding of what digital well-being is [6] and what does it consist of. Hence, the objective of this study is to deepen our understanding of how digital service users' favorite digital services can contribute positively to their digital well-being. In this study, the concept of 'favorite digital services' refers to services that individuals actively use, prefer, and find valuable in their everyday lives. To offer services that genuinely promote user well-being, it is essential for both businesses and scholars to gain insights into how users perceive diverse digital services and the specific dimensions of well-being these services can effectively nurture.

The paper proceeds as follows. Firstly, we offer an exploration of digital services and digital well-being. Secondly, we detail the empirical approach that underpins our research. Lastly, we present the empirical outcomes, scrutinize their implications from both theoretical and practical perspectives, and propose avenues for further research.

2 Digital Services and Digital Well-Being

A digital service refers to a type of service primarily accessed through digital channels like the internet, or technology such as smartphones. This encompasses services such as e-books, music streaming, and online entertainment [7]. Digital services have been examined through various concepts including e-services [8] and online services [9], for instance.

Because the role of digital services as a part of people's everyday lives is constantly growing, recent studies have highlighted the potential impact of services on the well-being of their users and other stakeholders, such as employees and families. Chen et al. [10] note that a widely recognized measure of well-being revolves around subjective well-being (SWB) that encompasses users' emotional and cognitive evaluations of their lives and includes elements often referred to as happiness, tranquility, contentment, and overall life satisfaction [11]. In service research, the stream of transformative service research (TSR) has particularly delved into the well-being topic [1, 12, 13]. The goal of TSR is to understand how services can be transformed from solely transactional to more meaningful and impactful, leading to positive outcomes for all stakeholders involved, including users and employees' (micro-level); service organizations' (meso-level); and service industries' wellbeing (macro-level) [1].

Prior studies indicate that digital services can contribute to user well-being. It has been concluded that the impact can be either positive or negative: services can either enhance or diminish users' well-being [7]. For instance, Reinecke and Hofmann [6] found a positive association between users' enjoyment of digital media and their subjective well-being. Jin and Li [14] noted that video games incorporating social elements, like interactions with characters controlled by real individuals or nonplayer characters, can enhance players' well-being. Regarding the negative impacts Salo et al. [15] determined that social networking sites and services can trigger technostress, which can negatively impact user well-being. Likewise, Brooks [16] discovered that increased personal usage of social media elevated technostress levels and reduced happiness.

However, although the well-being topic has sparked interest across various research fields, deficiencies can also be identified within the studies. First, TSR mostly centers on services whose transformative nature is self-evident. Studies have focused to health-care, non-profit, and social services sectors, where the wellbeing of all stakeholders is the main concern and outcome [4]. In other research avenues, including Information Systems studies, the current body of literature concerning digital services and user well-being demonstrates a strong emphasis on service type and application specificity. Researchers have examined strategies to encourage well-being, placing particular emphasis on appli-cations or app features aimed at heightening users' consciousness of their smart device usage duration when engaging with social networks, for instance [17]. Themes such as the impacts of fitness app features on user well-being [6], well-being among mHealth users [7], users' well-being in virtual medical tourism communities [8], and Smart-phone use wellbeing [9] have been investigated. Due to these focuses, there has been limited scrutiny of digital services that are preferred and frequently used by individuals during daily life and their impact on digital well-being. In addition, as Vanden Adeele [10] notes, the understanding of what constitutes digital well-being and the methods to achieve it still lack clarity. Monge Roffarello et al. [5] support this idea by noting that a comprehensive definition of digital well-being has not been established and a broader understanding is needed. For instance, considering interactions that involve multiple devices and interactions across devices is essential for achieving a more holistic grasp of digital well-being [5]. To sum up, there exists a distinct need to explore the interplay between different everyday digital services and their contribution to users' well-being. The purpose of this study is therefore to enhance the understanding of how users' favorite digital services can contribute to their digital well-being in a positive manner.

3   Methodology

As the purpose of this study is to increase the understanding of how favorite digital services can contribute to users' digital well-being positively, the participants for the research were obtained through purposive sampling. Purposive sampling entails choos-ing participants who possess significant knowledge or experience related to a specific phenomenon of interest [18]. In our case, the focus was on digital services and their role in users' well-being. Purposive sampling is common in qualitative research, aiming to understand experiences, perspectives, and behaviors within a specific group.

The participants were recruited by using Instagram story posts, referrals, and direct contact with those meeting the criteria of significant experience with digital services. Young adults (22–31 years old) were recruited as they are known for their active digital service use and experience. All interviews were conducted in Finland by the second author. A sample size of 14 semi-structured one-on-one interviews (with 7 females & 7 males) was determined through analytic saturation. 13 interviews were carried out through Zoom video conferencing software and the remaining interview was conducted in person due to the interviewee's preference. The interviews lasted around 41 min on average. The interviewees' background information and the durations of the interviews can be found in the Appendix.

The interviews encompassed three primary themes. These themes centered on the utilization of digital services (behavior), the participants' perceptions regarding digital services (cognitive aspects), and the emotional responses evoked by using digital services (emotions). Instead of directly inquiring about the effects of digital services on well-being, we took an indirect approach by employing questions through which the participants could reflect on the topic freely, through their own perspectives and experiences. This approach enabled us to gain multifaceted insights into how digital services can influence users' well-being from various perspectives. The participants were encouraged to name and discuss their favored and most valued digital services without any specific suggestions from the interviewer, ensuring they freely shared the services they personally deemed most significant. The participants' preferred digital services spanned various categories, primarily including streaming, gaming, instant messaging, and social media services (for more details, see Appendix).

The transcribed written data underwent analysis using a thematic approach, following the framework presented by Braun and Clarke [19]. Thematic analysis involves recognizing, scrutinizing, and presenting recurring themes that hold relevance to the research objectives. To achieve a user-centric comprehension, an inductive method was employed [20]. This approach involves transforming data into categories without any theoretical pre-assumptions, which are then structured into a framework to capture crucial themes deemed significant by the researcher [20].

The initial phase involved thoroughly reading and acquainting ourselves with the interview transcripts to develop a comprehensive grasp of the data. Subsequently, the data was imported into the NVivo qualitative data analysis software for further analysis. The generation of initial codes marked the second step. Initial codes were determined by pinpointing all relevant phrases and statements in which the participants discussed the positive aspects of digital services. Tentative labels were assigned to these statements. The subsequent step encompassed categorizing the statements into main themes and subthemes based on their content and significance. Ultimately, our analysis identified three primary well-being levels (psychological, social, and cognitive) that digital services can positively contribute to, each featuring distinctive subthemes. These themes are discussed next in more detail.

4 Findings

The findings of this study demonstrate that favorite digital services can enhance their users' digital well-being at psychological, social, and cognitive levels. Digital services can contribute to psychological well-being by providing retreat, serenity, and enhancement to daily tasks. Digital services also contribute to social well-being by fostering connectedness and sense of unity. Finally, digital services enhance cognitive well-being by promoting knowledge and understanding and offering inspiration.

4.1 Psychological Level

Digital services support users' digital well-being at psychological level by providing retreat, serenity, and enhancement to daily tasks. Hence, the psychological level here refers to the well-being of the user's mind, and comfort and enjoyment during everyday life. It encompasses factors that are related to stress management, coping with challenges, and maintaining a positive psychological state during everyday chores and tasks in the physical world.

Retreat. Digital services act as a retreat by providing users with a virtual space where they can temporarily escape from the physical world. Particularly, the favored digital services were perceived as avenues for breaking away from the routine monotony of daily life and moments of waiting. Digital services divert attention from unfavorable physical situations, and offer entertainment for moments where the physical world cannot provide sufficiently interesting content.

> "I make quick visits to them. [...] If you have two minutes of waiting, like if you're waiting in a doctor's reception area, well then you naturally start browsing, because there's always something to read and watch in them." -P2

> "To eliminate boredom... As soon as you get bored or have nothing to do, then you will probably browse." -P9

Serenity. Digital services provide serenity by providing users with a virtual space where they can escape from their everyday routines and find relaxation. The findings indicate that digital services can function as a mechanism for coping, empowering users to adeptly handle their well-being. Activities like movie-watching, music-listening, or gaming have the potential to alleviate stress and pressure, affording a respite and recovery from the challenges and obligations of daily life.

> "Streaming services offer a means to reset your mind. [...] A chance to clear your thoughts. [...] In the evenings, when exhaustion from work sets in and I find myself passive and unable to undertake anything else, I turn to browsing Instagram." -P4

> "Netflix is a perfect sleep aid. You watch it and fall asleep." -P14

Enhancement. Digital services make many obligatory daily tasks easier (e.g., banking, shopping) and thereby enhance the quality of life and well-being. In addition, digital services add an element of enjoyment and pleasure to routine tasks, making them more engaging and satisfying. Digital services have the capacity to elevate one's mood, frequently acting as a source of upliftment during daily responsibilities. Notably, music streaming services played a vital role in this context. Engaging with music can foster positive emotions even during less desirable chores like cleaning, and it can heighten the enjoyment of physical exercise, for example. Through playlists, various moods can be created in the physical world. Additionally, participants regarded streaming services and audiobooks as valuable additions to their everyday lives; also these services inject flavor and enrichment into the physical space and chores.

"The (digital service) must be useful, using it should not be waste of time. And it should solve a problem or at least make something easier. [...] There has to be some aspect that improves or enhances the quality of life. For example, online banking eases things at the mobile app level, making usage much simpler without having to wait for the computer to start up or search for credentials. Online banking in a browser also makes it easier, so you don't have to go to the bank." -P14

"I have a lot of playlists. There are dozens of them. They're for different situations—there's basic Finnish pop music, workout music, running music, and so on. I think there are dozens of hours of playlists." -P3

"I'm not really that attached to Netflix, but it's still nice to have as background entertainment and background noise to put on if I'm working, having something playing so there's a bit of noise." -P2

4.2 Social Level

The findings reveal that digital services support digital well-being by enhancing connectedness and sense of unity. The social level here refers to the individual's social connections, relationships, and interactions within their community and the extent to which a person feels connected, valued, and integrated into their social environment.

Connectedness. Digital services offer a chance to enhance connectedness and engaging with others, regardless of physical barriers. Digital services like WhatsApp, Snapchat, Instagram, and Tinder play a pivotal role in enabling social interaction and communication and facilitate the establishment and maintenance of connections and relationships. Maintaining connections with those who may not be very familiar, and with whom one might not necessarily engage in real-world conversations, is also facilitated. For instance, Instagram stories provide an easy and accessible way to 'get to know' or 'further acquaint oneself' with acquaintances or friends of friends, expanding one's social circle and engaging in social interactions.

"Almost all communication goes through WhatsApp. Having those friend groups there is really important, and then there's a separate group for family discussions." -P2

"I can comment on what others are doing, even if we would never call each other. The threshold to keep in touch with different people is lower." -P6

Sense of Unity. Digital services facilitate shared experiences that contribute to a feeling of unity and connection among people. This involves establishing a sensation of connection, a shared community, and mutual encounters that users can enjoy via digital platforms. The participants observed that digital services nurture the sense of community by attracting numerous like-minded users and acquaintances to the same platforms. For instance, a communication platform Discord was recognized as a significant catalyst for community bonds, as it enables people to congregate and engage socially. Similar to how users gather in physical places for social interaction, Discord offers a digital equivalent where users can convene and participate in a communal setting. Sharing one's thoughts and perspectives with like-minded individuals also emerged as an important theme in terms of digital well-being. Through digital services, it is easy to find like-minded individuals and even friends, and engage in discussions on topics of shared interest among participants.

> "Discord is a bit like a marketplace; you can find people and different groups where you enjoy hanging out. So, it's indeed a very important tool for socializing at the moment." -P1

> "You can share your own things there and everyone else is also there; you get a sense of community through having friends there." -P5

4.3　Cognitive Level

The findings indicate that digital services contribute to digital well-being by fostering knowledge and understanding and by offering inspiration. Hence, cognitive level refers to the state of an individual's engagement with cognitive pursuits and curiosity.

Knowledge and Understanding. Digital services enhance knowledge and understanding through various tools, resources, and platforms that provide engaging and interactive educational content. Digital services enable users to access up-to-date information, news, and insights, supporting their quest for broadening one's knowledge. For instance, participants highlighted the value of news outlets and search engines in keeping them informed about global events. Moreover, digital services aid critical thinking by offering diverse perspectives and resources for work and educational tasks. Participants emphasized the utility of platforms like YouTube for learning complex subjects.

> "YouTube is a platform with meaningful content. For example, you can find help for schoolwork to some extent, especially when there are quite difficult topics...explanations." -P7

On the other hand, through digital services, one can learn about any personal area of interest, not just through educational, written, or news content. Digital services offer a snapshot of the current trends and materials that users can reflect on and further analyze themselves. For example, whereas advertising might appear as an annoying aspect of digital services to many users, it can also provide educational and interesting content to follow.

> "Because I'm interested in advertising, I gladly look at the types of ads that have been created, and on Instagram, I follow different brands that do good advertising. I voluntarily follow advertisements." -P1

Inspiration. Digital services serve as sources of inspiration; discovery of ideas, creativity, or a sense of motivation. For instance, participants recognized that observing the activities and posts of others, including dance videos or Facebook and Instagram posts, serve as a well of inspiration where one can get tips for one's everyday life in different areas. Digital services and the content presented in them have the potential to encourage individuals to explore new activities or acquire skills they might not have been inclined to pursue otherwise.

"Especially from those Facebook groups, you can get ideas and inspiration [...] You see what others are doing and what else I could do myself." -P5

"Inspiration, there's entertaining content, even humor. You see different people and their ways of behaving. In general, I really enjoy videos where, for example, there's dancing or something related to photography or other things like that." -P6

5 Discussion and Conclusions

The role of digital services in people's lives, and consequently their significance in individuals' well-being, has grown significantly during the past decade. Consequently, digital well-being has been increasingly studied [4, 5, 17]. However, despite the growing focus across various research areas in recent years, there are identifiable shortcomings within these studies. For instance, the stream of transformative service research has predominantly focused on services characterized by inherently transformative qualities. In the field of Information Systems research, there has been a notable emphasis on specific service types and applications. Because of these emphases, everyday digital services and their role in digital well-being has received limited examination. In addition, it has been noted that the understanding of what constitutes digital well-being still lacks clarity [21] and there is also need for a broader understanding of the digital well-being concept [5].

Because of the above viewpoints, the objective of this study was to deepen the understanding of how users' favorite digital services can enhance their digital well-being. This study contributes to the existing literature by exploring the different levels of well-being favorite digital services can contribute to. The findings demonstrate that these digital services can enhance users' digital well-being in psychological, social, and cognitive levels.

First, the findings indicate that digital services can enhance users' digital well-being at the psychological level by acting as a retreat and letting their users briefly escape their physical surroundings. Preferred services offer an escape from daily routines and moments of waiting, providing entertainment and stimulus for unstimulating moments and environments. Digital services also offer serenity, allowing users to find relaxation. They serve as coping mechanisms, relieving stress through activities such as movie-watching, music-listening, and gaming. Digital services also enhance routine tasks with enjoyment. They can uplift their users' mood and engagement in daily responsibilities. Prior studies have also presented similar findings in different service contexts. Lukoff et al. [21], for instance, discovered that individuals utilized their mobile phones as a type of micro-escape, using them to shift their focus away from difficult emotions or situations they were encountering. Also, an increasing volume of research has showcased the effectiveness of media consumption as a viable recovery strategy that is capable of evoking a comprehensive range of recovery experiences [6]. As an example, Reinecke and Hofmann [4] found a positive association between users' enjoyment of digital media and their subjective well-being. Considering digital service management and design, and future research, these discoveries prompt contemplation on how digital services can be made even more immersive and more supportive of psychological well-being, blurring the boundaries between the physical and digital worlds. Technological advancements will undoubtedly bring new and intriguing applications related to this theme in the near future. Already now devices such as virtual reality (VR) glasses have the potential to offer immersive encounters, creating a sensation of actually being present within a simulated environment. An important question, of course, is also the threats and potential illbeing the immersion of the physical and digital worlds can cause.

Second, the findings indicate that digital services can enhance users' digital well-being at the social level by facilitating connections and by transcending physical barriers, enabling collaboration and interaction. Digital services foster relationships with both familiar and unfamiliar individuals. In addition, digital services can foster unity and connection by creating a sense of community and shared encounters. Digital services draw like-minded users together and nurture community feel. The social dimension of digital services has also been identified in previous studies. It has been noted, as an example, that video games incorporating social elements, like interactions with characters or other players, can enhance players' well-being [14, 22]. Altogether, these findings emphasize the importance of the social aspects of digital services. People are spending increasingly more time with digital services, and in some cases, digital services can even provide the only means of connecting with others. Therefore, the potential of digital services to facilitate social interaction is a crucial question from the perspective of digital well-being. An important question for digital service researchers and practitioners is how to make these services more interactive and how to incorporate elements that enable them to offer similar interaction to real-world situations.

Finally, the findings indicate that digital services can enhance users' digital well-being at the cognitive level by offering chances for both knowledge and understanding and inspiration. Digital services offer convenient access to up-to-date information for various knowledge and understanding purposes, such as keeping oneself informed about the latest developments in a particular field or area of interest. Also, by exposing users to diverse content and providing external stimuli, digital services can become valuable sources of inspiration; they arouse creative thinking, new ideas and motivation, and thus, fuel personal growth and development. In previous studies, the potential of everyday digital services from the perspective of knowledge, understanding and inspiration has not been extensively examined. Previous research has primarily investigated the utilization of digital tools within educational contexts. These studies have demonstrated that digital services have the potential to establish inspiring learning settings for students. For instance, TikTok can be used as an educational instrument that boosts creativity and nurtures curiosity [23], and Twitter can provide a wealth of opportunities for learning, engagement, discovery, and creativity [24]. In the future, based on these viewpoints, it is good to consider how mundane digital services could be developed to better support users' learning and inspiration – for example, what kind of user interfaces or presentation tactics best enhance users' cognitive pursuits.

To sum up, the study's findings underscore the need for further research into the dimensions of digital well-being to establish more consistent and reliable definitions and measures for these dimensions. The findings of this study are based on interviews with 14 young Finnish adults. The utilization of purposive sampling within the technologically proficient age group of 22–31 may not yield findings directly applicable to other age groups. Hence, further research related to the topic is needed with larger sample sizes, different kinds of users and various research methods. Also, users' favorite digital services including different types of services were examined in this study. In the future, digital well-being could be examined focusing on specific types of preferred services, such as gaming or streaming services. An interesting research topic is, for example, what kind of well-being certain types of digital services promote or what kind of well-being differences can be identified between different types of digital services.

In addition, as the usage of digital services continues to increase and digital and physical environments are constantly becoming more unified "metaverses", it is imperative to examine the impact of digital services on users' well-being from diverse perspectives. Paying attention to digital well-being is important from the perspective of individuals but also on a broader societal scale, in order to prevent the potential adverse effects of digital services and promote the effects that can enhance well-being. From a practical standpoint, the insights offered by this study prompt digital service providers to consider the different well-being aspects of their services. Given that digital services can influence their users' well-being positively but also negatively, it becomes crucial for service providers to assess the effects of their service and deliberate on the desired associations. From an ethical and sustainable business standpoint, promoting the positive outcomes and mitigating the negative ones is, of course, advisable. Achieving this requires a comprehensive evaluation of both intended and unintended consequences arising from the design choices of digital services.

Appendix. Participants' Background Information

	Age	Gender	Status	Favorite digital services	Interview Duration (min)
P1	22	Male	Student	Spotify, Discord, YouTube	42
P2	29	Female	Employee	Netflix, Spotify, Microsoft Gamepass	38
P3	26	Female	Student	Instagram, Spotify, WhatsApp	41
P4	30	Female	Student	Spotify, Instagram	52
P5	27	Female	Employee	Facebook, Instagram, Netflix, WhatsApp	51
P6	26	Female	Employee	Instagram, OP-mobile, WhatsApp	40
P7	25	Male	Student	Snapchat, WhatsApp, YouTube	23
P8	31	Male	Employee	Spotify, BookBeat, Netflix	57

(continued)

(continued)

	Age	Gender	Status	Favorite digital services	Interview Duration (min)
P9	25	Male	Student	Reddit, YouTube, Twitch	34
P10	24	Male	Student	Spotify, Discord, Twitch	42
P11	23	Female	Employee	YouTube, Discord, Counter-Strike: Global Offensive	35
P12	26	Male	Entrepreneur	Google, Gmail, Counter-Strike: Global Offensive	31
P13	31	Female	Student	Instagram, Spotify, WhatsApp	37
P14	28	Male	Employee	WhatsApp, Netflix	53

References

1. Anderson, A., Ostrom, L.: Transformative service research: advancing our knowledge about service and well-being. J. Serv. Res. **18**, 243–249 (2015). https://doi.org/10.1177/10946705155913
2. Kuss, D.J., Griffiths, M.D.: Social networking sites and addiction: ten lessons learned. Int. J. Environ. Res. Public Health **14**, 311 (2017). https://doi.org/10.3390/ijerph14030311
3. Hou, Y., Xiong, D., Jiang, T., Song, L., Wang, Q.: Social media addiction: Its impact, mediation, and intervention. Cyberpsychology J. Psychosoc. Res. Cyberspace (2019). https://doi.org/10.5817/CP2019-1-4
4. Burr, C., Taddeo, M., Floridi, L.: The ethics of digital well-being: a thematic review. Sci. Eng. Ethics **26**, 2313–2343 (2020). https://doi.org/10.1007/s11948-020-00175-8
5. Roffarello, A.M., De Russis, L., Lottridge, D., Cecchinato, M.E.: Understanding digital well-being within complex technological contexts. Int. J. Hum.-Comput. Stud. **175**, 103034 (2023). https://doi.org/10.1016/j.ijhcs.2023.103034
6. Reinecke, L., Hofmann, W.: Slacking off or winding down? an experience sampling study on the drivers and consequences of media use for recovery versus procrastination. Hum. Commun. Res. **42**, 441–461 (2016). https://doi.org/10.1111/hcre.12082
7. Rosenbaum, M.S., Russell-Bennett, R.: Editorial: when service technologies and human experiences intersect. J. Serv. Mark. **35**, 261–264 (2021). https://doi.org/10.1108/JSM-03-2021-0096
8. Kalia, P., Arora, D.R., Kumalo, S.: E-service quality, consumer satisfaction and future purchase intentions in e-retail. E-Serv. J. **10**, 24–41 (2016). https://doi.org/10.2979/eservicej.10.1.02
9. Adil, M., Sadiq, M., Jebarajakirthy, C., Maseeh, H.I., Sangroya, D., Bharti, K.: Online service failure: antecedents, moderators and consequences. J. Serv. Theory Pract. **32**, 797–842 (2022). https://doi.org/10.1108/JSTP-01-2022-0019
10. Chen, X.M.S., Schuster, L., Luck, E.: The well-being outcomes of multi-actor inter-organisational value co-creation and co-destruction within a service ecosystem. J. Serv. Mark. ahead-of-print (2023). https://doi.org/10.1108/JSM-03-2022-0082

11. Diener, E., Oishi, S., Lucas, R.E.: Personality, culture, and subjective well-being: emotional and cognitive evaluations of life. Annu. Rev. Psychol. **54**, 403–425 (2003). https://doi.org/10.1146/annurev.psych.54.101601.145056

12. Anderson, S., Nasr, L., Rayburn, S.W.: Transformative service research and service design: synergistic effects in healthcare. Serv. Ind. J. **38**, 99–113 (2018). https://doi.org/10.1080/02642069.2017.1404579

13. Parkinson, J., Mulcahy, R.F., Schuster, L., Taiminen, H.: A transformative value co-creation framework for online services. J. Serv. Theory Pract. **29**, 353–374 (2019). https://doi.org/10.1108/JSTP-04-2018-0098

14. Jin, Y., Li, J.: When newbies and veterans play together: the effect of video game content, context and experience on cooperation. Comput. Hum. Behav. **68**, 556–563 (2017). https://doi.org/10.1016/j.chb.2016.11.059

15. Salo, M., Pirkkalainen, H., Koskelainen, T.: Technostress and social networking services: uncovering strains and their underlying stressors. In: Stigberg, S., Karlsen, J., Holone, H., Linnes, C. (eds.) Nordic Contributions in IS Research, pp. 41–53. Springer International Publishing, Cham (2017). https://doi.org/10.1007/978-3-319-64695-4_4

16. Brooks, S.: Does personal social media usage affect efficiency and well-being? Comput. Hum. Behav. **46**, 26–37 (2015). https://doi.org/10.1016/j.chb.2014.12.053

17. Gennari, R., Matera, M., Morra, D., Melonio, A., Rizvi, M.: Design for social digital well-being with young generations: engage them and make them reflect. Int. J. Hum.-Comput. Stud. **173**, 103006 (2023). https://doi.org/10.1016/j.ijhcs.2023.103006

18. Etikan, I.: Comparison of convenience sampling and purposive sampling. Am. J. Theoret. Appl. Stat. **5**(1), 1 (2016). https://doi.org/10.11648/j.ajtas.20160501.11

19. Braun, V., Clarke, V.: Using thematic analysis in psychology. Qual. Res. Psychol. **3**, 77–101 (2006). https://doi.org/10.1191/1478088706qp063oa

20. Thomas, D.R.: A general inductive approach for analyzing qualitative evaluation data. Am. J. Eval. **27**, 237–246 (2006). https://doi.org/10.1177/1098214005283748

21. Lukoff, K., Yu, C., Kientz, J., Hiniker, A.: What makes smartphone use meaningful or meaningless? Proc. ACM Interact. Mobile Wearable and Ubiquitous Technol. **2**(1), 1–26 (2018). https://doi.org/10.1145/3191754

22. Halbrook, Y.J., O'Donnell, A.T., Msetfi, R.M.: When and how video games can be good: a review of the positive effects of video games on well-being. Perspect. Psychol. Sci. **14**, 1096–1104 (2019). https://doi.org/10.1177/1745691619863807

23. Escamilla-Fajardo, P., Alguacil, M., López-Carril, S.: Incorporating TikTok in higher education: Pedagogical perspectives from a corporal expression sport sciences course. J. Hosp. Leis. Sport Tour. Educ. **28**, 100302 (2021). https://doi.org/10.1016/j.jhlste.2021.100302

24. Marr, J., DeWaele, C.S.: Incorporating twitter within the sport management classroom: rules and uses for effective practical application. J. Hosp. Leis. Sport Tour. Educ. **17**, 1–4 (2015). https://doi.org/10.1016/j.jhlste.2015.05.001

Utilizing Degree Centrality Measures for Product Advertisement in Social Networks

Manoj Kumar Srivastav[1], Somsubhra Gupta[1], V. M. Priyadharshini[2], Subhranil Som[3], Biswaranjan Acharya[4], Vassilis C. Gerogiannis[5], Andreas Kanavos[6], and Ioannis Karamitsos[7](✉)

[1] School of Computer Science, Swami Vivekananda University, Barrackpore, West Bengal, India
[2] Department of IT, BIT Campus, Anna University, Trichirappalli, Tami Nadu, India
[3] Department of Computer Science, Bhairab Ganguly College, Kolkata, West Bengal, India
[4] Department of Computer Engineering, AI and BDA, Marwadi University, Rajkot, Gujarat, India
biswaranjan.acharya@marwadieducation.edu.in
[5] Department of Digital Systems, University of Thessaly, 41500 Larissa, Greece
[6] Department of Informatics, Ionian University, Corfu, Greece
akanavos@ionio.gr
[7] Graduate Programs and Research, Rochester Institute of Technology, 341055 Dubai, United Arab Emirates
ixkcad1@rit.edu

Abstract. Social networks, as abstract representations of relationships between entities, play a pivotal role in connecting individuals in the digital age. This paper delves into the realm of social network analysis (SNA), a method rooted in graph theory that explores the dynamics of social relationships within communities. One of the key objectives of SNA is to identify the most influential actors within a social network, a task often achieved by calculating various centrality metrics. These metrics, such as degree centrality, allow to quantify the significance and impact of individual nodes within a social network. In the context of marketing and brand promotion, these metrics are particularly relevant and useful. Leveraging social networks for marketing endeavors can enhance brand recognition and foster customer loyalty. When promoting products within social networks, targeting the "most important" members (i.e., those with higher centrality metrics) can exponentially increase the reach and impact of the marketing campaign. In this paper, an approach is suggested for supporting social network marketing by employing binary logistic regression analysis. Logistic regression is a valuable tool in case of models where the dependent variable is dichotomous, and it can be an ideal method for predicting node behavior in the context of a product promotion campaign. By analyzing the actions of the "most important" nodes within social networks, we can predict which nodes are likely to purchase a marketed product based on their interactions and centrality.

Keywords: Social Networks · Social Network Analysis (SNA) · Online Communities · Centrality Measures · Influence Analysis · Logistic Regression

© The Author(s), under exclusive license to Springer Nature Switzerland AG 2024
M. Papadaki et al. (Eds.): EMCIS 2023, LNBIP 502, pp. 77–91, 2024.
https://doi.org/10.1007/978-3-031-56481-9_6

1 Introduction

Social Network Analysis (SNA) is the study of social structures through the lens of a social network, consisting of actors and their relationships [21]. With the growing user base on social networks, individuals form various communities, and some even assume administrative roles within them [11, 12]. These communities vary in terms of governance, with administrators controlling some or allowing member participation.

In the realm of social networks, communication patterns differ—some communities enable unrestricted communication among members, while others impose restrictions. Messages exchanged within these communities may receive responses or remain unanswered, potentially influencing a member's standing within the social network [20]. Understanding members' positions within these networks is vital for harnessing influence [20].

The underlying foundation of social networks lies in graph theory, where nodes represent entities, and edges denote their connections [8, 13]. A network is defined as an ordered pair (V, E), where V represents nodes (or entities), and E represents edges (or relationships). Depending on the nature of these relationships, networks can be directed or undirected, with edges possibly carrying associated weights. In the context of social networks, we can define them as collections of socially connected elements, represented as the set S = social elements: social elements are connected. Within a social network S, let's consider two nodes: N_1 and N_2. We can introduce a function f: $N_1 \rightarrow N_2$ to represent message exchange between them, illustrating the set of messages sent by N_1 and received by N_2 [18].

Communities within a graph are subsets of nodes that exhibit denser connections among themselves than with the rest of the network. This concept parallels social circles formed within social networks like Facebook, Instagram, and Twitter, where members aim to connect and expand their social circles [20].

Logistic regression, a predictive analysis technique, plays a crucial role in understanding collaboration dynamics in product promotion. Businesses leverage social network analysis (SNA) to identify effective promotion strategies, often targeting members with the most followers to maximize their reach [17]. SNA represents networks through graphs, where actors are nodes connected by edges. Graph theory quantifies distances between nodes by measuring the number of edges traversed to reach one from another. Identifying the most influential nodes is a primary application of graph theory in social network analysis.

Social networking serves various purposes, including social and commercial motives. Marketers utilize these platforms to engage with customers, promote products and services, address inquiries, and bolster brand recognition. The largest or most influential nodes often occupy strategic positions within the network. In particular, the centrality degree of each node is crucial, signifying the node's central role within the network. Highly central nodes tend to be more active due to their numerous connections [14].

Our work contributes to a nuanced understanding of the pivotal role played by social networks in shaping modern marketing strategies. Furthermore, we propose potential avenues for extending this research, encompassing the exploration of more intricate relationships between multiple independent variables and dichotomous dependent variables.

This paper, therefore, serves as a valuable resource for professionals and researchers seeking to navigate the evolving landscape of social network-driven marketing strategies.

This paper presents a comprehensive exploration of the intricate relationship between social network dynamics and marketing outcomes. Section 2 reviews existing research on social network marketing strategies and the role of degree centrality in amplifying brand values. Building on this foundation, Sect. 3 delves deeper into the significance of degree centrality and other centrality measures, establishing their critical relevance in the context of social network advertising. Sect. 4 demonstrates the effectiveness of logistic regression in dissecting the intricate relationships between the dynamics of social network marketing and the outcomes for business entities. Section 5 offers a detailed analysis of our findings, showcasing how the binary logistic regression model performs in practical scenarios. Finally, in Sect. 6, we consolidate our findings and outline future research directions.

2 Related Work

Centrality in social networks serves as a quantitative measure to unveil the importance of network nodes. A centrality measure is a real-valued function applied to nodes within a graph, providing insights into their significance. Generally, centrality metrics help determine an individual's influence within a social network, whereas some centrality measures consider the importance of nodes based on geometric factors, such as their distance within the network.

In the field of media advertising and social media, researchers have explored various aspects of advertising effectiveness. Authors in [16] delve into the factors contributing to media advertising and explore its associated benefits. Similarly, a method that assesses effectiveness based on four critical dimensions: empathy, persuasion, impact, and communication is proposed in [19].

In terms of social media advertising appeals, the work in [22] investigates the effects of different appeals (negative, positive, and coactive) on online engagement and prosocial behavior. In the realm of social network analysis, authors in [10] present an automated method for analyzing Twitter user profiles within specific communities.

Moreover, a model that integrates business intelligence with social media networks is explored in [3]. This model is designed to analyze the substantial data provided by social network platforms, assisting marketers in making informed decisions about their advertisements. Another research on enhancing brand value based on social networks focuses on the promotion path of brand value within social network services [23].

These studies collectively contribute to our understanding of media advertising, social media ads, social advertising appeals, social network analysis, and the integration of business intelligence with social networks.

Additionally, in the context of social network analysis, various centrality measures are commonly employed to identify key players within social networks [4]. These measures aid in recognizing the most influential nodes, which are pivotal in the evolving marketing strategy of collaboration between brands and influential nodes. These influential nodes, often possessing significant online followings, have the potential to influence their target market's purchasing decisions, making them valuable assets for businesses.

Furthermore, logistic regression, a statistical analysis method frequently used in machine learning, plays a crucial role in predicting outcomes when the dependent variable is binary or dichotomous. For instance, logistic regression can be employed to predict whether a product will be purchased after being advertised through a social network platform.

3   Centrality Measures in Social Networks

3.1   Graph Representation

A graph is formally represented as a pair of sets (V, E), where V denotes the set of vertices, and E represents the multiset of edges connecting pairs of vertices. Typically, a graph is denoted as $G(V, E)$ or $G = (V, E)$ [8].

3.2   Degree

Degree, denoted as $deg(v)$, characterizes the number of connections a node v has in an undirected and unweighted network G. In simpler terms, it signifies the number of edges linked to a node in such a network [8]. For example, your degree in a social network like Facebook corresponds to the number of friends you have. If two nodes in an undirected network share an edge, they are referred to as neighbors. Therefore, the degree of a node in an undirected and unweighted network equals the count of its neighbors.

3.3   Centrality

Centrality measures the level of connectedness of a node within a network and quantifies its direct influence within its local neighborhood [7].

3.4   Degree Centrality

Degree centrality is pivotal because it emphasizes that the most active nodes must possess the most connections within the network [15].

The degree centrality of a node n_i is mathematically expressed as:

$$CD(n_i) = \frac{d(n_i)}{g - 1} \tag{1}$$

where $CD(n_i)$ represents the degree centrality of node n_i, $d(n_i)$ is the number of connections of node n_i and g signifies the total number of nodes in the network. A higher degree centrality implies greater activity within the network, granting the node the power to influence other nodes or acquire additional information.

3.5 Closeness Centrality

Closeness centrality measures how close a particular node is to the rest of the network. It quantifies this by calculating the reciprocal of the mean shortest path from the node to all other connected nodes in the network. Mathematically, the closeness centrality of a node u in a network $G(V, E)$ is defined as:

$$Cc(v) \frac{|V| - 1}{\sum d_{u,v} \quad \text{where } u \in V, u|=v} \tag{2}$$

where $d_{u,v}$ indicates the length of the shortest path between node u and v, and $|V|$ represents the number of nodes in the graph [14].

Closeness centrality is based on the proximity of an actor to all other actors in the set. If an actor can rapidly interact with all other actors, it possesses a high degree centrality and is less reliant on others for information relay.

3.6 Betweenness Centrality

Betweenness centrality measures how central a node is in terms of bridging paths within the network. It quantifies how many shortest paths pass through a particular node [14]. The betweenness centrality of node u in a network $G(V, E)$ is defined as:

$$Cb(v) = \sum_{x,y \in V, x \neq v \neq y} \frac{\sigma_{xy}(v)}{\sigma_{xy}} \tag{3}$$

where σ_{xy} denotes the number of shortest paths between nodes x and y in the network, and $\sigma_{xy}(v)$ denotes the number of shortest paths between nodes x and y that pass through node v. If $x = y$, then $\sigma_{xy} = 1$.

4 Proposed Method

4.1 Utilizing Degree Centrality in Social Networks for Product Advertisement: A Practical Approach

In today's digital era, social networks have emerged as powerful platforms for businesses to reach their target audience and promote their products. Understanding the structure and dynamics of social networks is crucial for effective advertising campaigns. This paper proposes a practical approach to utilize degree centrality in social networks for product advertisement. By identifying key individuals with high degree centrality, businesses can strategically target their promotional efforts to maximize their advertising impact.

Algorithm 1 Degree Centrality-Based Advertising in Social Networks

---

1: Input:
 - G: Main network with nodes $V = \{N_1, N_2, \ldots, N_m\}$ and edges E
 - $Data$: Data for calculating $p(x)$
2: Output:
 - $d(N)$: Node with the maximum degree centrality
 - $p(x)$: Probability calculated using the given formula
3: Calculate the degree centrality for each node in the main network G
4: **for** each node N_i in V **do**
5:     Calculate the degree centrality $d(N_i)$ of node N_i in G
6: **end for**
7: Find the node N with the maximum degree centrality
8: Set $d(N)$ to the maximum value among all $d(N_i)$ for $i = 1$ to m
9: Choose node N to make an advertisement in social networks
10: Calculate $p(x)$ using the given formula:

$$p(x) = \frac{1}{1 + e^{-(\beta_0 + \beta_1 x)}}$$

11: Use the provided data to compute the value of $p(x)$ using the formula.
12: Output the result: Print or store the value of $p(x)$ as the calculated probability.

The number of users on social networks is growing rapidly in the current decade, and social networks have become a platform for advertising products to reach the audience. Advertisement via social networks is useful for connecting with consumers and boosting commercial marketing. An organization should create a marketing strategy to reach consumers on the online platform of social networks. Many companies are entering social media advertising on the premise of advertising, but they may not fully understand the strategies required. Therefore, businesses need to develop a thoughtful and prudent social media marketing strategy for managing their corporate social accounts. Utilizing individuals with high degree centrality, such as celebrities with a massive following, can significantly impact the effectiveness of advertising campaigns.

Generally, companies have tried to make advertisements with celebrities because they have a substantial following. When celebrities endorse a product, their influence on consumers is more significant. The interaction can be classified as social networks (virtual) and real-life (physical) interaction [5]. The impact of communication gives a rough indication of the social power of a node based on how well they "connect" the network [18]. The aim of this type of connection is to prepare the relationship between maximum nodes. It is the decision or choice taken by the members of social networks to develop their relationship. Nodes in social networks have their personal views or opinions on social advertisements. The chances of a response on each option are equally likely.

For example, Instagram is one of the social media platforms that marketers use the most. Because a single post can reach millions of people, nearly all businesses and marketers have worked with or plan to work with celebrities who will serve as influencers at some point. Consequently, the majority of influencer marketing platforms use Instagram as their primary social media platform. Most people prefer or intend to work with influencers who are more authentic and affordable. The concept related to degree centrality in social networks for product advertisement exists. Therefore, the degree centrality concept can be used to make advertisements through social network platforms.

Node with the maximum degree centrality among celebrities (after removing the Instagram row): Cristiano Ronaldo, with a degree centrality value of 0.08695652173913043.

The tables provided above serve as examples to demonstrate that there are celebrities with massive followers on social networks, making them influential figures for advertising products.

4.2 Assumption

In this paper, we assume that an organization has appointed a celebrity based on their degree centrality within social networks. Furthermore, it is presumed that a raw dataset obtained from the "Social networking ads" dataset on Kaggle is utilized for analysis. Under this assumption, we consider that the most influential person within the social network has featured in the advertisement, and we examine the following results.

Please note that the above assumption is purely hypothetical and does not reflect any specific real-world scenario or finding. It is intended solely for the purpose of illustrating the utilization of degree centrality in selecting a celebrity endorser and analyzing the outcome of an advertisement based on this assumption.

Let us consider that a group of 400 people (members) in the social network is taken, whose income lies between Rs. 10,000/- to Rs. 160,000/-, and it is taken for analysis to study the relationship between income and the purchase of a product.

The dataset used in this analysis, containing information on individuals' income and purchase behavior, can be found in [1] (Table 1).

Table 1. Top 25 Instagram Users by Followers [2]

#	Profile	Followers	Following
1	Instagram	568,257,669	151
2	Cristiano Ronaldo	497,212,743	523
3	Leo Messi	373,589,990	302
4	Kylie	372,285,918	91
5	Selena Gomez	356,652,607	232
6	Dwayne Johnson	347,866,817	618
7	Ariana Grande	339,853,608	575
8	Kim Kardashian	333,806,602	206
9	Beyonce	282,459,320	0

(*continued*)

Table 1. (*continued*)

#	Profile	Followers	Following
10	Khloe Kardashian	279,951,226	103
11	Justin Bieber	264,964,050	730
12	Kendall	263,176,256	227
13	Nike	249,429,575	147
14	National Geographic	245,290,334	140
15	Taylor Swift	232,040,528	0
16	Jennifer Lopez	226,675,889	1,448
17	Virat Kohli	223,644,379	261
18	Barbie	205,413,638	640
19	Kourtney Kardashian Barker	203,994,880	145
20	Miley Cyrus	187,456,632	297
21	NJ	181,956,713	1,690
22	KATY PERRY	179,660,219	738
23	Zendaya	157,868,583	1,771
24	Kevin Hart	157,824,166	941
25	Demi Lovato	142,949,585	844

4.3 Logistic Regression

Logistic regression is a statistical technique used for modeling and analyzing the relationship between a binary dependent variable and one or more independent variables. It is a fundamental tool in the field of machine learning, particularly for binary classification problems where the goal is to predict one of two possible outcomes based on a set of predictor variables. When the dependent variable is dichotomous, as in our case, logistic regression is the appropriate choice. The key challenge in logistic regression is to model and predict probabilities that naturally fall in the range of 0 to 1, given a set of independent variables. To achieve this, the logistic function (also known as the sigmoid function) is utilized, mapping the linear combination of the independent variables to a value within the 0 to 1 range, thereby facilitating the modeling of probabilities.

The logistic function ensures that the output is confined to the valid range of probabilities, making it a powerful method for understanding and predicting binary outcomes. Furthermore, the log-odds transformation helps represent the relationship between the independent variables and the probability of the event occurring in a linear manner. This combination of the logistic function and log-odds transformation is fundamental to logistic regression [6, 24].

5 Results

In our analysis, we observed a total of 143 successful outcomes and 257 failures. The plot in Fig. 1 visually represents the relationship between the model and the actual data.

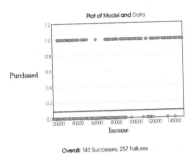

Fig. 1. Plot of model and data

Table 2 presents the results of the chi-squared test, an important statistical test for model fit. The chi-squared statistic is 53.8468 with 1 degree of freedom, and the p-value is extremely low, indicating that the model fits the data well.

Table 2. Chi-Squared Test of Model Fit

Parameter	Value
Chi-Squared (χ^2)	53.8468
Degrees of Freedom (df)	1
p-value	2.1675×10^{-13}

Table 3 provides detailed statistics for the logistic regression model [9]. It includes coefficients, odds ratios, means, standard deviations, counts, standard errors, p-values, and 95% confidence intervals. Notably, the p-values for both variables are very close to zero, suggesting their strong influence on the model. Table 4 presents predicted probabilities generated by the logistic regression model for various income levels, along with their corresponding 95% confidence intervals [9]. These predicted probabilities are essential for understanding how the model classifies individuals based on their income.

Figure 2 graphically illustrates the predicted probabilities generated by the model, showing how they change with different income levels.

Another analysis concerns the binary classification approach where the predicted probability $P > 0.5$ is taken as output 1, else 0. In this analysis, we employ a binary classification approach based on the predicted probabilities generated by our model. The model calculates the likelihood of an instance belonging to one of two classes: 1 (indicating a purchase) or 0 (indicating no purchase). If the predicted probability exceeds 0.5, it is classified as 1; otherwise, it is classified as 0. This threshold of 0.5 serves as the decision boundary.

Table 3. Model Statistics

Variable	Intercept	Dataset 2 (x)
Coefficient	−2.3227	0
Odds Ratio	0.098009	1
Mean	0.3575	69742.5
Standard Deviation (σ)	0.47926	34054.31
Count	400	400
Standard Error	0.2855	0
p-value	4.1396×10^{-16}	1.12×10^{-11}
95% Confidence Interval	–	(1, 1)

Table 4. Probabilities Predicted by The Model with respect to given dataset and 95% confidence interval using Logistic regression calculator – stats.blue

Income	Purchased	Predicted Probability (P)	Confidence Interval
19000	0	0.1336	(0.0902, 0.1936)
20000	0	0.1364	(0.0927, 0.1964)
43000	0	0.2148	(0.1672, 0.2716)
57000	0	0.2765	(0.2292, 0.3294)
76000	0	0.3756	(0.3257, 0.4284)
58000	0	0.2813	(0.2340, 0.3340)
84000	0	0.4214	(0.3668, 0.4779)
150000	1	0.7788	(0.6671, 0.8608)
...	...	...	...
59000	0	0.2862	(0.2389, 0.3386)
86000	0	0.4330	(0.3770, 0.4909)
149000	1	0.7746	(0.6632, 0.8571)
21000	0	0.1393	(0.0952, 0.1992)
72000	0	0.3535	(0.3049, 0.4053)
...	...	...	...

Table 5 presents a confusion matrix, which is a fundamental tool for evaluating the performance of binary classification models. It provides a clear picture of the model's strengths and weaknesses. It allows us to assess its ability to correctly classify instances, particularly in terms of its ability to identify actual purchases (true positives) and accurately recognize non-purchases (true negatives). It also highlights where the model tends to make errors, such as false positives and false negatives.

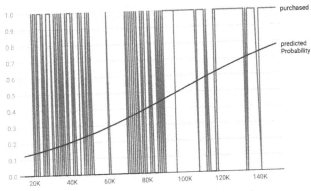

Fig. 2. Predicted Probability Representation

Table 5. Confusion Matrix for the Purchased/Not Purchased Buy Prediction

Actual value	Predicted value
0	0 (True Negatives = 247)
0	1 (False Positives = 10)
1	0 (False Negatives = 80)
1	1 (True Positives = 63)

True Negatives (TN) represent the cases where both the actual value and the predicted value are 0. In this specific analysis, there are 247 instances correctly classified as "Not Purchased". False Positives (FP) correspond to the cases where the actual value is 0 (Not Purchased), but the model predicted 1 (Purchased). There are 10 instances where the model made false positive predictions, meaning it incorrectly indicated a purchase. False Negatives (FN) represent the cases where the actual value is 1 (Purchased), but the model predicted 0 (Not Purchased). In this analysis, there are 80 instances where the model made false negative predictions, meaning it failed to recognize actual purchases. True Positives (TP) correspond to the cases where both the actual value and the predicted value are 1. There are 63 instances correctly classified as "Purchased".

Table 6 offers a comprehensive set of metrics that are commonly used to evaluate the performance of logistic regression models in the context of purchased/not purchased prediction. These metrics provide valuable insights into how well the model distinguishes between instances that represent purchases and those that do not.

The key metrics presented in the table can be analyzed as:

Sensitivity (True Positive Rate TPR): This metric measures the model's ability to correctly identify instances that belong to the "Purchased" class. In this analysis, the sensitivity is approximately 0.4406, indicating that the model correctly identifies about 44.06% of actual purchases.

Specificity (True Negative Rate – SPC): Specificity assesses the model's ability to correctly classify instances that belong to the "Not Purchased" class. In this case, the specificity is approximately 0.9611, implying that the model accurately identifies about 96.11% of non-purchases.

Table 6. Different Metrics (to evaluate logistic regression for purchased/not purchased prediction)

Measure	Value	Derivations
Sensitivity	0.4406	TPR = TP / (TP + FN)
Specificity	0.9611	SPC = TN / (FP + TN)
Precision	0.863	PPV = TP / (TP + FP)
Negative Predictive Value	0.7554	NPV = TN / (TN + FN)
False Positive Rate	0.0389	FPR = FP / (FP + TN)
False Discovery Rate	0.137	FDR = FP / (FP + TP)
False Negative Rate	0.5594	FNR = FN / (FN + TP)
F1 Score	0.5833	F1 = 2TP / (2TP + FP + FN)
Accuracy	0.7750	ACC = (TP + TN) / (TP + TN + FP + FN)

Precision (Positive Predictive Value – PPV): Precision quantifies the accuracy of positive predictions made by the model. The precision in this analysis is approximately 0.863, suggesting that when the model predicts a purchase, it is correct about 86.3% of the time.

Negative Predictive Value (NPV): NPV gauges the accuracy of negative predictions made by the model. The NPV is approximately 0.7554, indicating that when the model predicts no purchase, it is correct approximately 75.54% of the time.

False Positive Rate (FPR): FPR calculates the proportion of false positive predictions made by the model. The FPR here is approximately 0.0389, suggesting a relatively low rate of false positives.

False Discovery Rate (FDR): FDR measures the proportion of incorrect positive predictions relatlive to all positive predictions. The FDR in this analysis is approximately 0.137, indicating that about 13.7% of positive predictions are incorrect.

False Negative Rate (FNR): FNR quantifies the proportion of false negative predictions made by the model. The FNR is approximately 0.5594, suggesting that the model misses approximately 55.94% of actual purchases.

F1 Score: The F1 score is the harmonic mean of precision and sensitivity, providing a balanced measure of the model's overall performance. The F1 score in this analysis is approximately 0.5833.

Accuracy (ACC): Accuracy represents the proportion of correct predictions made by the model. The accuracy here is approximately 0.7750, indicating that the model is correct in its predictions about 77.5% of the time.

These metrics collectively provide a comprehensive assessment of the logistic regression model's performance in distinguishing between purchased and not purchased instances. They offer valuable insights into the model's strengths and weaknesses, helping stakeholders understand its reliability and potential areas for improvement.

Figure 3 provides a visual representation of the different evaluation metrics, helping to compare and understand the model's overall performance.

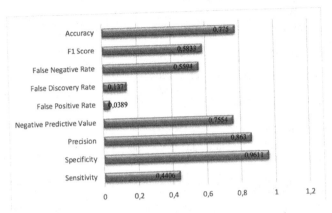

Fig. 3. Graphical Representation of Different Results

6 Conclusions and Future Work

In conclusion, the power of social networks as potent marketing tools cannot be underestimated in the contemporary business landscape. Organizations that harness platforms like Facebook, Instagram, and LinkedIn stand to gain significant advantages in their marketing strategies. Social networks, especially when leveraged through celebrity endorsements, can greatly enhance brand visibility and success. The degree centrality of network members emerges as a pivotal factor in augmenting a product's brand value.

Looking ahead, future research could delve into conducting comparative analyses of brand value when the same network member advertises multiple products on social network platforms. Additionally, exploring alternative methodologies such as kNN, Decision Trees, and SVM could offer valuable comparisons and advancements in predictive modeling within the social network marketing domain. Future work in these directions has the potential to deepen our understanding of the intricate dynamics between social networks and business outcomes.

References

1. Social networking ads. https://www.kaggle.com/datasets/sonalisingh1411/-social-networking-ads/data. Accessed on 10 Oct 2023
2. Top 100 instagram users by followers. https://www.socialtracker.io/toplists/top-100-instagram-users-by-followers/. Accessed 10 Oct 2023

3. Allaymoun, M.H., Hamid, O.A.H.: Business intelligence model to analyze social network advertising. In: IEEE International Conference on Information Technology (ICIT), pp. 326–330 (2021)
4. Arroyo, D.O.: Discovering sets of key players in social networks. In: Computational Social Network Analysis – Trends. Tools and Research Advances, pp. 27–47. Springer, Computer Communications and Networks (2010)
5. Bhadoria, R.S., Bhoj, N., Srivastav, M.K., Kumar, R., Raman, B.: A machine learning framework for security and privacy issues in building trust for social networking. Cluster Comput. **26**(6), 3907–3930 (2022)
6. Czepiel, S.A.: Maximum likelihood estimation of logistic regression models: Theory and implementation. czep.net/stat/mlelr.pdf **83** (2002)
7. Das, K., Samanta, S., Pal, M.: Study on centrality measures in social networks: a survey. Soc. Netw. Anal. Min. **8**, 1–11 (2018)
8. Deo, N.: Graph Theory with Applications to Engineering and Computer Science. Courier Dover Publications (2017)
9. Hosmer, D.W., Lemeshow, S., Sturdivant, R.X.: Applied Logistic Regression, vol. 398. John Wiley & Sons (2013)
10. Iglesias, J.A., Garcia-Cuerva, A., Ledezma, A., Sanchis, A.: Social network analysis: Evolving twitter mining. In: IEEE International Conference on Systems, Man, and Cybernetics (SMC), pp. 1809–1814 (2016)
11. Kafeza, E., Kanavos, A., Makris, C., Pispirigos, G., Vikatos, P.: T-PCCE: twitter personality based communicative communities extraction system for big data. IEEE Trans. Knowl. Data Eng. **32**(8), 1625–1638 (2020)
12. Kafeza, E., Kanavos, A., Makris, C., Vikatos, P.: T-PICE: twitter personality based influential communities extraction system. In: IEEE International Congress on Big Data, pp. 212–219 (2014)
13. Kanavos, A., Vonitsanos, G., Mylonas, P.: Clustering high-dimensional social media datasets utilizing graph mining. In: IEEE International Conference on Big Data, pp. 3871–3880 (2022)
14. Knoke, D.: Origins of social network analysis. In: Encyclopedia of Social Network Analysis and Mining, 2nd edn. Springer (2018)
15. Knoke, D., Yang, S.: Social Network Analysis. SAGE Publications (2019)
16. Lidwina, G.J., Vaidegi, T., Hemalatha, S.: Media advertising: a study on advertising on social media. Int. J. Sales Market. Manag. Res. Dev. (IJSMMRD) **10**(1), 93–104 (2020)
17. Mitchell, T.M.: Machine learning, International Edition. McGraw-Hill Series in Computer Science, McGraw-Hill (1997)
18. Srivastav, M.K., Nath, A.: Study on mathematical modeling of social networks. Int. J. Emerg. Technol. Adv. Eng. **5**(3), 611–618 (2015)
19. Tripiawan, W., Amani, H., Wijaya, A.T.: Effectiveness analysis of social media ads as a promotional media (case study: Instagram Taya.Id). IOP Conf. Ser.: Mater. Sci. Eng. **505**(1), 012095 (2019)
20. Tsvetovat, M., Kouznetsov, A.: Social Network Analysis for Startups: Finding Connections on the Social Web. O'Reilly Media, Inc. (2011)
21. Wasserman, S., Faust, K.: Social Network Analysis: Methods and Applications. Cambridge University Press (1994)

22. Yousef, M., Dietrich, T., Rundle-Thiele, S.: Social advertising effectiveness in driving action: a study of positive, negative and coactive appeals on social media. Int. J. Env. Res. Public Health **18**(11), 5954 (2021)
23. Zhao, C.: Research on the path to enhance the brand value based on social network. In: 2nd IEEE International Conference on E-Commerce and Internet Technology (ECIT), pp. 130–133 (2021)
24. Zheng, A., Casari, A.: Feature Engineering for Machine Learning: Principles and Techniques for Data Scientists. O'Reilly Media, Inc. (2018)

Innovative Research Projects

Business Models for Mobility Data Sharing Platforms: Stakeholders' Perceptions

Louis-David Benyayer[1]([⊠]) and Markus Bick[2]

[1] ESCP Business School, Paris, France
lbenyayer@escp.eu
[2] ESCP Business School, Berlin, Germany

Abstract. Potential economic benefits of data sharing have been estimated and described in many reports and research. However, data sharing initiatives are limited by a lack of alignment on the business models for data sharing among stakeholders. In this research we aim specifically to investigate the perceptions of the stakeholders of such data sharing platforms on business models for data sharing to identify ways to foster alignment on business models. In our qualitative exploratory analysis (interviews and focus groups with stakeholders involved in a EU-funded project, MobiDataLab) we identify 6 criteria used by the stakeholders to compare business models and rank 5 archetypal business models according to them. We conclude with insights useful for business model design for mobility data sharing platforms and possible future research.

Keywords: platforms · business models · data sharing

1 Introduction

Economic benefits of data sharing have been described and measured by several reports over the last years [1, 2]. Some estimate its impact to be between 1% and 2.5% of OECD countries´ GDP [3] and several regulations, laws and initiatives have been implemented at the European level to foster its development (Data Act [4] and Data Governance Act [5] in particular). Economic benefits of data sharing include transparency, accountability and empowerment of users, business opportunities for data intermediaries and start-ups, co-operation and competition across sectors and countries, crowdsourcing new insights and user-driven innovation, Increased efficiency across society through data linkage and integration [3]. E.g., the re-use of Transport for London (TfL) open data was generating annual economic benefits and savings of up to GBP 130 million a year for TfL customers, road users, London, and TfL itself [6].

Among the challenges which hamper the development of data sharing, are listed the coherent incentive mechanisms and sustainable business models [3]. The difficulties associated with business models for data sharing between organizations include the unclear ownership and decision rights [7] and the difficulty to settle a price for data [8].

In order for data sharing to scale, alignment is required between the business models of the various participants of a data sharing business ecosystem: data providers, data consumers, intermediary platforms, software and services providers. Various research has shown this alignment is lacking in many data ecosystems because of different perceptions between participants on the value of data [9–11] and the revenue sharing models [12, 13].

However, little is known yet on the criteria used by the participants to compare possible business models and which they favor. Hence, the research question we investigate is twofold: *What are the criteria used by stakeholders to compare business models? (RQ1) And how specific business models are evaluated according to these criteria? (RQ2).*

The main objective of this research is to get a better understanding of the perceptions of business models that stakeholders involved in data sharing have. We hence apply a qualitative research approach mixing semi-structured interviews and focus groups with stakeholders involved in a EU-funded project, MobiDataLab. Our study contributes to the body of knowledge of platform business models by offering new insights on their key success factors and in particular how to spur generative interactions with the surrounding ecosystem [14, 15]. This study also brings a new perspective on the role of data in platform business models [16]. Last, by studying the perceptions of business models in the context of environment challenges we contribute to improve our understanding of how firms could play a bigger role in addressing environmental challenges through their business models [17].

In the following, we present an overview of the research and frameworks for business models for data sharing (Sect. 2), then we describe the methodology (Sect. 3) and discuss our main findings (Sect. 4); last, we address the limitations and future research (Sect. 5).

2   Theoretical Background

Data sharing platforms have emerged in many contexts (e.g., NxtPort for port data [18] or Space data Marketplace [19] for space data), and we chose mobility data sharing platforms as subject of our analysis. With the development of connected mobility, mobility data are extensively produced by many different stakeholders and after a first analysis, it became apparent that the problems identified in the literature regarding business models were also present in mobility data sharing platforms. Limiting our analysis to this scope helps us getting a better understanding of our two research questions. This section presents an analysis of the existing literature of mobility data sharing and the associated business models. Initially, the analysis is focused on the importance of data sharing and on the related barriers. Next, a review of existing business models for data sharing platforms is presented.

2.1   Mobility Data Sharing

As they are more and more connected with sensors and GPS devices, public and private transport vehicles generate significant data on their use, location and functioning. In addition, data is generated by the individuals through their smartphone sensors when they are (or not) using a transport vehicle. Beyond data about mobile vehicles, mobility data might also include data on the road and transport infrastructures (for example bus stops and metro stations). In a nutshell, the mobility data landscape is composed of many different data types, provided by different systems which often are not interfaced. There are three types of stakeholders in mobility data sharing [20]: data generators (public and private transport operators), data aggregators (platforms which gather and synchronize data from various sources) and data analytics providers (service companies which offer tailor made digital services based on data, such as journey planners).

Even though the value of sharing has been described by many researchers [21, 22], a lot of research concludes that it is difficult to implement data sharing at scale. Among the barriers to share data, strategical reasons are regularly mentioned. The fear of a loss of control when data is reused by third parties and the uncertainty about a possible harm of their business interest are aspects that prevent the development of data sharing [23].

When it comes to the motivations to share, companies accept to share data if individual benefits exceed search and transaction costs [24]. Companies tend to follow open approaches if they have a strong interest in data-reuse (e.g., if they benefit from third party services built on these data). Another reason to provide business data on a larger scale for free is to serve the public interest, but this data philanthropy remains an exception [23]. Sometimes, sharing data is a way to be distinct from competition and win a contract. For example, Via [25], provides digital infrastructure for public transport since early on to establish the sharing of key data indicators with public transport partners as a core element of its business model. Via has gained a competitive advantage in being selected by Los Angeles Metro as provider of the on-demand ride-sharing service (after Los Angeles Metro first partnered with another provider without reaching satisfactory agreements on trip data sharing) [20]. Similarly, companies may be forced to share data when getting a contract from a public entity [26].

This has several implications. First, a data sharing solution should either increase the benefits or reduce the transaction costs in order to increase the likelihood of sharing. Second, not all data sharing models are commercial models, and some exchanges are not directly monetized, they are settled through cooperation or barter.

2.2 Business Models

According to [27], a business model is "designed to create value through the exploitation of business opportunities". A business model is formed by the value proposition, the value network and the revenue-cost model. Many researchers and practitioners have proposed descriptions of the components of a business model, e.g., [28, 29] or [30].

Companies can use data to support organizational value creation and capture in three ways [31]: improving internal processes, enriching products, services or experiences, monetizing their internal data by selling them to external stakeholders. Consistently, companies can select or combine three different roles in the value chain of value creation with data [32]: data users who leverage data for internal purposes, data suppliers which aim to market data and data facilitators which provide infrastructure, software and platform solutions.

We analyzed the most prominent business models on mobility data sharing with the aim to identify business model archetypes to be tested with our respondents. Some studies focus on describing the business model for accessing the data. UITP [20], for example, concludes with three data sharing models: *open access, bilateral restricted models* and *multilayered restricted models*. When revenue is involved, the same study identifies six models: *Free-to-all*, also known as *open data*; *Freemium*, free access to limited data or data services (access to higher quality data or specific services is available at a cost to the data consumer); *Licensing*, data is shared based on licensing agreements; *Sponsorship/branded advertisement*, free access to data because cost of data is subsidized by sponsors or advertisers; *Demand-oriented*, cost of data depends on availability and complexity and 6) *Barter system*, data consumer gain access to data based on the exchange of data with data provider.

Other studies focus on the business model of the platform enabling data sharing. In the literature, there are various contributions on platform business models [33, 34] or [35]. Mallon [36], for example, concludes on the impossibility to identify a fixed list of business models but rather names categories and dimensions or components and identifies 109 business model components of digital platform business models. Business models for data ecosystems vary according to the type of revenue or cost models, the level of collaboration and the level of openness [37]. Considering the main objective of this paper we refer to Schweiger et al. [38] who differentiate between three categories of data sharing platforms: *integrator platform, product platform* and *two-sided platform*. In terms of revenue models, two broad categories are identified generally [39]: the *service-fee-based* revenue model and the *advertising-based* revenue model. The service-fee-based model is intended to provide consumers with products and services and collect purchase fees (i.e., service fees). By comparison, revenue is directly raised from third party advertisers instead of users in the advertising-based revenue model. More precisely [40] identifies four possible revenue models in digital platforms: Commissions, Subscription, Advertising, Service sales.

Based on this analysis of existing data sharing and associated business models, we synthesize 5 business model archetypes for a mobility data sharing platform: Contributive and free (BM1), Two-sided (BM2), Marketplace (BM3), Software-as-a-service (BM4) and Barter (BM5). The 5 business model archetypes have been discussed with our respondents and are described in the Table 1 below.

Table 1. Archetypal business models (literature synthesis)

Business model	Description
BM 1 Contributive and free	This model blends open data and open source. All data is available without compensation. Software is maintained by platform stakeholder donations. Like other open-source projects, contributors can voluntarily join to contribute to maintenance. Generally, data and service providers are not compensated financially. Data users do not pay for the platform. The incentive for stakeholders to allocate resources that everyone benefits from is to profit from data and platform while investing limited resources

(continued)

Table 1. (*continued*)

Business model	Description
BM 2 Two-sided (advertising, sponsoring)	Like social networks and search engines, access is free for users. The platform is financed by ads or sponsorships. In this model, data consumers access data for free. Data and service providers are compensated financially
BM 3 Marketplace	A data marketplace allows buying and selling data. If data supports a transaction, like selling a ticket, the platform takes a commission. In this model, platform costs are balanced by commissions on transactions between data providers and sellers. Data consumers pay data producers. Analytics service providers are paid for services
BM 4 Software-as-a-service	In this model, users are charged fees based on usage. Servers, databases, and tools enabling access and use are controlled by the provider. Data and analytics providers are financially compensated. Consumers pay a subscription for data and platform access. The difference to the marketplace model is a single subscription model for all consumers, rather than independent data transactions
BM 5 Barter	Here organizations provide data in exchange for platform operation rights. It is highly unusual and often used with other models. Consumers access data for free and providers are not compensated. Platform costs are funded through subscriptions paid by consumers or providers

3   Methodology

Following our previous analysis, we can see that the criteria used by stakeholders to compare business models is still under-researched. Besides it remains unclear how business models are evaluated according to these criteria. To our knowledge, no prior studies investigated these perceptions about business models. Consequently, we chose a qualitative methodology to gather rich information about these perceptions or criteria which could be further tested within future quantitative studies.

Our qualitative methodology consisted of semi-structured interviews and focus groups. A qualitative approach helps to broadly explore the various criteria in place. Considering RQ1, we interviewed different stakeholders involved in mobility data sharing. Such interviews facilitate gathering insider insights. The open-ended format provides time to delve deeper into practical knowledge compared to other forms of data collection [41]. We then presented and discussed our findings in focus groups. Thereby we leveraged various benefits of focus groups, including observing interactions, quick turnarounds, and identifying issues or new topics, as group discussion often reveals new topics not previously considered. Finally, focus groups are particularly useful for diagnosing success factors and generating impressions of services [42].

3.1   Data Collection

We collected data from multiple sources and data collection took place between September 2021 and April 2022 during the initial phase of the MobiDataLab project [43].

MobiDataLab is an EU-funded lab for prototyping new mobility data sharing solutions. The consortium is composed of 10 partners from industry, research, academia, consultancy and governance sectors, located in 7 different countries.

Various stakeholders involved in the project (data generators, data aggregators, or data analytics service providers) got engaged in the two data collection steps. First, we conducted 11 interviews (Table 1) and second, we organized two focus groups with 16 participants in total (Table 2). During the interviews we used a semi-structured interview guideline based on our previous results. The interviews aimed at evaluating the typology of 5 archetypal business models and to have the opinion of each interviewee on the acceptance of the possible business models. During the interviews the participants were asked to express their views for each business model on their respective advantages, drawbacks, pre-requisites, attractiveness and feasibility. All interviews took place via video conferencing systems and lasted between 33 to 45 min. Extensive notes of the conversations were taken and syntheses of the discussions were sent to the respective participants for review.

Table 2. Overview of interviewees

Interviewee	Position	Type of Company
IP 1	Operations manager	Transport Authority (Belgium)
IP 2	Head of international affairs	Transport Authority (Italy)
IP 3	Operations manager	Transport Authority (UK)
IP 4	Digital mobility advisor	Transport Authority (Germany)
IP 5	Advisor on digitization and Smart City	City council (Romania)
IP 6	Sales Director	Digital service company (France)
IP 7	Strategy manager	Transportation technology provider
IP 8	Data and platform Strategy manager	City council (The Netherlands)
IP 9	Smart mobility coordinator	City council (Belgium)
IP 10	Project management officer	Transport Authority (Germany)
IP 11	Head of public Policy	Shared mobility solution provider

The two focus groups (Table 3) were asked to elaborate on the interview results and helped to triangulate the data obtained during the interviews. Focus groups encourage participants to share perceptions and points of view without pressuring them to vote or reach consensus, which allows to gather rich information [42]. Another advantage of focus groups is that they enable participants to debate distinct points of view, build their viewpoints on the individual contributions [44]. In the focus groups, the archetypal business models were presented as well as a synthesis of the interviews about each business model (advantages, drawbacks, pre-requisites, attractiveness and feasibility). In the focus groups, participants were asked to add and comment on the content provided. The first focus group was composed of researchers and experts in the domain and the second focus group was composed of potential users of the data sharing platform. The participants of the workshop were also asked to rank the archetypal business models by order of preference.

Table 3. Overview of focus group participants

Interviewee	Position	Type of Company
FG1 1	Researcher	Research center (Italy)
FG1 2	Consultant	Consulting company
FG1 3	Researcher	Research center (Belgium)
FG1 4	Project manager	Consulting company (UK)
FG1 5	Project manager	Digital service company
FG1 6	Researcher	Research center (Italy)
FG1 7	Consultant	Consulting company
FG1 8	Operations manager	Digital service company
FG1 9	Project manager	Consulting company
FG2 1	Strategy manager	Transportation technology provider
FG2 2	Project manager	City Council (Greece)
FG2 3	Digital mobility advisor	Transport Authority (Germany)
FG2 4	Advisor on digitization and Smart City	City council (Romania)
FG2 5	Head of international affairs	Transport Authority (Italy)
FG2 6	Scientific project manager	City Council (Greece)
FG2 7	Transport expert	Public agency

3.2 Data Analysis

After the interviews and the two focus groups, we had triangulated data for each business model on their respective advantages, drawbacks, pre-requisites, attractiveness and feasibility. These data were then analyzed and synthesized. Data on advantages, drawbacks, pre-requisites, attractiveness and feasibility of each model were coded. Open coding was applied and the codes were clustered to identify categories of evaluation criteria used to compare business models. This led to the identification of 6 criteria. In a second step, interviews and workshop results were analyzed to measure how each business model scored on each criterion previously identified.

4 Findings and Discussion

In the following, we present our main findings on the evaluation of criteria used to compare and assess business models for data sharing and the ranking of such business models on these criteria.

4.1   Evaluation Criteria

Based on our analysis we were able to distinguish between 6 criteria used by the various stakeholders to assess business models for data sharing.

Scalability. Scalability refers to the ability of the business model to handle increasing volumes of data and users. This is particularly important in mobility data sharing, given the vast amounts of data generated by various sources, such as GPS devices, sensors, and mobile applications. A scalable model can accommodate growth without significant increases in cost or decreases in performance.

Implementation Speed and Cost. This criterion evaluates how quickly and cost-effectively the business model can be put into practice. It takes into consideration the resources needed to establish the data sharing infrastructure, including both technological and human resources. Factors such as the need for new hardware, software, skills, or processes, as well as the time required to integrate with existing systems, can influence the speed and cost of implementation.

Openness. Openness refers to the degree of accessibility and transparency in the data sharing process. An open model would ensure that data is freely available to all interested parties, fostering innovation and collaboration. It also implies transparent governance and decision-making processes, where stakeholders have insight into who is accessing the data, how it's being used, and how decisions about the data are made.

Financial Incentives for Private Companies. Private companies will assess the business model for its potential to generate financial returns. This could come directly from selling data or providing data-based services, or indirectly from the efficiencies and improvements that the data enables. The model might also offer incentives such as access to new markets, improved customer relationships, or competitive advantages.

Attractiveness for Public Entities. Public entities will evaluate the business model based on its ability to serve the public interest. This could include improving transportation systems, supporting urban planning, enhancing environmental sustainability, or providing public access to data. The model should align with the public entity's mission and goals and provide tangible benefits for the communities they serve. Moreover, public entities would also be interested in models that foster collaboration, comply with regulations, and protect the public's privacy and security.

Technical Complexity. This criterion looks at the technological demands of implementing the business model. It includes the sophistication of the technology required, the technical skills needed to implement and manage it, and the complexity of integrating it with existing systems.

4.2   Ranking of Business Models

In a next step the archetypal business models were compared based on the previously explained six criteria (Table 4), according to a qualitative assessment (medium, high, low) derived from the views expressed during the interviews and the focus groups by the various stakeholders.

Table 4. Scores of each Business Model on each criterion

Criteria	BM1 Contributive and free	BM2 Two-sided (advertising)	BM3 Marketplace	BM4 Software-as-a-service	BM5 Barter
Scalability	Medium	High	High	Medium	Low
Implementation	Medium	High	Medium	Medium	Low
Openness	High	Medium	Medium	Medium	Medium
Financial incentives	Low	Medium	High	High	Low
Attractiveness	Medium	Low	Medium	Medium	High
Technical complexity	Low	High	High	Medium	Low

The comparison of business model archetypes illustrated in Table 4 shows that the choice of model might be influenced by implementation concerns. It also highlights possible paths: e.g., starting with a low resource model and then evolve towards a more resource intensive one. Lastly it may reveal possible combinations of business model archetypes which would offer symmetrical interests. In particular, several revenue models may coexist according to the granularity of data (e.g., different revenue models for real time and historical data, raw or aggregated data). Similarly, different models may be used for accessing the data and for using the solutions. According to the local characteristics, different business models may be implemented. Because of different local habits and local business organizations, as well as maturity level, some cities or regions may use different business models.

BM1 – Contributive and Free. This model excels in terms of openness and is then quite scalable. While there may be costs related to standardizing and cleaning the data for sharing, the open-source software can often be implemented at relatively low cost.. The data and source code are freely available for anyone to use and improve upon. Indirect benefits like innovation and problem-solving from the community can be substantial, however, direct financial incentives might be lower for private companies,. The model aligns well with the goals of many public entities, as it supports transparency, collaboration, and public access to data. Last, the complexity can vary. Open-source tools can range from very simple to highly complex, depending on their design and purpose.

BM2 – Two-Sided (advertising). This model might require a considerable investment to set up the infrastructure and attract advertisers but might scale well once initial investments are made. While user data might be closely guarded due to privacy concerns, the platform itself might be open to any advertiser that meets its criteria. Financial incentives can be significant as companies can generate revenue from advertisers, however, the attractiveness might be lower for public entities unless the advertising supports public goals or services. Given the size, scope and feature list, developing the solution can prove to be technically complex.

BM3 – Marketplace. The model is scalable as it benefits from network effects: more participants lead to more transactions and more valuable data. Creating a marketplace requires a considerable investment, both in terms of platform development and in attracting initial users and providers. Financial incentives can be substantial as companies can generate revenue from transactions, data, or services. This can be attractive for public entities if it helps to facilitate valuable services, though they may have concerns about control and fairness. Building and maintaining a marketplace can be technically complex, as it involves managing transactions, users, and data.

BM4 – Software as a Service. Cloud-based services can be easily expanded or contracted based on demand and can often be implemented quickly and at relatively low cost. The model tends to be more closed as the service provider controls the software and the data. The recurring revenue from subscription fees can provide significant financial incentives for private companies. SaaS can be attractive for public entities though there may be concerns about control and data privacy. The complexity can be relatively low, as the service provider handles the technical aspects.

BM5 – Barter. Barter systems can be scalable, however, managing many-to-many exchanges can be challenging. Barter systems can be fairly open, but they often require a certain level of trust between parties. The financial incentives are indirect, as companies gain value from the exchange rather than through monetary payment. Barter systems can be attractive for public entities if they support community and collaboration, but ensuring fair and equitable exchanges can be a challenge. The technical complexity can vary, from simple direct exchanges to complex multi-party barter systems.

Table 5. Ranking results of the two focus groups

	BM1 Contributive and free	BM2 Two-sided (advertising)	BM3 Marketplace	BM4 Software-as-a-service	BM5 Barter
Focus group 1 (researchers and experts	3,2	2,2	4,3	3,6	1,7
Focus group 2 (potential users	4,4	1,9	3,2	3,4	2,3

As shown in Table 5, the ranking results of the two focus groups are consistent regarding the least preferred business models: BM2 Two-sided and BM5 Barter rank at the lowest position in both cases. However, considering the most favored business models researchers and experts (FG1) tend to favor BM3 Marketplace whereas potential users (FG2) tend to favor BM1 Contributive and free and rank BM3 Marketplace number 3.

5 Conclusion

Contribution to Theory. The findings of this research contribute to better understand the perceptions of data sharing stakeholders on archetypal business models. We provide new insights on platform business models´ key success factors and on the role of data in those models, in particular. We offer insights on how to reduce the barrier to share mobility data.

Contribution to Practice. We conclude with four main insights which can be useful for organizations thinking about joining a data ecosystem or launching their own data platform: 1) data sharing business models are dynamic and agile. The business model for the launch of a platform is not necessarily the same as the business model for scaling. 2) the business model is likely to be a mix between several archetypal business models. In particular, several revenue models may coexist according to the granularity and origin of data (e.g.: different revenue models for real time and historical data, raw or aggregated data). Similarly, different models may be used for accessing the data and for using the solution. 3) according to local specificities, different business models may be implemented. Because of different local habits and local business organizations, as well as maturity level, some cities or regions may use different business models. 4) business models are perceived differently according to stakeholders. During the interviews as well as the focus groups we observed that potential users tend to favor BM1 (contributive and free) and BM5 (barter), whereas researchers and experts tend to favor BM3 (marketplace) and BM4 (software-as-a-service). In a nutshell, the development of data sharing platform business models should be designed in collaboration with the various stakeholders to ensure alignment of interests.

Of course, this research shows some limitations. On the one hand, the research has been conducted in the context of mobility data and the findings might not apply to other types of data (for example energy data or manufacturing equipment data). On the other hand, being exploratory and relying on a limited set of stakeholders interviewed, findings presented here can be confirmed through quantitative surveys. Finally, we studied the European context solely which is specific in terms of business organization and data regulation. Future studies investigating other geographical contexts would help to get a more global view on the subject.

Acknowledgements. This work was supported by MobiDataLab project co-funded by the European Commission, Horizon 2020, under grant agreement No. 101006879 (Research Innovation Action).

References

1. Manyika, J., et al.: Open data: unlocking innovation and performance with liquid information. McKinsey Global Institute (2013)
2. Open Data Institute: Understanding the social and economic value of sharing data. ODI (2023)
3. OECD: Enhancing Access to and Sharing of Data : Reconciling Risks and Benefits for Data Re-use across Societies. Editions OCDE, Paris (2019)

4. European Commission: Data Act: Proposal for a Regulation on harmonised rules on fair access to and use of data (2022)
5. European Commission: European data governance act. (2022)
6. Deloitte: Assessing the value of TfL's open data and digital partnerships (2017)
7. Eckartz, S.M., Hofman, W.J., Van Veenstra, A.F.: A decision model for data sharing. In: Janssen, M., Scholl, H.J., Wimmer, M.A., Bannister, F. (eds.) Electronic Government, pp. 253–264. Springer, Berlin, Heidelberg (2014). https://doi.org/10.1007/978-3-662-44426-9_21
8. Abraham, R., Schneider, J., vom Brocke, J.: Data governance: a conceptual framework, structured review, and research agenda. Int. J. Info. Manage. 49, 424–438 (2019). https://doi.org/10.1016/j.ijinfomgt.2019.07.008
9. Spiekermann, M.: Data marketplaces: trends and monetisation of data goods. Intereconomics. 54, 208–216 (2019). https://doi.org/10.1007/s10272-019-0826-z
10. Spiekermann, M., Wenzel, S., Otto, B.: A conceptual model of benchmarking data and its implications for data mapping in the data economy. In: Proceedings of the 2018 Multikonferenz Wirtschaftsinformatik (2018)
11. Khatri, V., Brown, C.V.: Designing data governance. Commun. ACM. 53, 148–152 (2010). https://doi.org/10.1145/1629175.1629210
12. Fox, K.: The Illusion of Inclusion — The "All of Us" Research Program and Indigenous Peoples' DNA. New England Journal of Medicine 383, 411–413 (2020). https://doi.org/10.1056/NEJMp1915987
13. Kembro, J., Selviaridis, K.: Exploring information sharing in the extended supply chain: an interdependence perspective. Supply Chain Manage. Int. J. 20, 455–470 (2015). https://doi.org/10.1108/SCM-07-2014-0252
14. Cozzolino, A., Verona, G., Rothaermel, F.T.: Unpacking the disruption process: new technology, business models, and incumbent adaptation. J. Manage. Stud. 55, 1166–1202 (2018). https://doi.org/10.1111/joms.12352
15. Zeng, J., Yang, Y., Lee, S.H.: Resource orchestration and scaling-up of platform-based entrepreneurial firms: the logic of dialectic tuning. J. Manage Stud. 60, 605–638 (2023). https://doi.org/10.1111/joms.12854
16. Markman, G.D., Lieberman, M., Leiblein, M., Wei, L.-Q., Wang, Y.: The distinctive domain of the sharing economy: definitions, value creation, and implications for research. J. Manage. Stud. 58, 927–948 (2021). https://doi.org/10.1111/joms.12707
17. Snihur, Y., Markman, G.: Business Model Research: Past, Present, and Future. J. Manage. Stud. n/a. https://doi.org/10.1111/joms.12928.
18. NxtPort, https://nxtport.com/.
19. Space Data Marketplace, https://www.space-data-marketplace.eu/en/.
20. UITP: Sharing of data in public transport (2020)
21. Prajogo, D., Olhager, J.: Supply chain integration and performance: the effects of long-term relationships, information technology and sharing, and logistics integration. Int. J. Prod. Econ. 135, 514–522 (2012). https://doi.org/10.1016/j.ijpe.2011.09.001
22. Kembro, J., Näslund, D., Olhager, J.: Information sharing across multiple supply chain tiers: A Delphi study on antecedents. Int. J. Prod. Econ. 193, 77–86 (2017). https://doi.org/10.1016/j.ijpe.2017.06.032
23. Richter, H., Slowinski, P.R.: The data sharing economy: on the emergence of new intermediaries. IIC. 50, 4–29 (2019). https://doi.org/10.1007/s40319-018-00777-7
24. Wysel, M., Baker, D., Billingsley, W.: Data sharing platforms: How value is created from agricultural data. Agricultural Systems. 193, 103241 (2021). https://doi.org/10.1016/j.agsy.2021.103241
25. Via, https://ridewithvia.com/.
26. Micheli, M.: Accessing privately held data: Public/private sector relations in twelve European cities. In: Data for Policy Conference Proceedings (2020)

27. Zott, C., Amit, R.: Business model design: an activity system perspective. Long Range Plann. **43**, 216–226 (2010). https://doi.org/10.1016/j.lrp.2009.07.004
28. Osterwalder, A., Pigneur, Y.: Business model generation: a handbook for visionaries, game changers, and challengers. John Wiley & Sons (2010)
29. Demil, B., Lecocq, X.: Business model evolution: in search of dynamic consistency. Long Range Planning **43**, 227–246 (2010). https://doi.org/10.1016/j.lrp.2010.02.004
30. Al-Debei, M.M., Avison, D.: Developing a unified framework of the business model concept. European J. Info. Sys. **19**, 359–376 (2010). https://doi.org/10.1057/ejis.2010.21
31. Wixom, B.H., Ross, J.W.: How to monetize your data. MIT Sloan Management Review **58** (2017)
32. Wiener, M., Saunders, C., Marabelli, M.: Big-data business models: a critical literature review and multiperspective research framework. J. Info. Technol. **35**, 66–91 (2020). https://doi.org/10.1177/0268396219896811
33. Tidhar, R., Eisenhardt, K.M.: Get rich or die trying… finding revenue model fit using machine learning and multiple cases. Strategic Management Journal **41**, 1245–1273 (2020). https://doi.org/10.1002/smj.3142
34. Tan, B., Anderson, E.G., Parker, G.G.: Platform pricing and investment to drive third-party value creation in two-sided networks. Inf. Sys. Res. **31**, 217–239 (2020). https://doi.org/10.1287/isre.2019.0882
35. Chatterjee, P., Zhou, B.: Sponsored content advertising in a two-sided market. Manage. Sci. **67**, 7560–7574 (2021). https://doi.org/10.1287/mnsc.2020.3873
36. Mallon, D.: A systematic literature review of digital platform business models. In: Ahlemann, F., Schütte, R., Stieglitz, S. (eds.) Innovation Through Information Systems, pp. 389–403. Springer International Publishing, Cham (2021). https://doi.org/10.1007/978-3-030-86800-0_27
37. D'Hauwers, R., Walravens, N., Ballon, P.: Data ecosystem business models: value and control in data ecosystems. J. Bus. Models **10**, 1–30 (2022). https://doi.org/10.54337/jbm.v10i2.6946.
38. Schweiger, A., Nagel, J., Böhm, M., Krcmar, H.: Platform business models. Digital Mobility Platforms and Ecosystems 66 (2016)
39. Su, Y., Jin, L.: The impact of online platforms' revenue model on consumers' ethical inferences. J. Bus. Ethics. **178**, 555–569 (2022). https://doi.org/10.1007/s10551-021-04798-0
40. Staub, N., Haki, K., Aier, S., Winter, R.: Taxonomy of digital platforms: a business model perspective. Hawaii International Conference on System Sciences 2021 (HICSS-54) (2021)
41. Bogner, A., Littig, B., Menz, W.: Interviewing experts. Palgrave Macmillan (2009)
42. Krueger, R.A.: Focus groups: A practical guide for applied research. Sage publications, Thousand Oaks, CA (2015)
43. Mobidatalab, https://mobidatalab.eu/.
44. Stewart, D.W., Shamdasani, P.N.: Focus groups: Theory and practice. Sage publications, Newbury Park (2010)

Innovation Process Integration Challenges and Their Influence on Project Management Performance in Oil and Gas Field- A Conceptual Framework Development

Ahmed Basioni[✉]

Faculty of Business and Law, The British University, Dubai, United Arab Emirates
22000662@student.buid.ac.ae

Abstract. Purpose – This paper aims to develop a comprehensive conceptual framework that elucidates the multifaceted factors affecting the dissemination of innovation processes within the oil and gas sector and how they interplay to impact project management, ultimately leading to improved project performance

Methodology – This study employs a literature synthesis method to build the conceptual framework.

Findings – The forthcoming study outcomes will help pinpoint the principal barriers to and catalysts for implementing innovation systems within the oil and gas industry's ecosystem.

Implications – Conducting subsequent research on integrating the concept of "green innovation" is pivotal for confirming the feasibility of constructing a conceptual framework encompassing critical variables that impact the adoption of innovation processes within the energy sector.

Originality – Exploring the interplay between the innovation process, project management, and performance contributes to a more comprehensive comprehension of a sustainable innovation process that bolsters the expansion of the oil and gas sector.

Keywords: Innovation · Process · organization · Oil · Performance

1 Introduction

Project managers often gauge the success of their projects by evaluating whether they have completed them within the designated timeframe, adhered to the allocated budget, and met the predefined scope and quality requirements. This perspective is commonly referred to as the "project triple constraints," encompassing Time, Cost, and Scope (Shenhar and Dvir, 2007), as depicted in Fig. 1.

Defining "Innovation" as a managerial competence is a challenging endeavor because, in business contexts, innovation entails translating opportunities into tangible and achievable products. Traditional project management models prioritize project delivery and the attainment of specified deliverables through well-defined and measurable execution strategies and success criteria. Regrettably, these models often lack

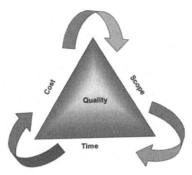

Fig. 1. Project Triple Constraints (Shenhar and Dvir, 2007)

emphasis on creative thinking, which plays a pivotal role in identifying opportunities for innovation, typically involving reflexive problem-solving. For instance, during the execution phase, when confronted with risks, project managers generate mitigation plans to either prevent or minimize the adverse effects of these risks (Gallagher, 2015). This paper aims to build a conceptual framework to explore the obstacles and facilitators involved in integrating innovation processes and performance metrics within the realm of project management, particularly in the energy sector. The research seeks to address the following query:

"How does the contribution of the Innovation Process impact the performance and efficiency of project management, while considering the challenges and facilitators for embracing innovation within the oil and gas sector?".

In subsequent sections, this study will delve into two critical areas: the theoretical foundation and research framework. The theoretical foundation encompasses a review of the literature on innovation theories, concepts and models of the innovation process, theories of individual acceptance, project management principles, and critical facilitators. The research framework will build upon this theoretical foundation, highlighting prior research efforts that have examined the challenges and opportunities associated with implementing innovation systems within the oil and gas sector. Finally, we will propose a conceptual framework tailored to the context of the oil and gas sector, incorporating specific variables and dimensions based on previous research findings.

2   Theoretical Background

The key focal points of this study involve evaluating the primary dimensions that characterize the significant hurdles encountered during the integration of the "Innovation Process" within the energy sector's ecosystem. Additionally, it seeks to investigate the relationship between the adoption of "Innovation Processes" and the acceptance of these processes by individuals, within the context of their contribution to project management.

2.1   Literature Review Background

To undertake this research, the chosen methodology for synthesizing the literature review involved initially defining the problem statement and specifying the topic for exploration within the realm of innovation practices in the oil and gas sector and its impact on the project management value chain. The next step was to conduct a comprehensive literature review, necessitating an extensive search process using various survey engines. The British University in Dubai (BUiD) library database served as a primary source, housing a wealth of articles, journals, papers, and e-books pertinent to risk subjects in project management and my specific focus on organizational structure, challenges, barriers, and enablers in the oil and gas industry. Access to the PMI database and articles provided additional valuable sources, complemented by targeted searches on Google for specific topics and supporting materials. The search criteria included keywords such as organizational structure, oil and gas, theory, challenges, barriers, energy, enablers, focusing on the past 10 years and preceding years for theoretical background. This comprehensive approach yielded approximately 98 publications. Subsequent to extensive readings of the content, including abstracts, introductions, literature reviews, findings, and conclusions, a screening process was initiated to exclude non-relevant topics and articles, resulting in around 27 selected references. The literature reviews in these articles were predominantly based on detailed analyses of case studies, qualitative and quantitative assessments, questionnaire surveys, and reviews on organizational structure theories, styles, merits, culture, as well as associated risks and mitigation strategies in the oil and gas business.

2.2 Innovation Concept and Process

Innovation terminology involves the creation of something distinctive and valuable, ultimately delivering fresh value to end users. It's important to note that innovation differs from invention, which pertains to the generation of new products, while innovation centers on generating value for customers. An interesting analogy between the two can be drawn: invention represents the transformation of financial resources into opportunities, whereas innovation involves converting ideas into revenue (Burmester et al., 2005).

The innovation process primarily encompasses three key phases: the initial phase of innovation, product development, and implementation. The initial phase is a comprehensive stage that delves into various components such as problem definition, idea exploration, scope delineation, marketing assessment, and culminates in the creation of a viable business case. This front-end phase is widely regarded as the pivotal aspect of the innovation process, as it lays the foundation for options to be further evaluated and considered in subsequent stages of development and commercialization.

2.3 Innovation Theories and Models

The realm of innovation management is expansive and exhibits numerous aspects. Despite the perception of innovation as a somewhat elusive concept, it is, in fact, a multifaceted subject of inquiry, encompassing various models, theories, and frameworks. One approach to categorize innovation involves dividing it into two categories based on its target market and the technology it employs. The innovation matrix model provides a framework for identifying the most prevalent forms of innovation, as depicted in Fig. 2.

Incremental innovation primarily entails making minor adjustments to an existing product formulation or service delivery strategy, resulting in marginal improvements over the previous iteration of the product or service. In contrast, disruptive innovation is a strategy frequently adopted by diverse industries. It involves the introduction of a concept, product, or service that establishes a new value network, either by disrupting an established market or creating an entirely new one (Christensen et al., 2013). On the other hand, sustaining innovation is market-oriented and, instead of forging new value networks through meeting consumer demands, it enhances and expands existing ones. Finally, radical innovation is a relatively rare occurrence as it shares characteristics with disruptive innovation but distinguishes itself by simultaneously utilizing a novel business model and groundbreaking technology (Linton, 2009).

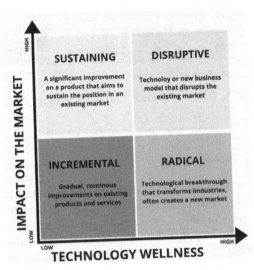

Fig. 2. Innovation Matrix (Linton, 2009)

Another pivotal theory that serves as the foundation for constructing the conceptual framework of this paper is the theory of the diffusion of innovations (as illustrated in Fig. 3). This theory gauges the circulation of newly generated opportunities within an existing ecosystem.

Initial research showed that the diffusion of ideas was widespread in tight-knit social groups like families and communities. These early studies primarily aimed to promote personal innovation by focusing on how people adopt new behaviors. Subsequently, researchers expanded their focus to examine how more complex technological innovations spread on a larger scale, encompassing areas such as business processes and information systems in sectors like healthcare and education (Ven et al., 1999).

Studies have argued that the adoption of innovation is a more intricate process within organizations compared to individuals. The innovation process is typically divided into five key phases, recognized across two primary stages: initiation, which involves outlining, and implementation, which encompasses redefining, reorganizing, and clarifying. Building on this theory, recent directives have emphasized the adaptation of innovation tools to align with the organizational context, fostering the development of novel innovation practices that enhance the organization's value and contribute to its success.

However, contemporary research has extended the scope of the innovation process beyond the boundaries of individual firms, encompassing a broader range of external domains, including projects, markets, and the overall business environment (Rogers, 2010).

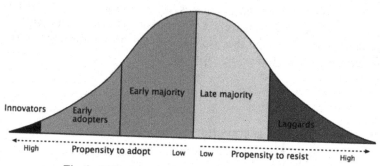

Fig. 3. Diffusion of Innovation Model (Rogers, 2010)

2.4 Individual Behavior Acceptance

The Theory of Planned Behavior (TPB), developed approximately two decades ago, has emerged as a potent framework for understanding human behavior. It has proven effective in explaining a wide array of behaviors. Interestingly, TPB has seen limited application in elucidating particularly intricate behaviors, such as managerial decision-making.

Creative thinking has garnered significant recognition as a pivotal element in driving employee success and ensuring an organization's sustainability and competitive edge in dynamic and demanding business environments (Anderson et al., 2014). The team's climate for innovation is acknowledged as a crucial determinant of employee creativity, underscoring the importance of fostering and rewarding collective efforts within the team. Nevertheless, prior research warrants further attention when it comes to investigating the role of personality in influencing the team's climate for innovation.

3   Research Framing

3.1   Theoretical Framework

In various literature contexts, the focus on process innovation can be categorized into two principal domains: Technological and Organizational innovation. Technological process innovations manifest as tangible outcomes, such as products enriched through technical enhancements, offering additional functionality and benefits. Conversely, organizational process innovations pertain to intangible strategies concerning process establishment, workplace dynamics, cultural environments, and frameworks for fostering creative thinking (Edquist et al., 2000). The formulation of the theoretical framework was conducted through factor analysis and empirical research, drawing from three main theories that contribute to the impact of the Innovation process on project performance and success factors. These theories include Institutional theory, Behavioral theory, and Attribution theory. Institutional theory revolves around the fundamental concept that structures featuring formal procedures, standards, and regulations exert an authoritative influence on social behavior and interactions. Organizational processes can either facilitate or impede innovation within this framework (DiMaggio and Powell, 2012).

Behavioral theory emphasizes the development of stability by minimizing misalignments and conflicts when striving to achieve objectives through strategies that align with constraints. The reaction of companies to performance discrepancies is scrutinized through attribution theory (Ford, 1985), which highlights the relationship between the effort invested in problem-solving and the value and uncertainty associated with the outcomes (Weiner, 1986). Senior managers should consider innovation as a viable remedy in cases of organizational decline, contingent upon causal attributions concerning the stability and controllability of the root causes (Rahimi et al., 2018).

Throughout the project life cycle, from initiation to completion and handover, project managers continuously generate potential opportunities and ideas to enhance value, ensuring the attainment of predefined targets and objectives. Given the market's volatility, rapid technological advancements, population growth, and heightened resource demands, innovation has become an essential competency for project managers. On the other hand, organizations grapple with challenges when it comes to developing and implementing a robust innovation process within their ecosystem, as outlined by Gallagher (2015). These challenges include managing risk, resistance to embracing new ideas to mitigate organizational risks and maximize the utilization of proven practices, knowledge gaps hindering the generation of innovative ideas, time constraints that leave little room for creative thinking and root cause analysis, and the impact of organizational size and structure on innovation tolerance.

Innovation necessitates a culture of collaboration deeply ingrained in an organization's DNA, highlighting the significance of fostering creative thinking and prioritizing team skills and expertise over positional authority. Building trust and influence from leadership is pivotal in promoting the adoption of the innovation process within the organization (Larson, 2015). Figure 4 provides a summary of the theoretical framework.

Fig. 4. Theoretical Framework for Project Performance.

3.2 Conceptual Framework

In line with Rogers' theory of spreading innovations, five key factors, namely relative benefit, compatibility, complexity, trialability, and observability, play a role in shaping the rate of acceptance. In this research, we introduce an additional dimension that requires investigation: the impact of individual acceptance on the innovation process's contribution. Consequently, we identify the dependent variable as financial performance, while the independent variables encompass technological factors, organizational factors, and individual behavioral factors. We have formulated four central hypotheses as follows:

Technological Factor

The oil and gas industry relies on well-established technology for the monitoring and control of process facilities, as well as for the remote development of extensive database and security systems, as indicated by Lappi et al. (2019). Achieving a seamless and effective integration of technology that is robust, high-speed, reliable, user-friendly, and maintenance-free will enhance the acceptance of the innovation process among all stakeholders within an existing business model. This, in turn, is expected to lead to positive outcomes such as accelerated decision-making, faster completion of deliverables, improved communication efficiency, time savings, enhanced overall efficiency, and increased productivity, as suggested by Almarri K. in 2023. As a result, we formulate the first hypothesis as follows:

Hypothesis 1: The adoption of the innovation process will be impacted by technological factors.

Organisational Factor

Introducing an innovation process across most of an organization can yield a dual impact: it can foster a drive for innovation development, while also potentially sparking resistance within the business, leading to potential risks. Given the perpetual nature of ongoing innovative changes, organizations must continually engage all available resources in the innovation process. However, the scalability of these resources may need enhancement at times to ensure consistent innovative transformations within the economic system. This phenomenon gives rise to various stochastic processes that, owing to their unpredictability and limited controllability, can have adverse effects on innovation effectiveness, as noted by Zotov et al. in 2020. Consequently, we posit the second hypothesis as follows:

Hypothesis 2: The adoption of the innovation process will be influenced by organizational factors.

Individual Behavior Factor

Differences in personality play a role in shaping the adoption of the innovation process. An individual's openness to trying new approaches and their level of risk aversion have a bearing on the encouragement of creative thinking. A positive attitude contributes value to the business unit, company, or industry. Individual behavior encompasses a spectrum of factors, including motivation, trust, mindset, fear, adherence to social norms, self-perception, risk perception, familiarity, acceptance, agility, and more (Roberts and Flin, 2019). Consequently, we formulate the third hypothesis as follows:

Hypothesis 3: The adoption of the innovation process is influenced by individual behaviors.

Project Performance

The hypotheses are rooted in one of two intellectual streams: the essential stream and the available stream. The essential stream posits that organizations or projects with lower performance levels seek to implement innovation as a means to improve their challenging circumstances (Amabile and Conti, 1999). One of the primary performance indicators for projects is financial performance, measured in terms of return on assets (ROA) as a profitability ratio, which gauges financial success. ROA is particularly relevant as a profitability indicator, especially when compared to metrics like return on equity, because it highlights how assets or resources are utilized to generate revenue, a significant consideration in capital-intensive industries like oil and gas (Merrow, 2012). Hence, we formulate the fourth hypothesis as follows:

Hypothesis 4: The innovation process will have a beneficial impact on project performance (Fig. 5).

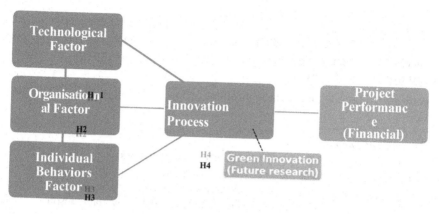

Fig. 5. Proposed Conceptual Framework

4   Discussion

Embracing positive disruption stands as a pivotal initiative for organizations, propelling them from a sluggish status quo into a transformative era characterized by heightened productivity and efficiency. Regardless of the size of the firm or organization, project practitioners serve as the driving force behind innovation. In many organizations, the ideation process is relatively straightforward. However, the true challenge lies in translating those ideas and opportunities into viable concepts and subsequently evolving them into finalized products or services.

To effectively manage breakthrough or innovative projects, it becomes imperative to blend fresh modes of thinking with familiar processes. In sectors like oil and gas, which are hubs for innovation, critical businesses seek secure and cost-effective solutions to align with global strategic growth plans. This research identifies four key stages for integrating the innovation process into significant projects:

1. *Bridging Stage*: This stage, which occurs during the initiation phase, involves preparation during front-end planning. It integrates innovative practices, technologies, and organizational processes to enhance project performance.
2. *Engaging Stage*: Focuses on establishing an innovative incentive scheme to engage stakeholders, simplify tendering and contracting procedures, and implement a reward system that encourages creative thinking, crowdsourcing, and innovative solutions.
3. *Leveraging Stage*: Occurs after contract award, involving the entire supply chain. It establishes a platform that brings together all stakeholders to develop new opportunities, enhance practices, and upgrade technologies.
4. *Exchanging Stage*: Takes place after project completion, facilitating the sharing of lessons learned and innovation history throughout the entire organizational ecosystem, thus enabling the exchange of experiences for future growth (Davies et al., 2014).

The literature review underscores a strong correlation between organization, technology, and innovation in driving digitalization and Artificial Intelligence as key contributors to improved firm performance. Given its iterative nature, the agile project management approach proves particularly suitable for creative projects characterized by a high degree of uncertainty.

Other theories accentuate the significance of human factors in the equation, emphasizing the level of acceptance and buy-in from individuals within the organization. This includes leadership support and employee engagement as critical elements for the successful adoption of innovation and technology within the organizational ecosystem, ultimately leading to value creation.

The ensuing relationships can be summarized as follows:

$$\text{Organisation} + \text{Innovative Technology} \rightarrow \text{Improved Performance} \qquad (1)$$

$$\text{Improved Performance} + \text{Enthusiastic Support by individual} \rightarrow \text{Value Creation} \quad (2)$$

One limitation/gap analysis in this study, which needs further development and exploration, pertains to the association between "Green Innovation" within the oil and gas sector and its influence on company performance. As environmental awareness has grown, the link between green innovation and the market for energy-related products has gained prominence. Research by Miao et al. (2017) suggests that green innovation has the potential to enhance energy efficiency, and the adoption of advanced technology, as indicated by Waheed et al. (2020), can be advantageous in reducing overall energy consumption and intensity.

5 Conclusion

In examining the outcomes of diffusion and implementation processes, this research promises to offer invaluable insights for both academia and practical application. While previous studies on organizational innovation have predominantly focused on bureaucratic firms operating within hierarchical structures and single market segments, this research ventures to broaden the perspective. It recognizes that an organization's innovation capacity is intricately linked to its organizational climate, culture, and various related factors. By exploring these dimensions, this research aims to illuminate the pathways to successful innovation. One of the primary contributions of this research lies in its recognition of the pivotal role of individuals in driving innovation, regardless of whether they hold managerial or professional positions. It underscores the critical importance of sustaining their motivation for innovation, which serves as a catalyst for organizational progress.

Moreover, this study acknowledges the profound influence of workplace culture on individual innovation. By delving into how employees' perceptions of the organizations support for innovation shape their inclination towards innovative behavior, this research strives to uncover the key drivers of a culture conducive to creativity.

However, it is essential to acknowledge certain limitations that may be encountered during the course of this research proposal. These limitations include potential challenges in obtaining data from diverse organizational settings, as well as the inherent complexities associated with assessing and quantifying organizational climate and culture. Despite these challenges, the anticipated benefits of this research for both academia and practical application make it a compelling endeavor with the potential to enrich our understanding of innovation dynamics within organizations (Prather, 2000).

References

Almarri, K., Boussabaine, H., Al Nauimi, H.: The influence of risks on the outturn cost of ICT infrastructure network projects. Constr. Innov. **23**(1), 85–104 (2023)

Amabile, T.M., Conti, R.: Changes in the work environment for creativity during downsizing. Academy of Management Journal (1999)

Anderson, N., Potočnik, K., Zhou, J.: Innovation and creativity in organizations: a state-of-the-science review, prospective commentary, and guiding framework. J. Manag. **40**(5), 1297–1333 (2014)

Aronson, Z., Shenhar, A., Reilly, R.: Project spirit: placing partakers' emotions, attitudes, and norms in the context of project vision, artifacts, leader values, contextual performance, and success. J. High Technol. Manage. Res. **21**(1), 2–13 (2010)

Burmester, M., Henry, P., Kermes, L.S.: Tracking cyberstalkers: a cryptographic approach. Acm Sigcas Computers and Society **35**(3), 2-es (2005)

Christensen, C., Raynor, M.E., McDonald, R.: Disruptive innovation. Harvard Business Review, Brighton, MA, USA (2013)

Davies, A., MacAulay, S., DeBarro, T., Thurston, M.: Making innovation happen in a megaproject: London's Crossrail suburban railway system. Proj. Manag. J. **45**(6), 25–37 (2014)

Deeb, G.: Reasons why big companies struggle with innovation. Forbes (2014). Retrieved from http://www.forbes.com/sites/georgedeeb/2014/01/08/the-five-reasons-big-companies-struggle-with-innovation/

DiMaggio, P.J., Powell, W.W.: The Iron Cage revisited: Institutional isomorphism and collective rationality in organizational fields [1983]. Contemporary sociological theory 175 (2012)

Edquist, C., Hommen, L., McKelvey, M.: Product versus process innovation: Implications for employment. In: Edquist, C., McKelvey, M. (eds.) Systems of innovation: Growth, competitiveness, and employment, pp. 377–400. Edward Elgar Publishing, Cheltenham, UK (2000)

Ford, J.D.: The effects of causal attributions on decision makers' responses to performance downturns. Acad. Manag. Rev. **10**(4), 770–786 (1985)

Gallagher, S.: PMs in the c-suite: The future of project leadership. Project Management Institute (2015)

Lappi, T.M., Aaltonen, K., Kujala, J.: The birth of an ICT project alliance. Int. J. Manag. Proj. Bus. **12**(2), 325–355 (2019)

Larson, E., Larson, R.: I still do not have time to innovate: I'm too busy doing business analysis. In: Paper presented at PMI® Global Congress 2015. Project Management Institute, North America, Orlando, FL. Newtown Square, PA (2015)

Linton, J.: De-babelizing the language of innovation. Technovation **29**, 729–737 (2009)

Merrow, E.W.: Oil and gas industry megaprojects: our recent track record. Oil and Gas Facilities **1**(02), 38–42 (2012)

Miao, C., Fang, D., Sun, L., Luo, Q.: Natural resources utilization efficiency under the influence of green technological innovation. Resour. Conserv. Recycl. **126**, 153–161 (2017)

Perrons, R.K.: How innovation and R&D happen in the upstream oil & gas industry: Insights from a global survey. J. Petrol. Sci. Eng. **124**, 301–312 (2014). Quarterly, 18, 105–120. https://doi.org/10.1016/j.leaqua.2007.01.002

Prather, C.W.: Keeping innovation alive after the consultants leave. Res. Technol. Manag. **43**(5), 17–22 (2000)

Rahimi, M., Kenworthy, T.P., Balakrishnan, J.: An analysis of innovation in oil and gas projects. Proj. Manag. J. **49**(5), 64–84 (2018)

Roberts, R., Flin, R.: Best practices for the introduction of new technologies: Investigating the psychological dimension (2019)

Rogers, E.M.: Diffusion of innovations. Simon and Schuster (2010)

Schneider, B., González-Romá, V., Ostroff, C., West, M.A.: Organizational climate and culture: Reflections on the history of the constructs in the Journal of Applied Psychology. J. Appl. Psychol. **102**(3), 468 (2017)

Shenhar, A.J., Dvir, D.: Reinventing project management: the diamond approach to successful growth and innovation. Harvard Business Review Press (2007)

Ven, A.H.V.D., Polley, D.E., Garud, R., Venkataraman, S.: The innovation journey. Oxford University Press (1999)

Waheed, R., Sarwar, S., Mighri, Z.: Role of high technology exports for energy efficiency: empirical evidence in the context of gulf cooperation council countries. Energy & Environment **32**(5), 803–819 (2021)

Weiner, B.: An attributional theory of achievement motivation and emotion. Psychol. Rev. **92**(4), 548–573 (1986)

Zotov, M.A., Ponikarova, A.S., Kadeeva, E.N.: Balanced management of innovative sustainable development of the petroleum and gas chemical complex. In: IOP Conference Series: Materials Science and Engineering, Vol. 971, No. 5, p. 052023. IOP Publishing (2020)

The Role of Adaptive and Transformational Leadership in the Successful Adoption and Implementation of New Technologies and Innovations in Organizations

Makram Abdel Malek[✉] and Khalid Almarri

The British University in Dubai, Dubai, United Arab Emirates
22000318@student.buid.ac.ae, khalid.almarri@buid.ac.ae

Abstract. With the increasing expansion and influence of technological innovations and breakthroughs, and as the world embarks into more digitization, automation, and man-machine integration, it becomes more vital than ever for businesses to identify and address any negative implications and drawbacks this advancement might have on the human factor in terms of people's acceptance and implementation of such innovations. One key theory that addresses this matter is the Unified Theory of Acceptance and Usage of Technology (UTAUT) which identifies and correlates variables that govern people's willingness and readiness to foster new technology. Furthermore, a major influencing factor of people's drives and attitudes in organizations is the leadership style and approach. Along those lines, the purpose of this research paper is to propose a conceptual framework to explore and study the role of Transformational and Adaptive leadership styles on the adoption of new technologies and innovations in organizations within the scope and frame of the UTAUT. The study follows a systematic literature synthesis methodology to gather and analyze information from multiple sources and provide a comprehensive overview of existing research on the topic, based on which an explanatory and guiding framework is proposed for enhancing people's successful adoption of new technologies.

Limitations to be highlighted are that the study bases its findings and analysis on previous research, and does not conduct primary research and data collection, no empirical findings are thus generated and presented to support or disprove the raised propositions. The study also addresses the topic conceptually and in general terms with no specific context, site, or sampling demographics, which are all influential factors that can alter findings. The paper however highlights the importance of sound and proactive leadership in enhancing organizational learning initiatives, innovation, and openness to technological and digital advancements, to maintain organizational sustainability and competitiveness, and thus sheds light on a critical management aspect in the digital and high-tech age which might have very harmful consequences if missed. It presents proposals on the influence of transformational and adaptive leaderships on adopting innovation and new technology, and calls for more study and primary research on the topic for further exploration and understanding.

Keywords: Transformational leadership · Adaptive leadership · Innovation · Agile strategies & structures

M. Papadaki et al. (Eds.): EMCIS 2023, LNBIP 502, pp. 120–132, 2024.
https://doi.org/10.1007/978-3-031-56481-9_9

1 Introduction

In the contemporary world, it is a rare occurrence for a day to transpire without the introduction of novel ideas or concepts within various fields, industries, or facets of life. The innate impetus of creativity and innovation has consistently propelled humankind forward, leading to groundbreaking advancements in technology and industry across all sectors. As individuals, we possess an inherent proclivity towards novelty; we seek diversity of options and crave innovation, whether in the form of new products or the revitalization of existing ones.

This reality holds profound significance for businesses and organizations on a broad scale. Consumers perpetually anticipate the evolution of goods and services, particularly in our present era characterized by rapid technological progress and heightened competition. Amidst these challenges and the mounting demand for fresh offerings, businesses must perpetually generate innovative ideas and concepts to not only survive but thrive in this fiercely competitive landscape [12].

Examples on this are countless, be it in the constant introduction of renovated consumer goods like vehicles or technological products, or in the rebranding that companies make every certain period, all for the purpose of sustaining and growing their client base.

Furthermore, and considering the modern digital business environment, it is not enough anymore for companies to renovate but it's rather becoming more vital to *innovate*. Companies need to think out of the box and create breakthrough innovations rather than stay satisfied with simple incremental modifications to the products they already have. With the massive advancement in telecommunication and digital solutions, disruptive innovations are finding their way into almost every industry. For that, companies need to transform their thinking and mindsets, and create new products and services that better meet customer requirements and needs, and hence enhance organizational performance and success [14].

Of the most important factors that would help organizations create a forward-thinking and progressive culture is a "Leadership" that believes in those values and perspectives and adopts them [14]. Research on the relationship between leadership and innovation shows clear correlation between the two, leaders have the authority and influence to affect organizations' practices, policies, and procedures; and the extent to which they adopt innovative strategies and cultures [12]. Thus, having the right kind of leadership becomes very vital to organizations' abilities to innovate. In this context, Transformational and Adaptive Leadership (TL and AL) are seen as the most suitable and relevant leadership styles to sponsor and drive creativity and innovation in organizations. Transformational leaders promote the excitement and momentum among their team members to think laterally and in non-conventional ways [12] while Adaptive leaders mobilize people to deal with changing circumstances and accompanying challenges through a shared learning and exploration process of both the leader and the followers together [6].

What makes leadership more particularly important at these times of strong innovative high-tech disruption, is the need to understand people's perceptions and feelings towards the sweeping wave of technological change for it is only the people in those organizations who can achieve the successful transformation to advanced technology utilization, and hence they need to accept, adapt, and adopt those technological solutions for them to succeed [11]. Another major reason for closer and stronger leadership support in this area is that most research studies address the technical features of high-tech industry implementation and very little attention is given to people's attitudes and acceptance levels of new technologies and this resembles a key and fundamental gap to address [17] for the sound and smooth organizational transition and sustainability.

1.1 Aim of the Research

Considering the above global business setting, and the importance of having the right direction and drive for organizations to continually develop and innovate; and knowing that people are generally reluctant to accept change, let alone create it! This paper attempts to address the identified gap, and to highlight and explore the role and influence of transformational and adaptive leaderships on people's attitudes, intentions, and actions to adopt and implement new innovations and solutions in their organizations. The study seeks to answer the following questions:

1.2 Research Questions

How do transformational and adaptive leadership styles help employees and organizations in successfully adopting and adapting to new innovations, technologies, and operational procedures? And how can such leaders help employees overcome obstacles and challenges that keep them from successfully accepting and embracing new technologies and innovations?

In pursuit of answering these questions, the study correlates and connects the attributes of Transformational and Adaptive leadership styles to the variables and components of the Unified Theory of Acceptance and Use of Technology (UTAUT) introducing a hypothetical analysis of how the characteristics of TL and AL can represent solutions to the different hindering factors identified in the UTAUT.

The following sections include a theoretical review and background setting of TL, AL, and the UTAUT; followed by an analytical correlation of those leadership styles' attributes to the variables that affect the intention and use of new technology as per the UTAUT. The study closes with a discussion of the research limitations and gaps and the room for more future study, followed by a general conclusion of the revised literature, the propositions made, and the relationship between the different variables involved.

2 Literature Review and Related Theories

2.1 Transformational Leadership

According to Transformational Leadership (TL) Theory, superior leadership occurs when leaders trigger employees' thinking into higher levels of reflection and awareness, strengthening their adoption of organizational goals, and raising their commitment to achieving them [3]. Transformational leaders are characterized by five major attributes which contribute to building and strengthening an innovation culture and mode of thinking [12], each in its own way. The first of these attributes is *Idealized Influence*, through which leaders act as role models influencing and inspiring followers through their behaviors and gaining their trust and respect. Leaders with idealized influence encourage risk taking and are reliable supporters to their followers in doing so. Idealized influence is also displayed through leaders inspiring a vision of self-aspiration and confidence, instilling ambition and pride in fulfilling higher potentials of self-actualization and achieving personal and organizational breakthroughs. Leaders also display idealized influence by showing their excitement about the new technology and their firm belief that it leads to a better organizational and professional future [4].

Strong charisma is the second key feature of TL through which leaders instill feelings of confidence and drive in their followers and teams; and raise their commitment to the leader's and the organization's vision and goals [12]. Charisma is also seen as the extent to which the leader's behaviors are admired by his/her subordinates and followers making them identify and relate to him/her mentally and psychologically [15] *Inspirational motivation* and *intellectual stimulation* are also attributes of transformational leaders that widen people's perspectives and broaden their thinking into solution-finding scopes. Inspirational motivation involves a leader's emotional characteristics and influence [4]. The Inspirational motivation attribute is a core characteristic of leadership at large because leadership in essence is all about connecting with people's core values and morale, having them close to heart and genuinely caring for them while challenging them positively and always extending a helping hand. According to [3], leaders through intellectual stimulation promote innovative thinking, and support their followers in challenging the status-quo and questioning current practices. They also encourage their team members to see challenges and problems from different perspectives and discover new ways and solutions for resolving them [16].

The fifth distinguishing feature of transformational leadership is *Individualized consideration,* by which the leaders provide individual and personal attention to each of their followers, giving them close mentorship and guidance and prioritizing their and the organization's development and interests through positively challenging and coaching them [3, 12]. According to Bilal [5] the role of TL in triggering and stimulating employees' creative behavior and thinking has always caught the attention and interest of researchers, however; not enough study has been conducted in this area [5], and while the relationship between individuals' perception of TL and their creativity has been studied, little awareness has been given to the influence of Transformational Leadership on Innovation as such [5]. This perspective highlights two main considerations. First, it acknowledges the significance and influence of Transformational Leadership on people's creativity and innovation, and second, it recognizes the need for further research on the importance and influence of TL in this regard. In that sense, the questions raised in this research paper align and relate to existing literature in its quest to further study the correlation between Transformational Leadership and employees' innovation and acceptance of new innovative ideas and applications.

2.1.1 Influencing Factors of Transformational Leadership on Technology Acceptance and Innovation

Literature identifies several possible causes behind people's reluctance to technological change, these include fear of losing one's job, fear of digital surveillance by the advanced systems and mechanisms, and the perception of lacking appreciation of their efforts by their managers, as well as feeling alienated and outdated by the overwhelming modern technologies. For all those reasons and to retain and sustain people's confidence and engagement, strong leadership support has become extremely important to create safe and secure working environments that foster people's growth and well-being. Transformational leaders' socio-emotional abilities and their close sensing of people's individual concerns, insecurities, as well as unique capabilities are now more significant than ever in driving people's acceptance of new technologies, and this has so far been very rarely and weakly addressed [17].

In looking closer at the components and constituents of TL and its impact on individuals' innovation in organizations, researchers identified and analyzed a number of factors that would mediate and moderate the intensity and level of influence of TL on people's creativity and innovativeness.

2.1.2 Mediating Variables

According to Bass 1999, *psychological empowerment* is a motivational factor that acts as a mediator to Transformational Leadership in its impact on innovation [5]. It is also believed that internal motivation which is created by transformational leaders through psychological empowerment is integral to innovation and creativity. In [7], researchers also studied mediating and moderating factors of Transformational Leadership on innovation at individual, team, and organizational levels, and they identify *self-efficacy*, which is a person's belief in their own capacity to meet objectives and achieve results, as a mediator to TL as well [7]. At the team level, they identify **team reflection** and **team heterogeneity** as mediators, where team reflection is defined as the common and collective consideration by the team members of the team's objectives, aims, and strategies, meaning that the stronger the members' commitment to the team's aims and goals, the stronger the transformational leadership's influence on the innovation initiatives of its members. *Team heterogeneity* is also identified as a mediator, whereby the more heterogeneity on a team, the more creative and of diverse knowledge its members are [7].

2.1.3 Moderating Variables

Factors considered by researchers to moderate the effect of TL on innovation include *self-construal* according to Gabriel et al., 2007 as cited in [5], which is the extent individuals see themselves as connected (dependent self-construal) or disconnected (independent self-construal) from fellow employees and the organizations at large. Also, according to Lord et al. (1999) and Lord and Brown (2004) and as depicted in [5], Transformational Leadership is more impactful when interdependent self-construal is stronger.

Organization self-esteem, which is how valuable to the organization an individual sees him/herself and *Self-presentation* are also perceived as moderators at the individual level according to [7].

2.2 Adaptive Leadership

While transformational leadership is based largely on the leader's influence on their followers' perceptions, attitudes and drives, Adaptive leadership is seen more as a process than individual ability of either the leader or the follower(s) [6]. Adaptive leadership involves the collaborative learning of both leader and followers of a new phenomenon or problem. It entails a shared responsibility on both followers and leaders to explore new happenings and circumstances and "adapt" to them in the most effective and efficient way. Though it is a shared process of learning between leaders and their followers, it is primarily the leader's responsibility and role to initiate and drive that process being in the more influential and responsible role and position. An adaptive leader is considered successful when their followers adopt new norms and practices and adjust to the "new" way of doing things. Adaptive leaders achieve that through a receptive and open-minded approach and encourage followers to be inquisitive and explorative of new modes and trends (Castillo, G.A. 2018). AL is mainly illustrated through three key sets of activities, 1) identifying and examining new developments and happenings taking place in organizational settings. These can include any changes in internal or external circumstances and situations, such as changes in market dynamics, changes in policies and procedures or public regulation, as well as any changes and advancements in applicable systems and technologies. 2) Analyzing these developments and happenings, and seeking to learn and understand their underlying and related features and consequential implications, and aiming to better correlate and adapt to these implications; and 3) taking action and initiative in addressing and meeting the desired and required adaptation according to Heifetz et al., 2009 as cited in [6]. In the context of introducing new technologies and innovations, AL would first acknowledge and accept the introduction of the new and developed technology, assess its implications and impact on the individual and organizational levels, and act towards better familiarization with it and usage of it. Adaptive leaders apply those activities and initiatives in a collective and collaborative manner. They provide coaching support, clear guidance and direction, and delegate responsibilities to their subordinates and teams throughout the process as a learning and exploration opportunity [6].

2.3 The Unified Theory of Acceptance and Use of Technology (UTAUT)

Another key theoretical reference to the study of Transformational and Adaptive Leadership influence on innovation is the Unified Theory of Acceptance and Use of Technology (UTAUT). UTAUT identifies aspects of people's reluctance to adopt new technology. People are generally hesitant when it comes to change, even though they desire new inventions and ways of doing things, when it comes to changing certain habits or adapting to new systems and procedures, people rather push-back and prefer to keep to ways they are used to and to familiar routines they feel comfortable with.

UTAUT is a combination of study frameworks and models aimed at studying individuals' acceptance and use of information technology. It was put together by Professor Viswanath Venkatesh, a prominent and reference scholar in Business Information Technology, along with other scholars. Venkatesh and his fellow researchers combined the Technology Acceptance Model (TAM) and other models based on the Theory of Planned Behavior (TPB) into the UTAUT [18]. The Unified Theory of Acceptance and Use of Technology presents four main components/variables that affect people's intention to use new technology. These are Performance expectancy, Effort expectancy, Social influence, and Facilitating conditions. Performance expectancy refers to how strongly an individual believes that the new technology or system at hand is likely to help improve his/her work experience and output. Effort expectancy refers to the amount of perceived effort needed to learn how to use the new system(s) and tools. Social influence is the impact of peer individuals and groups on subject individuals and those peer groups' belief that the concerned person should learn and use the new technology or system at hand [2]. Thus, individuals or organizations who are successful in using new technologies would have strong influence on those still attempting to do so as per Popovic et al., 2016 as cited in [2]. Lastly, facilitating conditions are the surrounding supporting circumstances and initiatives such as having suitable system infrastructures, proper technical support, and guidance; as well as relevant training and upskilling when it comes to the new systems as per Lee et al. 2018 as cited in [2].

The UTAUT theory also presents moderating variables including individual factors such as age, gender, level of experience and specialization, as well as the readiness and voluntariness to use new technology and systems. UTAUT also considers a person's level of technology anxiety and adaptability [18]. Having partially stemmed from the Technology Acceptance Model (TAM), the UTAUT shares basic features with TAM that have direct influence on people's tendency and readiness to use new technology. According to TAM, the level of usefulness, and ease of use as perceived by potential users have a direct impact on their willingness and intention to use new technology [15].

Here according to [15] leaders can significantly influence those perceptions whereby if a leader believes and communicates that the use of the new systems and technology will make it easier for individuals to achieve their goals and those of the organization, employees would be more likely and willing to try the new technology. Also, through their *Inspirational motivation* and *intellectual stimulation* attributes and attitude, transformational leaders would positively challenge and support their team members to explore and experiment with the new technology increasing chances of successfully adopting it.

Further research has also identified a positive correlation between TL and innovation in organizations. Transformational leadership is perceived to be among the leadership styles that most emphasize and promote innovation according to Church & Waclawski, 1998; and Howell & Higgins, 1990 as cited in [13]. Transformational leadership is also positively associated with increased R&D and innovation initiatives and performance, according to Keller, 1992, Waldman & Atwater, 1994 and others, as cited in [13].

From all the above cited research on the attributes and influences of TL and AL on working environments and employees' drives, and from the definition and highlights of the UTAUT, in addition to the numerous previous research identifying a positive relationship between TL and AL on one hand and employee innovation on the other, we propose a positive correlation between those constituents and suggest the following propositions:

Proposition 1: Transformational Leadership through its five key attributes of Idealized Influence, Strong charisma, Inspirational motivation, intellectual stimulation, and Individualized consideration will have a positive influence on the successful adoption of new technology and innovation in organizations through addressing the intention and usage variables of the UTAUT, Performance Expectancy, Effort Expectancy, Social Influence and Facilitating conditions.

Proposition 2: Adaptive leadership through its open-minded learning attitude and approach also mitigates the influence of innovation inhibiting factors and facilitates the acceptance and use of new technology and innovation.

Further elaboration and discussion on the correlation of the above interrelated variables and their graphic presentation is developed in the following section.

Framework

According to previous research, and as referred to above, transformational leadership is infused with motivating and inspiring influence and drive. Transformational leaders instill a sense of purpose and mission in their teams and followers, as well as high intellectual awareness of goals and aspirations [5]. Through those features and its five main characteristics, TL would counter and hinder any deterring effect to innovation imposed by the UTAUT's intention and usage variables. Through inspirational motivation and intellectual stimulation, transformative leaders can also help their team members face and overcome any difficulty the Effort Expectancy variable can have in their intention to adopt new technology. They would motivate their subordinates to engage in learning the new system and technology and exploring its features.

Similarly, characteristics of Adaptive leadership create "Cognitive and behavioral functional change" which Heifetz refers to as a mobilizing behavior that enhances curiosity and focus to learn and explore the features of change at hand. Adaptive Leaders exhibit a "learning by example" behavior which inspires followers to commit to progress and advancement initiatives through collaborative learning [6].

3   Conceptual Model

The following conceptual model presents the correlation among the related variables. The model illustrates how Transformational and Adaptive leadership styles connect to the framework components of the Unified Theory of Acceptance and Use of Technology (UTAUT). As per the UTAUT, and as explained earlier, there are four "intention and usage variables" for adopting and using new technology, these are: Performance expectancy, Effort expectancy, social influence and Facilitating conditions. These constructs have a direct effect on the "Behavioral intention" and the "Use Behavior" of new technology. In other words, these four constructs influence the extent to which people are inclined, and intend, to use new technology and consequently behave as such. In turn, and as this paper proposes, these "intention and usage variables" are strongly influenced by Transformational and Adaptive Leaderships, where through their attributes of Idealized influence, Strong charisma, Inspirational motivation, Intellectual stimulation and Individualized consideration, these leadership styles inspire and motivate employees to adopt and explore new technologies and innovations by instilling convictions that new technology helps improve performance and efficiency, and subsequently personal development and growth which drives people to endorse and foster their application.

Also, through their motivational inspiration, intellectual stimulation and individualized consideration, transformational leaders provide encouragement and solid social influence in favor of adopting new technologies. They associate positively to people's concerns and doubts about performance efficiency and project feelings of confidence and security that new technologies would make work more enjoyable and increase efficiency and effectiveness, rendering employees' work experience a more rewarding and fulfilling one and leading to the faithful adoption of that new technology. The same is true for the adaptive leadership feature of collaborative and engaged learning attitude and influence.

Transformational and Adaptive leaderships also instill reassurance when it comes to Effort expectancy, social influence, and Facilitating conditions showing people that the return on the invested costs and efforts in embracing and using new technologies would outweigh all the financial and non-financial costs and efforts. Transformational and Adaptive leaders also provide positive social influence by supporting their subordinates and creating healthy holding environments, in addition to securing the facilitating conditions that help people learn and adapt to the new techniques and systems, such as providing training and access to systems and information that help people succeed [17].

Subsequently, the positive influence on the "intention and usage variables" would impact the "Behavioral intention" and the "Use Behavior" dependent variables towards embracing and implementing new technologies and innovations.

Also as presented under influencing factors above, the relationship between TL and UTAUT's "intention and usage variables" is moderated by self-construal, Organizational self-esteem, and self-presentation, and mediated by psychological empowerment, self-efficacy and team reflection and heterogeneity. This reasoning is based on the understanding that self-construal, organizational self-esteem, and self-presentation provide the supporting environment that nurtures a positive correlation between the two sets of constructs. Similarly, Psychological empowerment, Self-efficacy and team reflection and heterogeneity are channels through which the correlation takes effect and gains positive impact.

Leadership is perceived through the literature as a major influencing factor on people's drive and engagement in professional as well as social settings. The proposed conceptual model details in a structured correlation how Transformational and Adaptive leaderships impact people's attitudes and drives towards adopting new technologies, through the structural framework of the UTAUT theory. This correlation represents a reasonably rational and scientific explanation of how the different variables interconnect as it refers to relevant research and theories in terms of people's motivation, drive, and inspiration to learning and development and to adapting and conforming to change.

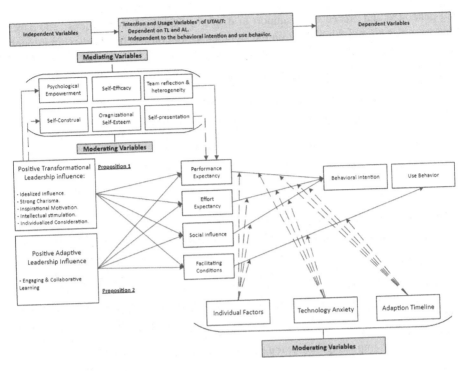

4 Methodology

The study follows a qualitative systematic review of pertinent literature to describe and correlate the included factors, Systematic literature reviews are practical tools that help in exploring research factors of the study and pinpoint associated gaps through the analytical exploration of concepts and themes pertinent to the subject. Qualitative content analysis was used to interpret text data contents through the structured categorization of the content into codes and the generation of themes and thought patterns [10]. Published literature was researched using "Google Scholar" and key terms used for the search included "Transformational Leadership impact on technology adoption", "Adaptive leadership and organizational innovation", and "Unified Theory of Acceptance and Usage of Technology".

In summary, the qualitative content analysis laid the foundation by systematically reviewing relevant literature, identifying key themes and concepts. The empirical analysis phase would involve further investigation, possibly through statistical methods, to validate and enhance the qualitative findings, ultimately contributing to a more robust and comprehensive understanding of the study's subject matter.

4.1 Limitations and Recommendation for Further Research

The study is limited to the conceptual review and analysis of related published literature with no primary research and data collection conducted, no empirical findings are generated and presented to support or disprove the raised propositions. The paper thus provides an initial highlight on the significance of constructive leadership through Transformative and Adaptive influence on new technology adoption, and this paves the way for further empirical research that would explore and assess the impact of these leadership styles within the framework of the UTAUT. This acquires particular importance since the topic is gaining increased importance and hopefully would receive more attention for enhanced academic and practical understandings and applications that can serve academy and practice in optimizing technology benefits both conceptually and practically.

5 Discussion

Despite the logical hypothetical correlation between Transformative and Adaptive leadership styles on one hand, and the positive behavioral intention to accept and adopt new technologies and innovations on the other, there is still good room for further empirical study and research of this topic. It would be wrong to generalize and consider the relationship to be affirmative and true under all circumstances and in all situations. Many studies have validated the relationship between the two sets of variables; however, many others still showed mixed results [13]. For example, while one study showed a positive relationship between TL and R&D teams' performance, it cannot be held as completely representative of impacting innovation per say of those teams (Keller, 1992; Waldman & Atwater, 1994 as cited in [13]). One study even found a negative correlation (Boerner, Eisenbeiss, & Griesse, 2007 as cited in [13]) and one did not find any shared influence at all [13]. Another study which assessed TL influence on smart government applications in the UAE concluded that TL did not have any important impact on the actual application of such smart systems [1], hence paving way to more research and implying the fact that this correlation is very contextual and needs to be addressed in light of particular settings, contexts and circumstances away from generalization. It is thus best to expand and diversify research in this field, and to generate and pursue research questions with more focused scope and more defined contextual parameters.

6 Conclusion

The developed conceptual framework assumes paramount importance when substantiating the aforementioned points with research evidence. In an era marked by relentless and accelerating technological advancements and innovations, businesses and organizations must fully grasp the gravity of this reality. Consequently, they must utilize every available means and resource to not only adapt but thrive in the face of perpetual change and heightened competition.

In this context, certain attributes take on paramount significance. Businesses must cultivate agile and dynamic strategies, underpinned by forward-thinking mindsets, in order to remain competitive. However, achieving such a state of readiness is contingent upon the presence of attentive and proactive leadership. Among the most widely acclaimed leadership styles in this context are Transformational and Adaptive leadership.

Research demonstrates that these leadership paradigms are revered for their substantial impact on fostering innovation and facilitating the adoption of technology. Numerous studies and empirical evidence consistently highlight their effectiveness in driving organizational growth and adaptability. While it is acknowledged that they may not be entirely devoid of imperfections, one undeniable truth remains: change will persist unabated. Therefore, organizations must continually embrace change with the utmost agility and flexibility to navigate the evolving landscape successfully.

At its core, this underscores the critical importance of the proposed conceptual framework, which aligns with established research findings and provides a robust foundation for organizations seeking to thrive amidst the relentless tide of technological transformation.

References

1. Al-Obthani, F., Ameen, A.: Association between Transformational Leadership and Smart Government among Employees in UAE Public Organizations. Int. J. Emerg. Technol. 10, 98–104 (2019). Available at: https://scholar.google.com/scholar
2. Baharuden, A.F., Isaac, O., Ameen, A.: Factors Influencing Big Data & Analytics (BD&A) Learning Intentions with Transformational Leadership as Moderator Variable: Malaysian SME Perspective. Int. J. Manage. Human Sci. (IJMHS 3(1), 10–20 (2019). Available at: https://ejournal.lucp.net/index.php/ijmhs/article/view/791/707
3. Bass, B.M.: From transactional to transformational leadership: Learning to share the Vision. Organizational Dynamics 18(3), 19–31 (1990). Available at: https://doi.org/10.1016/0090-2616(90)90061-s
4. Bednall, T.C., Rafferty, E., Shipton, A.H., Sanders, K., Jackson, J.C.: Innovative behaviour: how much transformational leadership do you need? British Journal of Management 29(4), 796–816 (2018). https://doi.org/10.1111/1467-8551.12275
5. Bilal, A.: Transformational leadership and Innovative Work Behavior. Academy of Management Proceedings 2019(1), 11866 (2019). Available at: https://doi.org/10.5465/ambpp.2019.11866abstract
6. Castillo, G.A.: The importance of adaptive leadership: management of change. Int. J. Novel Res. Edu. Learn. 5(2), 100–106 (2018). Available at: www.noveltyjournals.com
7. Denti, L.E.I.F., Hemlin, S.V.E.N.: Leadership and innovation in organizations: a systematic review of factors that mediate or moderate the relationship. Int. J. Innov. Manage. 16(03), 1240007 (2012). Available at: https://doi.org/10.1142/s1363919612400075
8. Gong, Y., Huang, J.-C., Farh, J.-L.: Employee learning orientation, transformational leadership, and employee creativity: the mediating role of employee creative self-efficacy. Academy of Management Journal 52(4), 765–778 (2009). Available at: https://doi.org/10.5465/amj.2009.43670890
9. Heifetz, R.A., Linsky, M., Grashow, A.: The practice of adaptive leadership: Tools and tactics for changing your organization and the world. Harvard Business Press, Boston, MA (2009)
10. Hsieh, H.-F., Shannon, S.E.: Three approaches to qualitative content analysis. Qualitative Health Research 15(9), 1277–1288 (2005). Available at: https://doi.org/10.1177/1049732305276687
11. Mikulić, I., Štefanić, A.: The Adoption of Modern Technology Specific to Industry 4.0 by Human Factor, p. 0941 (2018)
12. Mokhber, M., Khairuzzaman, W., Vakilbashi, A.: Leadership and innovation: the moderator role of organization support for innovative behaviors. J. Manage. Organiz. 24(1), 108–128 (2017). Available at: https://doi.org/10.1017/jmo.2017.26

13. Pieterse, A.N., et al.: Transformational and transactional leadership and innovative behavior: the moderating role of psychological empowerment. Journal of Organizational Behavior 31(4), 609–623 (2009). Available at: https://doi.org/10.1002/job.650
14. Schiuma, G., et al. The transformative leadership compass: Six competencies for digital transformation entrepreneurship. Int. J. Entreprene. Behav. Res. 28(5), 1273–1291 (2021). Available at: https://doi.org/10.1108/ijebr-01-2021-0087
15. Schepers, J., Wetzels, M., de Ruyter, K.: Leadership styles in technology acceptance: do followers practice what leaders preach?. Manag. Ser. Q. Int. J. 15(6), 496–508 (2005). Available at: https://doi.org/10.1108/09604520510633998
16. Tajasom, A., et al.: The role of transformational leadership in innovation performance of Malaysian smes. Asian J. Technol. Innovat. 23(2), 172–188 (2015). Available at: https://doi.org/10.1080/19761597.2015.1074513
17. van Dun, D. H., Kumar, M.: Social enablers of Industry 4.0 technology adoption: transformational leadership and emotional intelligence. Int. J. Operat. Prod. Manage. 43(13), 152–182 (2023). https://doi.org/10.1108/ijopm-06-2022-0370
18. Venkatesh, V., Thong, J., Xu, X.: Unified theory of acceptance and use of technology: a synthesis and the road ahead. J. Assoc. Info. Sys. 17(5), 328–376 (2016). Available at: https://doi.org/10.17705/1jais.00428

Managing Information Systems

Information and Communication Technology Enabled Collaboration: Understanding the Critical Role of Computer Collective-Efficacy

Andrew Hardin[✉] [ID]

University of Nevada, Las Vegas, NV 89154, USA
andrew.hardin@unlv.edu

Abstract. The organizational use of information and communication technologies (ICTs) presents sociotechnical challenges for organizational managers. One significant benefit of ICTs is facilitating the collaborative efforts of geographically dispersed teams of individuals with diverse skills. However, introducing ICT into these collaborative situations can compound the perceived barriers associated with these teams' typical virtuality constraints. Conversely, ICTs can facilitate better communication and performance despite these limitations. In the current study, we build on social cognitive theory propositions underlying the concept of collective efficacy to reconceptualize and re-operationalize the computer collective-efficacy construct. Collective efficacy has seen a recent surge of interest in ICT-mediated collaborative team environments because it strongly predicts performance. Although collective efficacy shows promise in these settings, it does not reference technology and, therefore, is not well-suited for gauging the sociotechnical aspects of using ICTs for collaborative purposes. Computer collective-efficacy, however, directly references ICT-mediated teams' abilities to utilize computer technology. Thus, researchers and practitioners should consider it an essential predictor of ICT-mediated team success. In this study, we discuss the theoretical framework, conceptualization, and operationalization of the computer collective-efficacy construct and provide a validated measure researchers can apply in future ICT-mediated collaborative research. We then conclude with significant theoretical and practical implications of this research.

Keywords: Computer collective-efficacy · technology-enabled collaboration · information and communication technologies · instrument development · structural equation modeling · bootstrapping

1 Introduction

Collaborating with individuals possessing diverse talents across different locations using information and communication technologies (ICTs) is essential for solving organizational and societal challenges [7, 23, 34]. The widespread deployment of geographically distributed ICT-mediated collaboration increased during the COVID-19 pandemic [8,

M. Papadaki et al. (Eds.): EMCIS 2023, LNBIP 502, pp. 135–147, 2024.
https://doi.org/10.1007/978-3-031-56481-9_10

42], and this trend is expected to continue, presenting new opportunities and challenges for managers [31].

Practitioners and researchers have intensified their efforts to understand the factors influencing the success of ICT-mediated, geographically distributed teams and to identify ways to address their challenges [13, 17, 30, 36, 45]. This form of collaboration is becoming more common due to digital advancements. Thus, further research is needed to explore new themes arising from this shift [43]. One contemporary concept in ICT collaboration research is collective efficacy (CE) [9, 18, 25, 32, 44, 48]. Researchers argue that CE, or a team's shared belief in its collective ability to work effectively, may help overcome the challenges of collaborating in geographically distributed ICT-mediated environments [32].

While the increasing attention to the impact of collective efficacy on ICT-mediated collaboration is notable, computer collective-efficacy (CCE), defined as a team's belief in its collective ability to utilize technology [14], is absent from this literature. Self-efficacy (SE) is an essential element within Bandura's (1997) social cognitive theory (SCT) and has been adapted for use within numerous research domains, including the application of computer self-efficacy (CSE) in the IS literature [22]. However, researchers have recognized that CSE is less useful in distributed team settings and thus have relied on CCE to predict performance [24] and virtual team efficacy (VTE) [14] in ICT-mediated environments.

Efforts to understand and apply CCE in ICT-mediated collaboration research are still evolving. A Google Scholar search using "computer collective efficacy" returns approximately 38 records but only four peer-reviewed studies directly focusing on CCE. One study captures software application-specific CCE [24], while the other three measure general computer skills CCE [14, 20, 21]. Providing evidence for the need to clarify the use of CCE, a recent study uses the term "collective technology efficacy" to reference the measure used by [20]. However, the authors did not directly adopt the measurement items or validate the indicators they eventually used to measure the adapted version of CCE [33]. Therefore, the operationalization of CCE requires additional clarification before researchers can effectively apply the measure during future studies of ICT-mediated collaboration.

Thus, this study contributes to the ICT literature by reconceptualizing a general CCE measure and discussing its operationalization at the collective level. Conducting empirical studies that advance CCE as a technology-focused conceptualization of CE not only addresses this need but also helps establish it as an essential component of IS scholarship that firmly embraces the sociotechnical tradition of IS research [40]. To achieve this outcome, we briefly introduce the critical theoretical foundation associated with the SCT framework and the general concept of SE. We then discuss collective efficacy (CE) and its relationship to SCT and SE. We next discuss CE in the computer domain by examining CCE as a general, collective-level computer efficacy measure. As a secondary objective, we address the advantages and disadvantages of current measurement approaches and develop and validate a general computer skills CCE measure. Finally, we discuss the theoretical and practical implications of this research.

2 Theory

2.1 Social Cognitive Theory

Bandura's development of social cognitive theory (SCT) [3] is based upon his earlier conceptions of social learning theory [2]. These theories address the short comings of personal and environmental determinism. Rather than grant that personal or environmental determinants drive human behavior or that a bi-directional interaction between the person and the environment causes behavior, [4] proposed a triadic reciprocal relationship between the person, the environment, and behavior. The concept of self-efficacy (SE) is central to SCT, which we will discuss next.

2.2 Self-efficacy

SE is "the belief in one's capabilities to organize and execute the course of action required to produce given attainments" ([4], p. 3). Efficacy beliefs form during a cognitive process that integrates enactive mastery, vicarious experience, verbal persuasion, and physiological and affective sources of information. *Enactive mastery* provides the most influential source of efficacy information and is based on knowledge gained from the performance of a given behavior. As an individual experiences success or failure performing a given behavior, their efficacy beliefs may increase or decrease depending on the interpretation of the performance. *Vicarious experience* provides a second source of efficacy information. People encounter vicarious experiences through visualizing themselves or observing others performing a given behavior. *Verbal persuasion* represents a third source of efficacy information. This source of efficacy generally comes in the form of performance feedback and, based on the valance of the input, can either raise or lower efficacy beliefs. *Physiological and affective states* provide the final source of efficacy information. Physiological information is most relevant during the performance of physical activities, with factors such as muscle soreness representing performance feedback. Affective information is associated with the influence of states of arousal, such as anxiety. The result of anxiety on cognition relates to cognitive capacity or the limited amount of attention an individual can focus on any task at any given time. Arousal states such as anxiety demand cognitive resources that individuals would otherwise use for attending to attentional processes [12].

3 Collective Efficacy

Bandura defines CE as "a group's shared belief in its conjoint capabilities to organize and execute the courses of action required to produce given levels of attainments" ([4], p. 477) and further proposes that CE operates within social cognitive theory like SE [5]. The following quote emphasizes this similarity: "Perceived personal and collective efficacy differs in the unit of the agency, but in both forms, efficacy beliefs have similar sources, serve similar functions, and operate through similar processes" ([4], p. 478).

Recent ICT-mediated distributed collaboration research establishes that CE moderates the influence of teamwork competency on satisfaction [1], is positively related to virtual team motivation [9], is predicted by trustworthiness [25], predicts virtual team performance over time [32], is an outcome of physiological synchronicity [39] and is positively related to team cohesion, social presence, and performance [48]. Finally, information systems researchers report that VTE, a geographically distributed ICT-mediated collaboration-specific measure of CE, predicts perceptual and objective performance in university and organizational settings [14, 20, 21, 41].

3.1 Collective Efficacy Assessment

Group members' aggregated SE beliefs predict the performance of groups completing tasks with limited interdependence [46]. Using group members' aggregated individual SE beliefs for predicting team performance makes sense conceptually, as the ability of the individual participants to complete their tasks independently is critical to the team's overall success. However, this method has seen limited use as group research has generally focused on the performance of groups completing interdependent tasks. Nonetheless, when used in this specific capacity, the predictive properties of the group-level aggregation of SE beliefs have been previously established [46].

Researchers also use non-aggregated CE beliefs of individual group members as an indicator of group efficacy [11, 27, 38, 49]. Researchers sometimes use this method to predict individual-level perceptions of job or team satisfaction [49]. In other words, these non-aggregated individual perceptions of CE predict perceptual outcomes rather than objective measures such as group performance. This method avoids the potential for cross-level analysis from the prediction of group-level performance variables using non-aggregated, individual-level perceptions of CE.

Aggregating team member group efficacy beliefs [33] predicts group-level performance variables and is the method that [4] recommends most. Researchers use this procedure to aggregate the group members' perceptions of the team's ability to predict group-level outcome variables. One complication of the aggregation method is that it necessitates establishing inter-rater agreement among the team members to ensure the data is meaningful at the group level of analysis [46]. Accepted methods for accomplishing this agreement include the rwg(j) coefficient and within and between analysis (WABA), with both having acknowledged strengths and weaknesses [26].

Finally, holistic group efficacy beliefs reached through group discussion also predict group performance [15, 19, 24, 29]. One obvious benefit of this approach is that it avoids the requirement for calculating inter-rater agreements. On the other hand, social persuasion by dominant group members may result in inflated CE responses that are less predictive of performance [21].

4 Computer Collective-Efficacy

To remain consistent with using *computer self-efficacy* (CSE) in reference to an individual's belief in their ability to use computers, we refer to the shared belief in a team's ability to use computers as *computer collective-efficacy* (CCE). we adapt [14]'s definition and define CCE as "a team's shared belief in its ability to use computers." Thus, researchers can use CCE to measure an ICT-mediated team's belief in its abilities regarding general computer use [14, 20, 21] or general or specific software use [24, 47].

4.1 Computer Collective-Efficacy – Nomological Network

CCE operates based upon similar socio-cognitive mechanism as CSE [20]. As articulated by [4], the central propositions of SCT include the four sources of efficacy discussed earlier and the four mediating processes responsible for the influence of efficacy beliefs on behavioral outcomes (see [4], Chapter 4). Thus, expectations are that these respective sources and mediating processes play a significant role in models involving CSE and CCE. The CE literature also reports that individual members' SE beliefs influence CE, and thus, CSE should similarly influence CCE. Researchers have also shown that national culture's individualist and collectivist components affect CSE, CCE, and VTE. Essentially, members of individualist cultures rate their self-efficacy higher than collective efficacy, and members of collectivist cultures rate their team's collective efficacy higher than their self-efficacy [20]. Finally, [14] shows that CCE influences VTE. Figure 1 depicts CCE's nomological network based on SCT and the current literature.

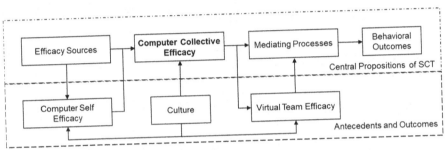

Fig. 1. CCE Nomological Network

4.2 Measuring Computer Collective-Efficacy

Bandura recommends measuring self- and collective efficacy utilizing a yes/no scale of 10 – not at all confident to 100 – totally confident [6]. A sample CCE item is: *we believe my group has the ability to explain why a program (software) will or will not run on a given computer.* The participant would first indicate if they believe their group has this ability, then rate their confidence on a 10 – not at all confident to 100 – totally confident scale. This example represents the most widely accepted method for measuring efficacy beliefs. In the case of CCE, [20] and [24] both utilized versions of this scale, while [33] did not report their scale anchors; however, the means and correlations table (Table 1) indicate that they may have used a 1–5 or 1–7 Likert-type scale.

As noted earlier, scholars have proposed three approaches for assessing collective-level efficacy beliefs when tasks are interdependent. The first is to gather individual perceptions of collective abilities and to use them to predict collective-level perceptual outcomes. The second is to obtain individual perceptions of collective abilities, aggregate them, and use them to predict collective-level perceptual and objective results. The final method is having members respond to the measurement items through discussion [4, 14, 16, 21, 24].

The CCE literature has used all three measurement methods. Some studies have utilized non-aggregated measures to be consistent with other individual-level variables [33]. Other studies have used non-aggregated and aggregated CCE measures to predict outcomes at both levels [14]. Finally, [24] and [47] measured CCE by having collective members respond to the measurement items through discussion. However, [4] cautions against this latter approach due to concerns of elevated collective beliefs resulting from social pressure [4]. Despite this warning, some researchers have recommended using the discussion approach [16], a suggestion that [24] and [47] followed. Shedding light on [4]'s conceptual arguments, there is empirical evidence establishing the superiority of the aggregated individual response approach. Consistent with [4]'s concerns, [21] found that teams' consensus-based CE responses in both collocated and virtual teams were higher than aggregated responses. Further, the authors found that aggregated CE beliefs in virtual teams significantly predicted outcomes, while consensus-derived beliefs did not.

Another consideration is the specificity of CCE. [24] measured general and specific CCE based on the software-focused measures they adapted from [10]. [47] similarly adapted a spreadsheet software measure proposed by [28] to a collective-level database context. On the other hand, [14, 20], and [21] utilized measures focused on general computer skills, representing a higher level of abstraction. [33] purportedly adopted the measures used by [20], but the final wording is imprecise.

The items used to measure CCE can reference computers in general or software. [4]'s recommendations and [21]'s findings suggest that researchers should measure CCE using individual members' aggregated perceptions of CCE. Like [14, 21], and [22], we thus investigate the factor structure for a CCE measure adapted from the general CSE measure proposed by [35]. To do so, we examine the factor structure at the unaggregated and aggregated level. Data for the analyses come from a digital collaboration involving 318 US and Hong Kong university students comprising 52 teams. We utilized graduate students from universities in metropolitan areas who worked and attended school and were from different countries and cultures with varying group and individual work norms. This helped increase the realism of working across various tasks, collaborative technologies, time differences, geographical separation, and cultural differences. Supporting the choice of sample and methodology further, prior virtual team research reveals that results from studies using similar student samples, tasks, and collaborative technologies [20, 21] generalized to organizational settings [41].

We relied on AMOS 26 to evaluate the CCE factor structure utilizing the non-aggregated data and SPSS 28 to assess the CCE factor structure employing the aggregated data. More precisely, we used covariance-based structural equation modeling (CBSEM) to analyze the non-aggregated data and principal axis factoring to analyze the aggregated data.

Following [14], we examined the two-level measure developed and validated by [35] utilizing the non-aggregated data (n = 318). However, unlike [14], we examined the full factor structure of the CCE measure in isolation. We utilized all ten original items from [35] and specified a second-order structure to account for the two levels. Bootstrap results from the maximum likelihood estimator (MLE) within AMOS showed that the initial model fit was unacceptable. Four items were associated with highly correlated error terms and subsequently deleted from the model. The final second-order variable consisted of three items for Level 1 and three for Level 2. Model fit for the final model was good, $\chi^2_8 = 18.475$; $\chi^2/df = 2.31$; CFI $= 0.99$; TLI $= 0.98$; RMSEA $= 0.06$, 90% CI [.025, .103]; SRMR $= 0.027$. All item loadings were above .7 and correlated higher with their corresponding lower-order variable, indicating convergent and discriminant validity. Table 1 depicts the factor loadings, and Table 2 depicts the correlations.

Table 1. Factor Loadings

Variable	CCE		CCEL1				CCEL1	
Item	CCEL1	CCEL2	CCE1	CCE2	CCE3	CCE4	CCE5	CCE6
Loading	0.75	0.94	0.84	0.81	0.84	0.91	0.83	0.75

Table 2. Item Correlations

	CCE	CCEL1	CCEL2	CCE 1	CCE 2	CCE 3	CCE 4	CCE 5	CCE 6
CCE	1								
CCEL1	0.69	1							
CCEL2	0.92	0.64	1						
CCE 1	0.58	0.84	0.54	1					
CCE 2	0.56	0.81	0.52	0.68	1				
CCE 3	0.59	0.85	0.54	0.71	0.69	1			
CCE 4	0.84	0.58	0.91	0.49	0.48	0.49	1		
CCE 5	0.77	0.53	0.83	0.44	0.43	0.48	0.76	1	
CCE 6	0.69	0.48	0.75	0.4	0.4	0.41	0.69	0.62	1

We ran an additional model utilizing the MLE estimator with bootstrapped coefficients to establish predictive validity. We specified CCE as an exogenous variable and group satisfaction (GS) as the endogenous variable, expecting that higher CCE teams would likely be more satisfied. Model fit for the final model was good, $\chi^2_{25} = 43.832$; $\chi^2/df = 1.75$; CFI $= 0.99$; TLI $= 0.99$; RMSEA $= 0.50$, 90% BSCI [.023, .072]; SRMR $= 0.043$. The analysis revealed a significant relationship between the second-order CCE and GS variables, $b = .068$, [.038, .100], $p = .002$, $\beta = .25$, establishing the predictive validity of the CCE measure. Table 3 in the Appendix lists the final six items.

We next investigated the factor structure of the CCE variable using the aggregated data. Because of the smaller sample size (n $= 52$), we utilized SPSS Statistics 28 and principal axis factoring with direct oblimin rotation. We used the same six items and specified the fixed number of factors to two but did not specify which factor the items should load on. Thus, the analysis is more consistent with exploratory factor analysis (EFA) than the confirmatory factor analysis (CFA) conducted using CBSEM. Confirming the CBSEM results, the principal axis factoring results revealed a similar two-factor structure, with the items loading on the same factors as specified in the CBSEM CFA. All item loadings were above .7 and loaded higher on their corresponding lower-order variable, indicating convergent and discriminant validity. Table 4 depicts the factor loadings using the aggregated data.

Table 4. Factor Loadings – Principal Axis Factoring

| | Factor | |
	CCEL1	CCEL2
CCE1	0.85	0.01
CCE2	0.92	−0.05
CCE3	0.83	0.11
CCE4	0.17	0.80
CCE5	0.06	0.83
CCE6	−0.10	0.90

To establish predictive validity, we conducted linear regression with bootstrapped coefficients. We created composite variables by averaging across the three items for each level. We then specified the CCEL1 and CCEL2 composites as hierarchical predictor variables and group satisfaction (GS) [20] as the criterion, which was also a composite based on averaging its three items. We then ran two separate regressions, one in which we entered CCEL1 first and a second where we entered CCEL2 first. In both cases, the variable entered first significantly predicted GS. However, multicollinearity exceeded acceptable limits in both cases, as expected. Thus, we created a third composite by averaging across all six items. The analysis revealed a significant relationship between the CCE composite and GS variables, $b = .044$, BSCI [.015, .077], $p = .024$, $\beta = .27$, once again establishing the predictive validity of the CCE measure.

5   General Discussion

Researchers have measured and applied CCE differently. In this study, we explained how SCT provides the theoretical framework underlying CCE, discussed the nomological network of CE and CCE, and then validated a measure of CCE at both the individual and collective levels. CBSEM and principal axis factoring results establish a two-factor structure for the CCE measure. CBSEM and regression analysis results further confirm the CCE measure's predictive validity. The following section briefly discusses this study's theoretical and practical implications.

6   Theoretical and Practical Implications

In the following sections, we discuss the theoretical and practical implications of our work. We begin with a discussion of the theoretical implications.

6.1 Theoretical Implications

Researchers have thoroughly investigated SE and CE. Bandura [4] confers how the concepts differ in terms of the level of analysis yet act through the same socio-cognitive processes. While there has been significant discussion of CSE as a context-specific measure of SE, information systems researchers have yet to investigate CCE as a context-specific measure of CE at a similar level of detail. Given the lack of attention to the concept, researchers have conceptualized and operationalized CCE inconsistently.

By addressing these shortcomings, this study contributes significantly to the ICT literature. First, the study offers a robust validation of the CCE construct; second, it elucidates the theoretical framework underlying CCE. Formally validating the CCE construct provides a robust general measure suitable for predicting individual and collective-level outcomes in ICT-mediated collaborative settings. Researchers can now confidently apply the CCE measure in collocated and geographically dispersed settings where technology is essential for team effectiveness. Using CCE in these settings can aid the advancement of theory on both collocated and virtual teams, as it will allow researchers to examine the interplay of CCE with other team success factors such as trust, cohesion, transactive memory, etc. Understanding how CCE performs in comprehensive models that include these established success factors can build more robust theoretical perspectives that advance research on both traditional and ICT-mediated teams.

Articulating the current nomological network of CCE provides a sound foundation for future ICT studies incorporating the concept. While some studies have included CCE, they have not articulated the complete nomological network of antecedents, moderators, and mediators that either influence the development of CCE or explain how CCE acts on outcomes. For example, research has mostly ignored the sources of information that SCT proposes are responsible for developing CCE. Understanding how these sources influence CCE's development is critical for building a sense of collective computer use capabilities in team settings. Further, recognizing the influence of CSE on CCE can help researchers confidently test theory-driven research models in which individual computer self-efficacy beliefs lead to computer-collective efficacy beliefs in technology-focused project settings. Understanding how cultural differences influence the development of CCE is particularly important for researchers interested in studying globally distributed, ICT-mediated teams. Finally, understanding how CCE affects outcomes through the mediating processes proposed by SCT and other essential variables, such as virtual team efficacy, is especially critical for advancing research on ICT-mediated teamwork.

6.2 Practical Implications

This research also contributes to practice. While the literature on CSE enjoys wide recognition in industry, much less is known about CCE. Thus, practitioners will be interested in a validated measure of CCE that they can apply in ICT-mediated collaborative settings. Given the growing reliance on remote workers who accomplish organizational work through teamwork [42], understanding factors that influence collective-level work is of increasing concern. The knowledge that CSE predicts CCE is also critical as practitioners can implement training initiatives that increase the CSE beliefs of individual team members to develop teams' CCE. Finally, given the cultural diversity of today's workforce, it is essential for practitioners to recognize that certain national cultures view CSE and CCE differently.

7 Conclusion

Self-efficacy (SE) is an essential element within Bandura's [4] social cognitive theory (SCT). SE has been adapted for use within numerous research domains, including the application of CSE in the seminal IS literature [10]. However, researchers have recognized that CSE is less useful for prediction at the collective level [14]. Since a substantial portion of today's organizational labor involves ICT-enabled collaboration [37], researchers must develop and validate collective-level measures such as CCE to better understand behaviors in these settings [24, 47]. This research takes a modest step towards formalizing the CCE concept by articulating its current theoretical framework and validating a general computer-use CCE measure.

Appendix

Table 3. Final Computer Collective-Efficacy Measurement Items

Computer Collective-Efficacy Level 1 (CCEL1)	
CCE1	I believe my group has the ability to use computers
CCE2	I believe my group has the ability to use the computer to write a report
CCE3	I believe my group has the ability to work with personal computers
Computer Collective-Efficacy Level 2 (CCEL2)	
CCE4	I believe my group has the ability to understand terms/words relating to computer hardware
CCE5	I believe my group has the ability to describe the function of computer hardware
CCE6	I believe my group has the ability to explain why a program (software) will or will not run

New Computer Collective-Efficacy Measurement Items for the Conference

CCEL1	
CCE1	I believe my group has the ability to use collaborative software (e.g., discussion boards, videoconferencing, etc.)
CCE2	I believe my group has the ability to use presentation software
CCE3	I believe my group has the ability to use word-processing software
CCE4	I believe my group has the ability to use chat software
CCEL2	
CCE5	I believe my group has the ability to write computer programs
CCE6	I believe my group has the ability to use database management software
CCE7	I believe my group has the ability to use web development software
CCE8	I believe my group has the ability to use project management software

References

1. Awuor, N.O., Weng, C., Piedad, E., Jr., Militar, R.: Teamwork competency and satisfaction in online group project-based engineering course: The cross-level moderating effect of collective efficacy and flipped instruction. Comput. Educ. **176**, 104357 (2022)
2. Bandura, A.: The self system in reciprocal determinism. Am. Psychol. **33**(4), 344–358 (1978)
3. Bandura, A.: Self-efficacy mechanisms in human agency. Am. Psychol. **37**(2), 122–147 (1982)
4. Bandura, A.: Self-efficacy: The Exercise of Control. W.H. Freeman/Times Books/ Henry Holt & Co., New York, NY (1997)
5. Bandura, A.: Exercise of human agency through collective efficacy. Curr. Dir. Psychol. Sci. **9**(3), 75–78 (2000)
6. Bandura, A., Guide for constructing self-efficacy scales. In: Pajares, F., Urdan, T. (eds.) Adolescence and education, Self-efficacy beliefs of adolescents, pp. 1–39. Information Age Publishing, Greenwich (2005)
7. Barlow, J., Dennis, A.: Not as smart as we think: a study of collective intelligence in virtual groups. J. Manag. Inf. Syst. **33**(3), 684–712 (2016)
8. Bilotta, I., et al.: Remote communication amid the coronavirus pandemic: optimizing interpersonal dynamics and team performance. Ind. Organ. Psychol. **14**(1–2), 36–40 (2021)
9. Blay, T., Froese, F.J.: How motivation fluctuates during virtual team work: the role of team characteristics and dynamics. In: Academy of Management. Seattle (2022)
10. Compeau, D.R., Higgins, C.A.: Computer self-efficacy - Development of a measure and initial test. MIS Q. **19**(2), 189–211 (1995)
11. Earley, P.C.: East meets West meets Mideast: Further explorations of collectivistic and individualistic work groups. Acad. Manag. J. **36**(2), 319–348 (1993)
12. Eysenck, H.J.: Is there a paradigm in personality research? J. Res. Pers. **17**(4), 369–397 (1983)
13. Feitosa, J., Salas, E.: Today's virtual teams: adapting lessons learned to the pandemic context. Organ. Dyn. **50**(1), 100777 (2021)
14. Fuller, M., Hardin, A., Davison, R.: Efficacy in technology-mediated distributed teams. J. Manag. Inf. Syst. **23**(3), 221–247 (2007)
15. Gibson, C.: Do they do what they believe they can? Group efficacy and group effectiveness across tasks and cultures. Acad. Manag. J. **42**(2), 138–152 (1999)
16. Gibson, C., Randel, A.E., Earley, P.C.: Understanding group efficacy: An empirical test of multiple assessment methods. Group Org. Manag. **25**(1), 67–97 (2000)
17. Gibson, C.B., Grushina, S.V.: A tale of two teams: next generation strategies for increasing the effectiveness of global virtual teams. Organ. Dyn. **50**(1), 100823 (2021)
18. Gruman, J.: How to lead virtual teams: a simple model can help you build and manage effective virtual teams. Career (2020)
19. Guzzo, R.A., Yost, P.R., Campbell, R.J., Shea, G.P.: Potency in groups: Articulating a construct. Br. J. Soc. Psychol. **32**(1), 87–106 (1993)
20. Hardin, A., Fuller, M., Davison, R.: I know I can, but can we? Culture and efficacy beliefs in global virtual teams. Small Group Research **38**(1), 130–156 (2007)
21. Hardin, A., Fuller, M., Valacich, J.: Measuring group efficacy in virtual teams: new questions in an old debate. Small Group Research **37**(1), 65–85 (2006)
22. Hardin, A., Looney, C., Fuller, M.: Computer based learning systems and the development of computer self-efficacy: are all sources of efficacy created equal? In: America's Conference For Information Systems. Acapulco, Mexico (2006)
23. Havakhor, T., Sabherwal, R.: Team processes in virtual knowledge teams: the effects of reputation signals and network density. J. Manag. Inf. Syst. **35**(1), 266–318 (2018)
24. Hsu, M.-H., Chen, I.Y.-L., Chiu, C.-M., Ju, T.L.: Exploring the antecedents of team performance in collaborative learning of computer software. Comput. Educ. **48**(4), 700–718 (2007)

25. Jacques, P.H., Garger, J., Filippelli-DiManna, L.P.: Antecedents to trustworthiness, satisfaction, and potency in ad hoc face-to-face and computer-mediated teams. J. Behav. Appl. Manag. 21(1), 1–16 (2020)
26. James, L.R., Demaree, R.G., Wolf, G.: Estimating within-group interrater reliability with and without response bias. J. Appl. Psychol. 69(1), 85–98 (1984)
27. Jex, S.M., Gudanowski, D.M.: Efficacy beliefs and work stress: An exploratory study. J. Organ. Behav. 13, 509–517 (1992)
28. Johnson, R.D., Marakas, G.M.: Research report: the role of behavioral modeling in computer skills acquisition: toward refinement of the model. Inf. Syst. Res. 11(4), 402–417 (2000)
29. Jung, D.I., Sosik, J.J.: Group potency and collective efficacy: Examining their predictive validity, level of analysis, and effects of performance feedback on future group performance. Group Org. Manag. 28(3), 366–391 (2003)
30. Klonek, F., Parker, S.K.: Designing SMART teamwork: how work design can boost performance in virtual teams. Organ. Dyn. 50(1), 100841 (2021)
31. Malhotra, A.: The Postpandemic Future of Work. J. Manag. 47(5), 1091–1102 (2021)
32. McLarnon, M.J.W., Woodley, H.J.R.: Collective efficacy in virtual teams: emergence, trajectory, and effectiveness implications. Can. J. Behav. Sci. 53(2), 187–199 (2021)
33. Mehta, N.K., Jha, S., Bhattacharyya, S.S.: Explicating collective technology efficacy in work from home context: study of employees with positive feelings. Business Perspectives and Research, p. 22785337221113165 (2022)
34. Morrison-Smith, S., Ruiz, J.: Challenges and barriers in virtual teams: a literature review. SN Applied Sciences 2(6), 1096 (2020)
35. Murphy, C.A., Coover, D., Owen, S.V.: Development and validation of the Computer Self-Efficacy Scale. Educ. Psychol. Meas. 49(4), 893–899 (1989)
36. Nordbäck, E.S., Espinosa, J.A.: Effective coordination of shared leadership in global virtual teams. J. Manag. Inf. Syst. 36(1), 321–350 (2019)
37. Ogbonnaya, C.: When teamwork is good for employees — and when it isn't. Harvard Business Review (2019)
38. Parker, L.E.: Working together: Perceived self- and collective-efficacy at the workplace. J. Appl. Soc. Psychol. 24(1), 43–59 (1994)
39. Peng, S.: Physiological synchrony in virtual teams: prediction of team emergent states. In: Academy of Management. Academy of Management, Seattle WA (2022)
40. Sarker, S., Chatterjee, S., Xiao, X., Elbanna, A.: The sociotechnical axis of cohesion for the IS discipline: Its historical legacy and its continued relevance. MIS Q. 43(3), 695–720 (2019)
41. Schepers, J., Jong, A.D., Ruyter, K.D., Wetzels, M.: Fields of gold: perceived efficacy in virtual teams of field service employees. Journal of Service Research 14(3), 372–389 (2011)
42. Schlaegel, C., Gunkel, M., Taras, V.: COVID-19 and individual performance in global virtual teams: The role of self-regulation and individual cultural value orientations. J. Organ. Behav. 44(1), 102–131 (2023)
43. Vuchkovski, D., Zalaznik, M., Mitręga, M., Pfajfar, G.: A look at the future of work: The digital transformation of teams from conventional to virtual. J. Bus. Res. 163, 113912 (2023)
44. Wei, L.H., Thurasamy, R., Popa, S.: Managing virtual teams for open innovation in Global Business Services industry. Manag. Decis. 56(6), 1285–1305 (2018)
45. Whillans, A., Perlow, L., Turek, A.: Experimenting during the shift to virtual team work: Learnings from how teams adapted their activities during the COVID-19 pandemic. Inf. Organ. 31(1), 100343 (2021)
46. Whiteoak, J.W., Chalip, L., Hort, L.K.: Assessing group efficacy: comparing three methods of measurement. Small Group Research 35(2), 158–173 (2004)
47. Wright, K.: Effects of self and collective efficacy perceptions on integrated information systems task performance, p. 208. The George Washington University, Ann Arbor (2005)

48. Yoon, P., Leem, J.: The influence of social presence in online classes using virtual conferencing: relationships between group cohesion, group efficacy, and academic performance. Sustainability **13**(4), 1988 (2021)
49. Zellars, K.L., et al.: Beyond self-efficacy: interactive effects of role conflict and perceived collective efficacy. J. Manag. Issues **13**(4), 483–499 (2001)

Communication is Key: A Systematic Literature Review of Transformation Competencies

Luca Laule(✉) [ID] and Markus Bick [ID]

ESCP Business School, Berlin, Germany
llaule@escp.eu

Abstract. Transformation competencies are essential for individuals tasked with executing organizational transformations, whom we refer to in this study as transformation managers. The transformation manager plays a crucial role as a mediator to successful organizational transformation. Consequently, it is critical to understand what competencies the transformation manager must possess. In this study, we conducted a systematic literature review of transformation competencies by screening an initial sample of 1,932 articles, resulting in a final sample of 11. We examined the characteristics of the final sample and subsequently interpreted the results. Our findings emphasize the novelty of the topic, given the limited extent of prior research in this domain. We further provide a list of competencies necessary for executing transformations where communication and competency in teamwork and collaboration were identified as the most important competencies for the transformation manager. Finally, we conclude with a comprehensive research agenda and provide future research directions.

Keywords: Systematic Literature Review · Transformation · Competencies · Transformation Management · Change Management

1 Introduction

Organizations are facing the need to make various transformations, and the ones responsible for these processes require a set of different competencies to carry them out successfully. Persons in charge of transformation processes are called transformation managers. Due to the different success factors relating to transformation processes, a transformation manager must be highly flexible in order to be universally deployable.

To achieve such flexibility, they need a set of competencies which we term *transformation competencies*. A competency, as described by Boyatzis [1], is *"essential to performing a job but is not causally related to superior job performance"* [1, p. 23]. Thus, identifying necessary competencies is fundamental for adequate job performance. Regarding theory, a transformation manager with the necessary competencies must be seen as a mediator of a successful transformation process.

Knowing what transformation competencies are, and which ones the most important are, is crucial and of high relevance. Organizations are currently operating in highly volatile times. Disruptive innovations like OpenAI's ChatGPT are driving transformation

© The Author(s), under exclusive license to Springer Nature Switzerland AG 2024
M. Papadaki et al. (Eds.): EMCIS 2023, LNBIP 502, pp. 148–163, 2024.
https://doi.org/10.1007/978-3-031-56481-9_11

because they are shaking up the way things have been done in the past. Authors are not the only ones who are writing books and musicians are not the only song writers anymore. Another big reason for organizational transformation is the climate change and the resulting need for organizations for transformation to become CO_2 neutral. The result of these various drivers of transformation is that going through a transformation is no longer a temporary condition but a constant one.

Competency requirements for transformation managers differ based on the type of transformation at hand. Some are related to specific transformations, like digital competencies for digital transformation. These are so called *domain-specific competencies* [2, 3] while others span different transformation types and are termed *domain-unspecific competencies*. This study solely focuses on the latter, aiming to identify universally relevant competencies for transformation managers regardless of the transformation's specific nature.

However, existing research hasn't adequately examined transformation competencies, especially the domain-unspecific ones. One research stream has focused on change processes [e.g., 4], which isn't sufficient as change processes differ significantly from transformation processes, although the terms are often mistakenly used synonymously. Change processes can be incremental in their scope [5, 6] in contrast to transformation processes, which by definition [7] have fundamental effects. Furthermore, change processes have a defined beginning and an end – as Lewin [8] already outlined in 1951 in his three-stage model of change by establishing the phases *"unfreeze, changing, and refreezing."* A transformation process, on the other hand, has an open end and can become a never-ending process [9]. Appelbaum and Wohl [10] differentiate transformation from change, stating that transformation doesn't enhance the current but creates the nonexistent. They describe the transformation process as crafting a new context, mandating shifts in underlying decision assumptions [10]. Thus, the competencies required for transformation processes may differ significantly from those required for change processes, in which case they need to be explored further.

Another research stream has investigated the competencies required for specific transformation processes, which is also doubtful. For example, Blanka et al. [11] explored digital transformation, whilst Schneider et al. [12] examined transformation towards sustainability. As the kinds of transformations organizations face are diverse, focusing on only one single type is not an adequate strategy.

The purpose of this paper is to provide a comprehensive summary of the research done so far on transformation competencies, as well as developing a detailed research agenda. Therefore, we conducted a systematic literature review by following the suggestions of vom Brocke et al. [13]. A second output of this study is a collection and tabular presentation of the identified transformation competencies.

Our paper makes several contributions. Firstly, we enhance the literature on organizational transformation and change theory. We offer a thorough overview of the current state of research on transformation competencies and provide insightful interpretations. Secondly, we improve the understanding of how a transformation process can succeed. Since, to the best of our knowledge, no other study has taken this approach, we highlight the importance of the topic and open up a possible new research stream.

The remainder of this study is structured as follows. After providing background information, we present our research design and our findings. Then we discuss these findings, conclude, and provide a future research agenda.

2 Theoretical Background

2.1 Organizational Transformation

The study of organizational transformation began in the mid-1980s [14] and is also termed as institutional transformation. The transformation of an organization *"is seen as an iterative, multilevel process, with outcomes emerging not merely as a product of rational or boundedly rational debates, but also shaped by the interests and commitments of individuals and groups, the forces of bureaucratic momentum, gross changes in the environment, and the manipulation of the structural context around decisions"* [15, p. 658]. It is a multidimensional process with diverse influences that require management by a transformation manager.

Organizations undergo transformation to remain competitive and respond to evolving business contexts. Orlikowski stated in 1996 that *"stability is out, change is in"* [16, p. 63] due to shifting economic, political, and technical conditions. This trend has only accelerated since 1996, compelling today's organizations to consistently adopt new technologies to stay competitive. Additionally, they must meet legal requirements for eco-friendly operations, becoming sustainable entities. Moreover, firms are doing well if they transform themselves in relation to ensuring a more diverse workforce to help develop more creative solutions [17]. The range of reasons for organizational transformation is extensive.

Theoretical descriptions of transformation processes exist, e.g., Gersick's [18] punctuated equilibrium model, which was tested by Romanelli and Tushman [19].

2.2 The Concept of Competency

Before defining the concept of competency, we must state that there is a difference between the often wrongly synonymously used terms *competence* and *competency*. Škrinjarić [20] highlighted that the origin of this uncertainty comes from the different origins of the terms, in that our understanding of competency comes from the US, while competence is rooted in the UK. Competencies pertain to an individual's attributes enabling task or job execution, whereas competences center around outcomes, actions, or behaviors [21, 22]. Competences are used when the context is not individual but organizational, for example by using competence models to manage human resources and to achieve corporate goals. Competencies, on the other hand, are used in the individual context.

McClelland [23] first called for applying competence tests instead of intelligence tests to decide if someone is suited to a school, a college, or a job because competency expresses appropriateness far more than an intelligence test. Boyatzis [1] introduced and discussed the modern concept of competency in his book *The Competent Manager: A Model for Effective Performance*. He stated that a competency has several dimensions that may be *"a motive, trait, skill, aspect of one's self-image or social role, or a body of knowledge"* [1, p. 21]. Furthermore, he clarified that a job competency signifies an ability, with a person's set of competencies reflecting their capability [1].

Spencer and Spencer [24] further defined competencies as consisting of five underlying characteristics, which in combination represent a competency: (1) motives, (2) traits, (3) self-concept, (4) knowledge and (5) skill. This is still the predominant understanding of the concept of competency [25].

Erpenbeck et al. [26] extended these definitions by emphasizing self-organization and creativity as fundamental to every competency. They further introduced four competency classes based on the research of Heinrich Roth [27]: (1) personal competencies, (2) activity- and implementation-oriented competencies, (3) social-communicative competencies and (4) professional-methodological competencies. These competency classes are commonly used to classify competencies [e.g., 3, 28].

3   Research Design

To comprehensively summarize the existing research on transformation competencies, we performed a systematic literature review, adhering to the guidelines proposed by vom Brocke et al. [13]. Our review considered studies identifying specific competencies required for effectively carrying out organizational transformations.

As a second step, we extracted single competencies by deductively conducting Mayring's [29] qualitative content analysis. Therefore, we used the four well-established categories of competencies provided by Erpenbeck et al. [26].

3.1   Systematic Literature Review

Our literature review design was sequential and exhaustive. Thus, we first collected relevant literature and then analyzed it in several steps. Hereby, we considered the entire body of literature available on the topic. To ensure we covered all relevant works, we used three databases: EBSCO (Business Source Complete, EconLit, APA PsycArticles, Psychology and Behavioral Sciences Collection, APA PsycInfo), JSTOR and ScienceDirect. We imposed no publication date limit to encompass all studies exploring transformation competencies up to our data collection date of September 2022. To ensure high quality, we considered only peer-reviewed articles, written in English and published in academic journals. We conducted a keyword search using the following search terms: EBSCO: TI(competenc* AND transformation*) OR SU(competenc* AND transformation*) OR AB(competenc*AND transformation*); JSTOR: Item Title: competenc* AND transformation* OR Abstract: competenc* AND transformation*; ScienceDirect: Title, Abstract, Keywords: (competence OR competences OR competency OR competencies) AND (transformation OR transformations OR transformational).

One might question why we didn't interchange terms like *transformation* with *change* or *competency* with *skill*. The answer can be found in theory. As we highlighted before, we see a significant difference between a transformation and a change process due to the much bigger scope [5–7] or mismatches in the open and closed ends [8, 9]. Furthermore, the terms *competency* and *skill* must be distinguished and cannot be used synonymously, since a skill is one part of a competency, but the competency cannot be reduced to a skill [24].

Our initial search yielded 1,932 articles. After applying the exclusion criteria *peer-reviewed* and *English language* as well as deleting doublings, the sample shrank to 1,674. In the next step, we screened the titles, abstracts, and keywords of the remaining works, which led to a sample of 43 possibly relevant articles. After reading the full texts, we ended up with a final sample of 11 articles which fulfilled our inclusion criterium: competencies for transformations are named. This process is displayed in Table 1.

To organize the final sample, we developed a data set in the form of an Excel file. Each article received a row, and details regarding the title, author(s), publication year, geographical location, journal, etc. was documented. An extract from this database, including the most important information, is displayed in Table 3.

Table 1. Process of literature identification and inclusion

	EBSCO	JSTOR	ScienceDirect	Total
Initial Search	n = 263	n = 402	n = 1,267	n = 1,932
Exclusion Criteria	n = 134	n = 274	n = 1,266	n = 1,674
Title, Abstract, Keywords	n = 17	n = 2	n = 24	n = 43
Full text	n = 5	n = 0	n = 6	**n = 11**

3.2 Identification of Transformation Competencies

One side outcome of our systematic literature review was a list of competencies most often mentioned in the identified articles. Therefore, we conducted qualitative content analysis [29] deductively and used QCAmap as a tool for coding. As Mayring [29] suggests, we evaluated the fit of the defined categories after coding the first studies. The codes are reported in Table 2 and refer to the four classes of competencies introduced by Erpenbeck et al. [26]. To ensure objectivity during the coding process, two researchers executed it independently. The correspondence rate reached 84%, and mismatches were discussed afterwards. The competency atlas, provided by Erpenbeck and Rosenstiel [30, p. 383], was used for assistance.

Table 2. Coding categories for the deductive, qualitative content analysis [26]

Category	Explanation	Anchor Example
(1) Personal competencies	The ability of someone to act reflexively and self-organized	Loyalty, credibility, humor
(2) Activity- and implementation-oriented competencies	The ability to act actively and holistically in a self-organized way towards specific goals or plans – either for oneself or for others or an organization	Vigor, consistence, decision-making
(3) Social-communicative competencies	The ability to act in a self-organized manner and in a communicative and cooperative way	Adaptability, capacity for teamwork, communication
(4) Professional-methodological competencies	The ability to act in a self-organized way while solving factual problems by using professional and instrumental knowledge, skills, and abilities	Market knowledge, analytical skills, project management

4   Findings

4.1   Sample Characteristics

The most relevant specifications of the final sample are documented in Table 3. The list of 11 studies consisted of eight articles published in academic journals and three articles from conference proceedings. The years of publication ranged from 2008 to 2022, with a strong weighting on the period 2018 to 2022.

Among these, three articles [11, 31, 32] were conceptual papers, five were case studies [3, 12, 33–35], and three were empirical papers [28, 36, 37]. Excluding the work of Busulwa et al. [32] from Australia, each study originated from researchers in Europe.

Nine of the studies covered a specific topic in the field of the digital transformation, while Schneider et al. [12] focused on transformation towards sustainability and Ormerod [31] examined consulting projects in the field of operational research.

Table 3. Overview of the results

Author	Title of publication	Year of public-ation	Focus on the identification of competencies	Data sources
Blanka et al.	The interplay of digital transformation and employee competency: A design science approach	2022	No	Literature
Busulwa et al.	Digital transformation and hospitality management competencies: Toward an integrative framework	2022	Yes	Literature
Butschan et al.	Tackling Hurdles to Digital Transformation - The Role of Competencies for Successful Industrial Internet of Things (IIoT) Implementation	2018	No	Literature
Gilli et al.	Leadership competencies for digital transformation: An exploratory content analysis of job advertisements	2022	Yes	Job Ads
Hulla et al.	A case study based digitalization training for learning factories	2019	No	Literature

(*continued*)

Table 3. (*continued*)

Author	Title of publication	Year of public-ation	Focus on the identification of competencies	Data sources
Hulla et al.	Towards digitalization in production in SMEs – A qualitative study of challenges, competencies and requirements for trainings	2021	Yes	Interviews
Imran et al.	Digital Transformation of Industrial Organizations: Toward an Integrated Framework	2021	No	Unknown
Lodgaard et al.	Future competence at shopfloor in the era of Industry 4.0 – A case study in Norwegian industry	2022	No	Literature
Ormerod	The Transformation Competence Perspective	2008	No	Unknown
Pihlajamaa et al.	Competence combination for digital transformation: a study of manufacturing companies in Finland	2021	Yes	Literature, Interviews
Schneider et al.	Transdisciplinary co-production of knowledge and sustainability transformations: Three generic mechanisms of impact generation	2019	No	Literature

4.2 Transformation Competencies

In addition to reviewing prior research, we extracted competencies mentioned in each study by deductively using qualitative content analysis by Mayring [29]. We identified a total of 235 competencies and grouped them into the competency classes introduced by Erpenbeck et al. [26].

Table 4 provides an overview of the competencies named by at least two studies. We excluded the category of professional-methodological competencies for two reasons. Firstly, our review focused on general, domain-unspecific competencies. Thus, competencies specific to certain fields like *information literacy* [32] or *systems thinking* [12] are not generalizable and thus beyond the scope of this study. Second, a huge amount of the competencies in this category are labelled competencies but are basically skills or knowledge. For example, *agile and scrum* [37], *install computer software* [32] or *knowledge of lean-management* [36] do not meet the definitions of competency as outlined by Spencer and Spencer [24] or Erpenbeck et al. [26].

Applying these criteria resulted in a list of 14 of the most important transformation competencies, based on research done so far.

Table 4. Overview of the most named competencies, grouped to the competency classes by Erpenbeck et al. [26]

Competency Class	Competency	Counts	Sources
(1) Personal competencies	Creativity	4	[3, 11, 34, 37]
	Personal flexibility	3	[3, 34, 35]
	Attitude	3	[33, 35, 37]
	Vision	2	[11, 33]
	Motivation and perseverance	2	[11, 35]
(2) Activity- and implementation-oriented competencies	Decision making	2	[3, 31]
(3) Social-communicative competencies	Communication	9	[3, 11, 28, 31, 33–37]
	Teamwork and collaboration	6	[3, 11, 28, 34, 35, 37]
	Problem solving	5	[3, 28, 34, 35, 37]
	Adaptability	3	[3, 33, 37]
	Mobilizing others	3	[3, 11, 33]
	Intercultural competency	3	[3, 34, 35]
	Relationship management	2	[12, 37]
	Customer orientation	2	[34, 37]

5 Discussion

5.1 Relevance of the Topic

The strong weighting of the years of publication between 2018 and 2022 might be an indicator of the rising relevance of the topic. In all, 90% of the articles of our final sample were published during that period with an increasing number from 2018 (1) to 2022 (4). There was no publication in 2020, which could be caused by the spread of COVID-19 in that year and the corresponding strong restrictions, but finally remains unclear. However, it should be noted that the final sample size of eleven of this study is relatively modest. This could be attributed to the increased demand for organizational transformation in recent years and the gradual onset of academic discussions in the field.

5.2 Research Focus of the Studies

One problematic characteristic of the sample was the research focus of the studies. Only four [32, 34, 36, 37] studies concentrated on identifying competencies, whilst the remainder listed only a set of competencies which they stated were necessary to get an organizational transformation done.

Another issue regarding the focus of the studies is that all of them investigated a specific kind of transformation. While the majority, nine of the eleven studies, examined a type of digital transformation, only Schneider et al. [12] focused on sustainability transformation, and Ormerod [31] on transformations in the operations research field.

5.3 Competency Data Sources

Data sources for identifying required competencies in transformations varied among the studies. While most drew from existing literature, Gilli et al. [37] analyzed job ads, which is a well-established procedure for the identification of competencies [e.g., 38] and a form of collecting new data. Hulla et al. [36] interviewed industry representatives, seeking competency needs for manufacturing and digitalization. Pihlajamaa et al. [34] used two methods to collect data by combining a referenced study with interviews. Imran et al. [33] described the digital transformation of four organizations and thereby provided a list of 12 competencies they identified in their data.

Data collection methods across the studies were used in appropriate ways, as far as we can evaluate, but as mentioned in the previous chapter, most of the studies did not explicitly focus on the identification of competencies. Especially those studies which used the literature as a source of data often neglected the initial context of the competencies they named, which is problematic [39].

5.4 Understanding of Competencies

The articles in our sample suggest that certain competencies are vital for successful transformations. However, not all articles accurately define competencies. For instance, Busulwa et al. [32] label skills and knowledge (e.g., *install computer software, use email systems*) as competencies. Ormerod [31] also confuses the terms *competence* and *competency* and equates competencies with skills.

However, some of the studies handled this issue in an appropriate way. Butschan et al. [28] referred to established experts like Erpenbeck and Rosenstiel, and Gilli et al. [37] drew from Boyatzis [1] to define competencies.

Our findings relating to the unclear and sometimes misunderstood term *competency* are in line with previous research [e.g., 20, 25].

5.5 Understanding of Transformation

While the majority of the studies handled the concept of transformation well, one case sticks out. Ormerod [31] wrote an article about consultancy projects in the field of operations research. He explained that the selection of the method to be used depends on two factors: the competence of the participants and the transformation to be achieved. This in turn implies that each OR consultancy project would be an organizational transformation. This might be questioned, however, by considering the definition of a transformation process compared to the definition of a change process.

5.6 Identified Transformation Competencies

Table 4 summarizes the transformation competencies mentioned most often across the sample. Competency in *communication* was mentioned the most by nine of the 11 studies and can thus be seen as the essential key to the success of a transformation effort. Competency regarding *teamwork and collaboration* was listed by six studies and ranked in second place. Having these two competencies in the first two places stresses the interdisciplinary role of the transformation manager. The competencies *problem solving*, and *creativity* are ranked third and fourth and are more oriented to the individual work of the transformation manager.

One interesting comparison of competencies might be between the change manager and the transformation manager. Therefore, we use the required competencies of a change manager identified by Crawford and Nahmias [40] across several studies. The comparison is displayed in Table 5. In order to compare the competencies to each other, we slightly adapted the list of change manager competencies by omitting those gathered in the fourth group of competencies by Erpenbeck and Rosenstiel [30], as we did with our sample. Such a comparison between similar and different competencies across job responsibilities has already been done by Crawford and Nahmias [40].

The comparison reveals both similarities and differences. A striking difference is that the change manager seems to be more of a leader than the transformation manager because he or she requires competencies in *leadership* and *influencing skills*. On the other hand, the transformation manager must be more of a *visionary* with a strong *attitude* and who might have more free space in regard to executing tasks where he can benefit from his *personal flexibility and adaptation*.

Table 5. Comparison between the competencies of change and transformation managers

Change Manager	Transformation Manager
Similar Competencies	
Creativity and challenge	Creativity
Communication	Communication
Decision-making and problem-solving	Decision-making
	Problem-solving
Team development	Teamwork and collaboration
Cross-cultural skills	Intercultural competency
Stakeholder management	Relationship management
Different Competencies	
Leadership	Personal flexibility
Facilitation skills	Attitude
Initiative and self- management	Vision
Influencing skills	Adaptability
Strategic thinking	Customer orientation
Learning and development	Motivation and perseverance
Action orientation	Mobilizing others

6   Conclusion, Limitations, and Research Agenda

This study examined the concept of transformation competencies and provided a summary and a presentation of the current state of the research by conducting a systematic literature review, following the suggestions of vom Brocke et al. [13]. We initially identified 1,932 scientific papers, which resulted in a final sample of 11 relevant articles. These articles were analyzed and discussed regarding different characteristics. Furthermore, we provided and discussed a list of competencies that were mentioned the most across the sample and can be seen as the competencies required to successfully conduct an organizational transformation, based on the research carried out to date. It is challenging to locate our results in the current literature, since we combined the topics *organizational transformation* and *competencies* without specifying to which kind of organizational transformation our findings refer. On the other hand, this might be one of the biggest advantages of our results, in that they are universally applicable to transformation managers and can be the basis for any organizational transformation.

We contribute to transformation and change theory by enriching the literature. We provide a comprehensive overview of the research done so far, including the approaches of different aspects of the concepts *transformation* and *competency*. As a second contribution to theory, we investigated transformation competencies that can be seen as mediators between the transformation manager and the success of a transformation project. The competencies we presented herein enable – and are a prerequisite to – organizational transformation. One practical implication is the use of our results by human resource managers in the employment process.

However, there are some limitations to our study, the biggest one of which is probably the database. All of the articles in our final sample handled a specific transformation, none of them investigated general transformation competencies. Furthermore, our sample was biased because nine of the eleven studies investigated digital transformation and it is obvious that there are more kinds of transformations. Another limitation is the lack of scientific rigor of some of the studies. While one of the authors framed consultancy projects in the field of operational research as transformations, some others did not paint a clear picture of competencies. One last major limitation is the sources of the competencies of the articles, as some of them neglected the context of the referenced studies.

Following the principles laid out by Webster & Watson [41], we understand that a systematic literature review paper entails more than just presenting existing knowledge and gaps in knowledge. However, since our topic of transformation competencies is still very new and lacks sufficient content to build a strong framework, we limit ourselves to two main points: First, we provide a clear distinction between change- and transformation competencies along the dimensions of Table 5. Second, we provide a comprehensive research agenda, displayed in Table 6. We recommend undertaking future research to address existing limitations and build that strong framework. Additionally, we put forth avenues for further investigation to advance the understanding and knowledge of transformation competencies.

Table 6. Research Agenda

Topic	Problem to be addressed	Research design
Enriching the database	The sample within this review encompasses nine studies in the realm of digital transformation, with merely two studies exploring other forms of transformations, which decreases the generalizability	Carry out research that (1): Explores competencies required for transformations beyond the digital. For instance, transformations towards diversity, sustainability, compliance with recent governmental regulations, or business model transformations. Alternatively, (2): Examine transformation competencies without a particular emphasis, mirroring the approach adopted in this study

(continued)

Table 6. (*continued*)

Topic	Problem to be addressed	Research design
Validation of the findings	We have identified competencies that we claim to be general transformation competencies. This generalization should be validated in future research to ensure that the competencies we have identified are indeed those necessary for each type of organizational transformation	To prove the universality of the identified competencies, a study on their generalizability should be conducted. This can take the form of a case study, which investigates the daily work of transformation managers. This study would observe individuals working on various types of transformations. Other research designs addressing this issue are also possible
Methodology	The process of identifying the competencies relied solely on secondary data due to the design of our chosen research approach	Gather primary data and use for instance the qualitative content analysis or grounded theory to develop a theory concerning the competencies of transformation managers
Methodology	The ranking of competency importance in this study is determined solely by the frequency of occurrence within the sample	Examine the practical significance of each transformation competency through a quantitative research design
Fundamental work	A particularly time-consuming challenge was establishing a distinct differentiation between *transformation* and *change*, grounded in scholarly literature	Precisely delineate the distinction between a *change* process and a *transformation* process and formulate a precise definition of the latter
Future direction	The data collection of this study occurred prior to the surge in interest surrounding large language models, which began in late 2022. Considering the significant transformative potential of this technology, it is necessary to explore potential competencies related to AI	Examination of the transformative impact of AI and the corresponding required competencies through an exploratory research design

References

1. Boyatzis, R.E.: The Competent Manager. John Wiley & Sons Inc, New York (1982)
2. Tittel, A., Terzidis, O.: Entrepreneurial competences revised: developing a consolidated and categorized list of entrepreneurial competences. Entrepr. Educ. **3**, 1–35 (2020). https://doi.org/10.1007/s41959-019-00021-4

3. Hulla, M., Hammer, M., Karre, H., Ramsauer, C.: A case study based digitalization training for learning factories. In: Procedia Manufacturing, pp. 169–174. Elsevier B.V. (2019)

4. Doyle, M.: Selecting managers for transformational change. Hum. Resour. Manag. J. **12**, 3–16 (2002). https://doi.org/10.1111/j.1748-8583.2002.tb00054.x

5. Kindler, H.S.: Two planning strategies: incremental change and transformational change. Group Organ. Stud. **4**, 476–484 (1979). https://doi.org/10.1177/105960117900400409

6. Dörfler, T., Gehring, T.: Analogy-based collective decision-making and incremental change in international organizations. Eur. J. Int. Relat. **27**, 753–778 (2021). https://doi.org/10.1177/1354066120987889

7. Combley, R.: Cambridge Business English Dictionary. Cambridge University Press, Cambridge (2011)

8. Lewin, K.: Field Theory in Social Science: Selected Theoretical Papers (Dorwin, C. ed.). Harpers, Oxford, England (1951)

9. Seliger, B.: Toward a more general theory of transformation. East Europ. Econ. **40**, 36–62 (2002)

10. Appelbaum, S.H., Wohl, L.: Transformation or change: some prescriptions for health care organizations. Manag. Serv. Qual. **10**, 279–298 (2000)

11. Blanka, C., Krumay, B., Rueckel, D.: The interplay of digital transformation and employee competency: a design science approach. Technol. Forecast Soc. Change **178**, 121575 (2022). https://doi.org/10.1016/j.techfore.2022.121575

12. Schneider, F., Giger, M., Harari, N., et al.: Transdisciplinary co-production of knowledge and sustainability transformations: three generic mechanisms of impact generation. Environ. Sci. Policy **102**, 26–35 (2019). https://doi.org/10.1016/j.envsci.2019.08.017

13. Vom Brocke, J., Simons, A., Niehaves, B., et al.: Reconstructing the giant: on the importance of Rigour in documenting the literature search process. In: ECIS 2009 Proceedings 161 (2009)

14. Bartunek, J.M., Ringuest, J.L.: Enacting new perspectives through work activities during organizational transformation. J. Manage. Stud. **26**, 541–560 (1989). https://doi.org/10.1111/j.1467-6486.1989.tb00744.x

15. Pettigrew, A.M.: Context and action in the transformation of the firm. J. Manage. Stud. **24**, 649–670 (1987). https://doi.org/10.1111/j.1467-6486.1987.tb00467.x

16. Orlikowski, W.J.: Improvising organizational transformation over time: a situated change perspective. Inf. Syst. Res. **7**, 63–92 (1996). https://doi.org/10.1287/isre.7.1.63

17. DiTomaso, N.: Why difference makes a difference: diversity, inequality, and institutionalization. J. Manage. Stud. **58**, 2024–2051 (2021). https://doi.org/10.1111/joms.12690

18. Gersick, C.J.G.: Revolutionary change theories: a multilevel exploration of the punctuated equilibrium paradigm. Acad. Manag. Rev. **16**, 10 (1991). https://doi.org/10.2307/258605

19. Romanelli, E., Tushman, M.L.: Organizational transformation as punctuated equilibrium: an empirical test. Acad. Manag. J. **37**, 1141–1166 (1994). https://doi.org/10.2307/256669

20. Škrinjarić, B.: Competence-based approaches in organizational and individual context. Humanit. Soc. Sci. Commun. **9**, 28 (2022). https://doi.org/10.1057/s41599-022-01047-1

21. Moore, D.R., Cheng, M., Dainty, A.R.J.: Competence, competency and competencies: performance assessment in organisations. Work Study **51**, 314–319 (2002). https://doi.org/10.1108/00438020210441876

22. Cheng, M.-I., Dainty, A.R.J., Moore, D.R.: What makes a good project manager? Hum. Resour. Manag. J. **15**, 25–37 (2005). https://doi.org/10.1111/j.1748-8583.2005.tb00138.x

23. McClelland, D.C.: Testing for competence rather than for "intelligence." Am. Psychol. **28**, 1–14 (1973). https://doi.org/10.1037/h0034092

24. Spencer, L.M., Spencer, S.M.: Competence at Work: Models for Superior Performance. Wiley John + Sons (1993)

25. Campion, M.A., Fink, A.A., Ruggeberg, B.J., et al.: Doing competencies well: best practices in competency modeling. Pers. Psychol. **64**, 225–262 (2011). https://doi.org/10.1111/j.1744-6570.2010.01207.x

26. Erpenbeck, J., Rosenstiel, L., Grote, S., Sauter, W.: Handbuch Kompetenzmessung. Schäffer-Poeschel (2017)

27. Roth, H.: Pädagogische Anthropologie. Band II: Entwicklung und Erziehung. Schroedel, Hannover (1971)

28. Butschan, J., Heidenreich, S., Weber, B., Kraemer, T.: Tackling hurdles to digital transformation - the role of competencies for successful industrial internet of things (IIoT) implementation. Int. J. Innov. Manag. **23**, 19500361 (2019). https://doi.org/10.1142/S13639196 19500361

29. Mayring, P.: Qualitative Content Analysis: Theoretical Foundation, Basic Procedures and Software Solution. Klagenfurt (2014)

30. Erpenbeck, J., von Rosenstiel, L.: Handbuch Kompetenzmessung. Schäffer-Poeschel, Stuttgart (2003)

31. Ormerod, R.J.: The transformation competence perspective. J. Oper. Res. Soc. **59**, 1435–1448 (2008). https://doi.org/10.1057/palgrave.jors.2602482

32. Busulwa, R., Pickering, M., Mao, I.: Digital transformation and hospitality management competencies: toward an integrative framework. Int. J. Hosp. Manag. **102**, 103132 (2022). https://doi.org/10.1016/j.ijhm.2021.103132

33. Imran, F., Shahzad, K., Butt, A., Kantola, J.: Digital transformation of industrial organizations: toward an integrated framework. J. Chang. Manag. **21**, 451–479 (2021). https://doi.org/10.1080/14697017.2021.1929406

34. Pihlajamaa, M., Malmelin, N., Wallin, A.: Competence combination for digital transformation: a study of manufacturing companies in Finland. Technol. Anal. Strateg. Manag. (2021). https://doi.org/10.1080/09537325.2021.2004111

35. Lodgaard, E., Torvatn, H., Sørumsbrenden, J.: Future competence at shopfloor in the era of Industry 4.0 – a case study in Norwegian industry. Proc. CIRP **107**, 961–965 (2022). https://doi.org/10.1016/j.procir.2022.05.092

36. Hulla, M., Herstätter, P., Wolf, M., Ramsauer, C.: Towards digitalization in production in SMEs – A qualitative study of challenges, competencies and requirements for trainings. In: Procedia CIRP. Elsevier B.V., pp. 887–892 (2021)

37. Gilli, K., Nippa, M., Knappstein, M.: Leadership competencies for digital transformation: an exploratory content analysis of job advertisements. German J. Hum. Resour. Manage. (2022). https://doi.org/10.1177/23970022221087252

38. Murawski, M., Bick, M.: Demanded and imparted big data competences: towards an integrative analysis. In: Proceedings of the 25th European Conference on Information Systems (ECIS), pp. 1375–1390 (2017)

39. Bamberger, P.: Beyond contextualization: using context theories to narrow the micro-macro gap in management research. Acad. Manag. J. **51**, 839–846 (2008)

40. Crawford, L., Nahmias, A.H.: Competencies for managing change. Int. J. Project Manage. **28**, 405–412 (2010). https://doi.org/10.1016/j.ijproman.2010.01.015

41. Webster, J., Watson, R.T.: Analyzing the past to prepare for the future: writing a literature review. MIS Q. **26**, xiii–xxiii (2002)

Evaluating the Success of Digital Transformation Strategy in Greek SMEs

Maria Kamariotou and Fotis Kitsios[✉]

Department of Applied Informatics, University of Macedonia, Thessaloniki, Greece
mkamariotou@uom.edu.gr, kitsios@uom.gr

Abstract. Prior studies have neglected the significance of IS strategy in choosing and implementing the suitable IS according to situation, instead concentrating on the impact of digital transformation methods on performance and crucial elements that influence IS adoption. The use of strategy-as-practice is useless because there are no formal procedures in place, and no strategic planning has been done. This makes it impossible for management to use IS planning to gain a sustainable competitive advantage. In this study, we look at how the performance of IS is impacted by strategic processes and practices. The survey's participants were 294 information technology managers from Greek SMEs. The results of the study suggest that top-level management should receive training on how to strategically apply IS planning to strengthen competitive advantage. In order to align the organization's strategy and structure, it is also the manager's duty to select the proper IT infrastructure. The findings aid managers in setting priorities for logistical tasks and in comprehending the importance of IS planning in this area.

Keyword: Information Systems strategizing · Success · SMEs · Digital transformation · Strategy-as-practice

1 Introduction

Small and Medium Enterprises (SMEs) operate in a modern and intricate financial environment. Increased complexity and drastic shifts are features of the modern world that impact businesses' operations and limit their responsiveness to the economic crisis [1–6]. In addition to budgetary constraints, a lack of technical, administrative, and human capacities, as well as a lack of strategic planning, may be impeding their ability to confront the financial crisis [7–9]. Managers at small and medium-sized enterprises can benefit from formal processes associated with strategic management and information handling if they implement them.

Adopting digital transformation appears to come with expenses, even if doing so has many advantages. There are a number of obstacles that must be considered when making this choice; all of them have to do with the introduction of Information Systems (IS). Thus, managers face a difficult decision when deciding whether or not to apply IS. Therefore, it is necessary to investigate the factors that influence the choice to adopt it [10–15]. An important issue for managers to invest in, digital investment not only affects

© The Author(s), under exclusive license to Springer Nature Switzerland AG 2024
M. Papadaki et al. (Eds.): EMCIS 2023, LNBIP 502, pp. 164–182, 2024.
https://doi.org/10.1007/978-3-031-56481-9_12

company performance but also helps managers align business strategy with corporate achievement. To accomplish environmental durability through abatement and financial effectiveness, businesses should develop organized processes in dynamic environments that apply consistent standards and protocols [16–18].

Unfortunately, IS strategy has been examined as a monolithic issue, which means that recent research has failed to define the appropriate actions and strategies based on the firm's capabilities pursued by executives during a crisis [19, 20]. While scholars [21–24] have researched the factors that influence digital transformation adoption, they have not yet investigated the ongoing processes and practices of strategy development, including IS and IT. However, there is a subset of businesses for which the constraints of small size and scarce resources have a disproportionately negative impact on alignment factors and organizational effectiveness [25]. Increasing acceptance and usage of technology in SMEs and deeper integration of technology into their processes and operations can be found in management literature [26, 27]. Alignment between business and IS strategy has a significant impact on firm outcomes, which should be recognized by both practitioners and scholars.

The idea of IS strategizing has evolved into what is now a well-established framework for how enterprises participate in the continual procedures and practices of strategy making involving IS and IT [28, 29]. This theory describes how enterprises engage in the on-going processes and practices of strategy-making involving IS and IT. The principles of strategy include a high-level and comprehensive understanding of how information systems strategy evolves through time as a dynamic, iterative, and knowing-and-learning-based set of behaviors, both formal and informal [30].

Scholars have pointed out that the focus of IS strategizing shouldn't be solely on the IT artifact but rather on how organizational actors are able to investigate and capitalize on possibilities as well as obstacles associated with IT [28, 31]. This is because the IT artifact is only one aspect of the equation. To this end, IS strategy involves a number of tensions [32]. Some examples of these disputes include those among formal and informal approaches, between human and IT aspects, and between standardized procedures such as business process "engineering" hand-in-hand with enterprise systems [33, 34] and flexible knowledge management systems [35]. IS planning also emphasizes dynamic activities that are carried out jointly by IT and business professionals [36]. This helps to reinforce or improve ongoing alignment [32].

Thus, in this study, we investigate how strategic processes and practices affect IS performance. 294 information technology managers from Greek SMEs participated in the survey. Data was analyzed using Factor Analysis.

The structure of this paper is as following: after a brief introduction to this field, the next section includes the literature review in order to highlight the issues which are discussed in this paper. Section 3 describes the methodology, while Sect. 4 shows the results of the survey. Finally, Sec. 5 discusses the results and Sect. 6 concludes the paper.

2 Theoretical Background

2.1 IS Strategizing and Organizational Issues

It has been a major concern for chief information officers (CIOs) for the better part of four decades [37–39], and strategizing through information technology has been a "hot" topic for quite some time.

IS planning "provides a shared understanding across the organization to guide subsequent IT investment and deployment decisions" [28]. This is not without its conflicts, a topic to which we will now turn. Recalling Earl's (1993) [40] "organizational" view of IS strategy and reflecting the continual assessment of business needs to promote the ability (or capability) to innovate [41, 42], this is not without its tensions.

The dynamic process of strategizing compares the intended strategy with its implementation while recognizing that certain features of the actual strategy are emergent. This suggests that there are tensions between utilizing current plans, ideas, and resources and investigating new and emerging approaches to attain organizational objectives. These tensions emerge because there is a tension between exploiting existing plans, ideas, and resources [28].

It is pertinent to reflect on the concept of alignment (in a dynamic and ongoing sense) between IT and business as a key aspect of strategizing [32, 41, 43]. The focus of this section is on how strategy is enacted in practice [44]. In addition, the planned strategy and the emerging strategy, in addition to the exploitation strategy and the exploratory strategy, are not seen as being sequential but rather as overlapping [45, 46]. Emerging practices are the result of the day-to-day "doings" of strategy, and they have the potential to alter the initial assumptions that were inherent in the planned strategy. This is because emerging practices are continuously refined and adapted to meet the requirements of new contexts, needs, and circumstances [47, 48].

In terms of the hierarchical aspects of strategy and its execution, strategizing acknowledges the limits of considering it as an exclusively "top-down" exercise, which is an exercise in which execution follows planning in a relatively straightforward manner [28]. Strategy is a considerably more chaotic and emergent phenomenon than this (emphasis added). For example, Newkirk et al. (2003) [34] highlight that an excessive amount of planning restricts flexible execution, which in turn inhibits creativity. On the other hand, an inadequate amount of planning is not managed since the strategy seems vague and confusing to those who seek to carry it out. This hints that there should be a balance struck between top-down planning and developing strategies, techniques, and practices in the workplace.

The focus on human resources is particularly essential from a strategic point of view because strategists are continually pushed by the opposing needs of using existing organizational knowledge to produce efficiency and, at the same time, exploring new knowledge in order to be innovative [49]. This presents a problem for strategists, who must find a balance between the two sets of demands in order to be successful. Therefore, even though it is necessary for strategists to implement "codified solutions" [32], they frequently must improvise because the planned strategy cannot be carried out in its entirety. In this way, for instance, the stages of growth framework provides a means of asking pertinent questions as regards the likely feasibility of strategies as they are being formulated on the basis of the various elements contained therein. On the other hand, feedback makes it easier for strategizing processes to involve review, ongoing learning, and revisions of current strategies in light of actual experience.

2.2 The Role of Practice in IS Strategizing

IS strategizing provides a contrast to the conventional "grand vision" of strategy (cf. Porter, Prahalad, and Hammer, for example) by implying that it is iterative and emergent in nature. Strategy-as-practice takes this concept one step further and emphasizes how crucial it is to take into account the routine activities of those who put strategy into action. Therefore, strategy-as-practice concentrates its attention primarily on the microscopic level [50–53].

That strategy is continually evolving in the flow of practice, which is comparable to what a number of IS strategizing researchers have explored over the years [30, 31, 44]. The literature on IS strategy, on the other hand, has a tendency to place an emphasis on emerging processes that strategists engage in when "the plan" (long-term, top-down) cannot be carried out in its entirety. To this purpose, as Chia and MacKay (2007) [54] point out, a significant amount of IS strategy research has recently moved its attention from a "strategy content" approach to a "strategy process" methodology. The latter captures the internal reality of organizations "in flight" [54] and highlights the relevance of reconfiguring processes (e.g., emerging processes) where circumstances change over time.

IS strategizing is therefore clearly positioned as a dynamic, ongoing construct that focuses on how organizations deal with the conflicting demands associated with exploiting existing assets such as infrastructure, resources, and expertise (based on previous experience), while also exploring new solutions and tasks through social interactions and creative thinking. IS strategy investigates these tensions not in terms of the results that are achieved but rather in terms of the methods that are used to achieve these results. However, the high-level viewpoint that defines the majority of studies on IS strategy, while beneficial in understanding such processes on a more holistic level, pays relatively little attention to how humans deal with day-to-day exigencies that vary and often disrupt the intended strategy. This is despite the fact that such studies are valuable in understanding such processes. As a consequence of this, we propose that research on IS strategizing should take into account the insights that come from strategy-as-practice theorizing [28].

In conclusion, the research on strategic ambidexterity gives valuable insights into how organizations deal with the opposing demands of exploitation (planned strategy, pursuit of efficiency) and exploration (emergence, flexibility, and agility). These insights pertain to how companies exploit planned strategies and pursue efficiency. On the other hand, particular practices have received a smaller share of attention up until this point. In addition, we have seen that power considerations have gotten scant attention in the IS strategizing literature, despite the fact that this literature has thoughtfully analyzed power considerations. According to Chia (2004) [55], the strategy-as-practice viewpoint investigates how power is formed and re-produced in the processes of strategy-making. Therefore, in this article, we suggest that, in order to do justice to the tensions that characterize IS strategizing processes, it is worthwhile to consider both hierarchical power, which focuses on the exploitation of existing resources and assets (for example, influence and "official" lines of command), and a more "practice-based" perspective, which sees power as an emerging accomplishment that occurs while strategy is "in the making." This brings the main tension between planned and emergent strategies to the surface once again. This clearly associates hierarchical power and the exploitation of existing knowledge and resources with that exercised by top management (e.g., in the setting up of long-term strategies). On the other hand, performative power and exploration of new knowledge and opportunities characterize the emergent aspect of IS strategizing, which is characterized by everyday practices. Below, we have a breakdown of our insights into an all-encompassing framework that deepens our comprehension of IS strategy while retaining its "core" traits.

IS strategizing and the strategy-as-practice literature have views that are comparable in the sense that they consider strategy-making to be dynamic and unpredictable. Despite the fact that planned (as in long-term planning) strategies need to be constantly updated in line with unique conditions and practices that develop, these two bodies of research have similar perspectives. As a result of this, both points of view steer clear of the mindless adoption of so-called "best practice solutions" [33, 34] and instead offer a method for thinking about the emergence of practices as well as a set of non-prescriptive principles that can be used to investigate IS strategizing activities in depth and in situ. For example, Galliers (2004) [30] suggests an IS strategizing framework that can be utilized as a "sense-making device, meant more as an aide memoir, to be used to raise questions and facilitate discussion concerning the strategizing elements and connections that may or may not be in place in any particular organization." However, strategy-as-practice and IS strategizing are not the same thing in a number of respects; for example, the literature on IS strategizing does not place a strong emphasis on individuals or the activities they engage in on a daily basis.

IS planning is typically geared toward the organizational (or SBU) level, whereas strategy-as-practice offers chances to take into consideration individuals and the behaviors they engage in [48]. Spee and Jarzabkowski (2009) [50] acknowledge that the emphasis placed on individuals is a point of strength but also a point of weakness in the strategy-as-practice theorizing. This is both a strength and a weakness of the theory. Their worry is connected to the possibility that placing an excessive amount of emphasis on individuals could cause one to ignore the interconnected nature of decision-making processes [56]. In addition, Jarzabkowski and Spee (2009) [57] state that the majority of strategy-as-practice studies do not combine multiple levels of analysis. Therefore, adopting this part of strategy-as-practice theorizing into IS strategic thinking would not only account for practice (at the micro-level), but it would also provide us with the possibility to expand on strategy-as-practice research into the meso- and macro-levels in addition.

2.3 IS Strategizing and Success

IS strategy is developed by a dynamic, iterative, and knowing/learning set of processes, both formal and informal, according to the principles of strategizing [58]. Rather than concentrating exclusively on the IT artifact, as some researchers have suggested [59], IS strategists should consider how organizational actors might best investigate and make use of IT's many advantages and disadvantages. Thus, IS planning necessitates resolving a number of conflicts, such as those between formal and informal methods, human and IT considerations, and standardized processes like business process 'engineering' and enterprise systems. IS planning also emphasizes collaborative, dynamic procedures between IT and business staff, which helps maintain or improve alignment [60, 61].

The IS strategizing procedure consists of five stages. Important planning issues, priorities, and goals must be established at the outset of the IS strategizing process, and the development team and executive sponsors must be chosen. Examination of existing organizational structures, business processes, and IS, along with examination of the external and internal IT environment, are the primary threats of the second stage of the process. In the third stage, CIOs and other IT executives establish pivotal goals, investigate potential sources of disruption, and develop comprehensive IT plans. Step four of developing an IS strategy is the following stage: In the fourth stage, IT managers will do activities like define new organizational procedures and IT architectures to help them reach their IT goals, as well as determine which new IT projects are the highest priority and how to organize them. Action plans and strategies for change management are defined in the fifth and final stage. Additionally, in this stage, IS executives evaluate the results of the IS strategizing process and determine if the goals established in the previous step have been met [7, 62–71].

IS managers have been seen to focus their efforts during this third step [7, 62–64, 70, 71]. More practical options may become apparent in Step 3 when they have been merged with Steps 2 and 3. Better options and decisions may sustain the plan and help the company continue to see positive results once it has established its IT goals. Two of the biggest problems that occurred during the IS were a lack of involvement from upper management and an inability to develop effective action strategies for carrying out the IS strategizing procedure. Without management buy-in, IT project teams are less likely to stick to development plans and are more likely to encounter difficulties when putting strategy into effect. If it isn't full or suitable for its strategic setting, IS fails. Therefore, managers should establish priorities that will aid in the effective implementation of their strategy in practice and the attainment of their objectives [72].

However, more comprehensive planning would be more effective since it would help planners comprehend and react to environmental impacts. Managers should be aware that if they keep pushing their teams to put in more and more effort, they run the risk of increasing tensions within the group. Because of these disagreements, progress may be slowed down. On the other side, wasting too much time might cause tensions to build among the team, which can ultimately slow down the project. Process evaluation, then, is a decision-making problem in which options must be determined. It's crucial since it gives managers the means to lessen the impact of these undesirable outcomes [65].

It has been determined through research that IT executives prioritize the third and fifth steps while ignoring the first and second. Therefore, current IS strategies are inefficient, ineffective, and fall short of IT objectives [66, 69]. Managers also prioritize cutting down on the project's overall timeline and budget. Managers that are solely concerned with the process's execution may speed up the implementation of the IS strategizing process, but this comes at the expense of the company's ability to achieve its IT goals [67, 68, 71].

3   Methodology

For IT managers, we designed a quantitative survey. Both IS strategizing and success were assessed using a 5-point Likert scale. The instrument was derived from studies that analyzed IS strategizing [7, 62, 65–71] and success [7, 62, 65–71].

Four IS executives were asked to participate in a pilot test. Each one completed the survey and commented on the contents, length, and overall appearance of the instrument. A sample of IS executives in Greece was selected from the icap list. SMEs which provided contact details were selected as the appropriate sample of the survey. The survey was sent to 1246 IS executives and a total of 294 returned the survey. Data analysis was implemented using Factor Analysis.

4   Results

The IS executive is typically seen as the most suitable person in the organization to provide data regarding IS strategizing and success as defined in this study [13]. Respondents in this study were employed in a variety of industries, well educated, and experienced. 16% of them worked in agriculture and food, 11.3% in business services, 10.6% in retail and the rest in other industries. 35.2% had some postgraduate studies and 44.7% had a degree. They also had 16–25 years of IS experience. Tables 1–5 show further respondent breakdown by industry, education and IS experience.

Table 1. Respondents' industry.

Primary Business Category	Respondents	Percentage
Agriculture & Food	47	16%
Business Services	33	11.3%
Chemicals, Pharmaceuticals & Plastics	18	6.1%
Construction	22	7.5%
Education	4	1.4%
Electrical	11	3.8%
Energy	8	2.7%
IT, Internet, R&D	24	8.2%
Leisure and Tourism	16	5.5%
Metals, Machinery & Engineering	28	9.6%
Minerals	3	1%
Paper, Printing, Publishing	14	4.8%
Retail & Traders	31	10.6%
Textiles, Clothing, Leather, Watchmaking, Jewellery	14	4.8%

(continued)

Table 1. (*continued*)

Primary Business Category	Respondents	Percentage
Transport & Logistics	20	6.8%
Total	294	100

Table 2. Education Level

Education Level	Respondents	Percentage
2 year college graduate	59	20.1%
4 year college graduate	132	44.7%
Post graduate degree	103	35.2%
Total	294	100

Table 3. Respondents' IS experience.

Years	Respondents	Percentage
0–5	22	7.5%
6–15	81	27.6%
16–25	118	39.9%
26–35	61	20.8%
Total	12	100

Table 4. IS employees.

Employees	Respondents	Percentage
0–5	261	89.1%
6–10	22	7.5%
11–20	3	1%
21–30	5	1.7%
31–40	0	0
41–50	0	0
>=51	2	0.7%
Total	294	100

Table 5. IS budges.

IS budget	Respondents	Percentage
0–50.000 €	175	59.7%
51.000–100.000 €	64	21.8%
101.000–150.000 €	17	5.8%
151.000–200.000 €	12	4.1%
>=201.000 €	25	8.5%
Total	294	100

The internal consistency, calculated via Cronbach's alpha, ranged from 0.774 to 0.980, exceeding the minimally required 0.70 level [13, 16]. Table 6 displays these results.

Table 6. Reliability of variables.

Variables	Cronbach a values
1st step of IS strategizing	0.937
2nd step of IS strategizing	0.944
3rd step of IS strategizing	0.941
4th step of IS strategizing	0.942
5th step of IS strategizing	0.943
Success	0.938

Factor analysis was implemented on the detailed items of the IS strategizing process and success constructs. Table 7 describes the reliability of IS strategizing steps and success constructs.

Table 7. Reliability analysis.

Variables	Mean	S.D
Determining key planning issues	3.820	.9190
Determining planning objectives	3.949	.9130
Organizing the planning team	3.752	1.0562
Obtaining top management commitment	3.830	1.0540
Analyzing current business systems	3.633	1.0258

(continued)

Table 7. (*continued*)

Variables	Mean	S.D
Analyzing current organizational systems	3.721	.9147
Analyzing current digital technologies	4.207	.7932
Analyzing the current external business environment	3.888	.9443
Analyzing the current external digital environment	4.082	.9384
Identifying major objectives for digital technologies	3.946	.8608
Identifying opportunities for improvement	3.823	.8952
Evaluating opportunities for improvement	3.714	.9420
Identifying high-level digital strategies	3.963	.8677
Identifying new business processes	3.748	.9299
Identifying new architectures for digital technologies	3.748	1.0920
Identifying specific new projects	3.701	.9486
Identifying priorities for new projects	3.776	.9035
Defining change management approaches	3.765	.9252
Defining action plans	3.704	1.0004
Evaluating action plans	3.415	1.0538
Defining follow-up and control procedures	3.565	.9953
Maintaining a mutual understanding with top management on the role of digital technologies in supporting strategy	3.993	.9347
Understanding the strategic priorities of top management	3.864	.9502
Identifying digital technologies-related opportunities to support the strategic direction of the firm	3.922	.9185
Aligning digital strategies with the strategic plan of the organization	3.854	.9469
Adapting the goals/objectives of digital technologies to changing goals/objectives of the organization	3.878	.9303
Educating top management on the importance of digital technologies	3.728	1.0552
Adapting technology to strategic change	3.827	.9350
Assessing the strategic importance of emerging technologies	3.793	.9462
Identifying opportunities for internal improvement in business processes through digital technologies	3.827	.8388
Maintaining an understanding of changing organizational processes and procedures	3.782	.8667
Generating new ideas to reengineer business processes through digital technologies	3.779	.9212
Understanding the information needs through subunits	3.707	.8721

(*continued*)

Table 7. (*continued*)

Variables	Mean	S.D
Understanding the dispersion of data, applications, and other technologies throughout the firm	3.806	.9740
Development of a "blueprint" which structures organizational processes	3.517	.9589
Improved understanding of how the organization actually operates	3.735	.8885
Monitoring of internal business needs and the capability of digital technologies to meet those needs	3.759	.8702
Developing clear guidelines of managerial responsibility for plan implementation	3.639	1.0349
Identifying and resolving potential sources of resistance to digital technologies' plans	3.415	.9188
Maintaining open lines of communication with other departments	3.687	.9621
Coordinating the development efforts of various organizational subunits	3.551	.9541
Establishing a uniform basis for prioritizing projects	3.622	,9654
Achieving a general level of agreement regarding the risks/tradeoffs among system projects	3.476	.9694
Avoiding the overlapping development of major systems	3.554	.9787
Ability to identify key problem areas	3.711	.9464
Ability to anticipate surprises and crises	3.571	.9564
Flexibility to adapt to unanticipated changes	3.670	.9365
Ability to gain cooperation among user groups for digital technologies' plans	3.840	.9871

Tables 8 and 9 present the principal component analysis using the Maximum Likelihood Estimate and the extraction of factors with Promax with Kaiser Normalization method. The factor loadings and cross loadings provide support for convergent and discriminant validity.

Table 8. Factor loadings for IS strategizing stages.

Factors	Items	Loadings
5th step of IS strategizing	Evaluating action plans	.929
	Defining action plans	.890
	Defining change management approaches	.661

(*continued*)

Table 8. (*continued*)

Factors	Items	Loadings
	Defining follow-up and control procedures	.595
1st step of IS strategizing	Analyzing current digital technologies	.878
	Analyzing the current external business environment	.728
	Analyzing the current external digital environment	.639
	Organizing the planning team	.635
	Determining planning objectives	.472
	Obtaining top management commitment	.329
3rd step of IS strategizing	Identifying opportunities for improvement	.925
	Identifying major objectives for digital technologies	.846
	Identifying high level digital strategies	.556
	Evaluating opportunities for improvement	.504
4th step of IS strategizing	Identifying new architectures for digital technologies	.906
	Identifying new business processes	.500
	Identifying priorities for new projects	.474
	Identifying specific new projects	.351
2nd step of IS strategizing	Analyzing current business systems	.895
	Analyzing current organizational systems	.711

Table 9. Factor loadings for success constructs.

Factors	Items	Loadings
Cooperation	Identifying and resolving potential sources of resistance to digital plans	.835
	Coordinating the development efforts of various organizational subunits	.814
	Achieving a general level of agreement regarding the risks/tradeoffs among digital projects	.800
	Establishing a uniform basis for prioritizing projects	.772
	Maintaining open lines of communication with other departments	.748
	Developing clear guidelines of managerial responsibility for plan implementation	.688

(*continued*)

Table 9. (*continued*)

Factors	Items	Loadings
	Ability to anticipate surprises and crises	.523
	Avoiding the overlapping development of major digital technologies	.480
	Ability to gain cooperation among user groups for digital technologies' plans	.471
Analysis	Maintaining an understanding of changing organizational processes and procedures	.795
	Generating new ideas to reengineer business processes through digital technologies	.774
	Improved understanding of how the organization actually operates	.773
	Understanding the information needs through subunits	.746
	Identifying opportunities for internal improvement in business processes through digital technologies	.738
	Monitoring of internal business needs and the capability of digital technologies to meet those needs	.599
	Understanding the dispersion of data, applications, and other technologies throughout the firm	.597
	Ability to identify key problem areas	.562
	Development of a "blueprint" which structures organizational processes	.522
	Assessing the strategic importance of emerging technologies	.425
	Flexibility to adapt to unanticipated changes	.323
Strategic Alignment	Understanding the dispersion of data, applications, and other technologies throughout the firm	.910
	Aligning digital strategies with the strategic plan of the organization	.707
	Identifying digital technologies-related opportunities to support the strategic direction of the firm	.393

(*continued*)

Table 9. (*continued*)

Factors	Items	Loadings
Managers' understanding of IS	Maintaining a mutual understanding with top management on the role of digital technologies in supporting strategy	.979
	Understanding the strategic priorities of top management	.677

5 Discussion

The results of previous studies [7, 62, 64–66, 70, 71] are corroborated by the data presented here. Managers in SMEs prioritize strategy execution, as seen in Table 8. IS executives also avoid spending time on the first two stages. Therefore, implementing an IS strategizing approach leads to IT strategies that are inefficient and don't help businesses achieve their goals. IT managers can only spend so much on IS development. This diverts their attention away from more pressing matters, such as determining the specific ways in which IS will boost business success. They were just concerned with speeding up the process and cutting costs as much as possible. Therefore, the most commonly encountered issues include the inability of IT projects to meet business needs, misalignment with existing systems, inflexibility of systems, and inadequate planning [7, 62, 70, 71].

A crucial aspect of IS strategizing is deciding which individuals will serve on the development team. Team members need to work together and hone their abilities in order to generate successful IT projects, so this exercise is crucial. Managers should support the planning team during IT project development to help businesses reach their objectives, fulfill their processes, and influence their own expansion. In addition, top-level management must lay out detailed instructions for carrying out IT initiatives.

Researchers came to the conclusion that IT managers typically skip Step 2 in favor of Steps 4 and 5. Because of this misalignment, neither the firm's strategic priorities nor the produced IS plans are enough to achieve the firm's IT goals [65, 66]. Additionally, executives exclusively focus on minimizing the project's implementation time and budget [70, 71].

Managers who put all of their energy into executing the IS strategizing process face significant challenges, including the possibility that they may devote less time to the process overall and the possibility that the strategic goals of the organization will not align with the goals of IT. According to these experts [7, 62] shifts in the internal IS context will call for a new approach to IS strategizing. Executives in charge of IS should keep an eye out for shifts in the external environment and internal organizational conditions that will create uncertainty and alter IS's function within enterprises. To better match the portfolio of IT projects undertaken by SMEs with their performance, the IS strategy should place an emphasis on environmental scanning and the strategic use of IT. Executives can utilize the results of an IT environment scan to adjust corporate strategy, introducing new or refocused strategic thrusts to boost innovation and SMEs' performance.

6 Conclusions

In this paper, we looked at how strategic processes and practices affect IS success. The study's primary findings suggest that IS strategizing is crucial for SMEs to succeed in the modern business world because it improves the efficiency with which IT projects are developed and implemented, ultimately leading to a larger proportion of the market. However, putting the approach into action is no simple task. Because of the complexity of modern business, managers must have a thorough familiarity with the company's goals and strategy. For SMEs to successfully complete the IS strategy, it is essential that they evaluate all relevant factors.

This paper's theoretical contribution is its explanation of IS strategy for executive use. It's crucial that everybody knows their roles and responsibilities and that nobody overlooks anything. IS managers may benefit from learning the processes because it will help them not only focus on the organization's goals but also appreciate the value of strategy as practice. This has led to more success in IS strategizing and fewer issues with IT initiatives. Otherwise, achieving both would be challenging, and the strategy as practice can always be better.

The paper's practical contribution is that it can serve as a roadmap for enhancing the quality of decision-making in SMEs. Better options and decisions can keep the plan going and produce the same or even better results; thus, it's important for the company to recognize its IT goals so it can establish its future IT and organizational goals. IT executives can get insight into the difficulties inherent in implementing IS projects and the bearing of IS strategy by fostering interaction between business and IT management in strategic IS planning. Executives in SMEs that follow the strategy-as-practice framework to the letter will have access to more up-to-date information regarding the external environment analysis of their businesses, allowing them to make better strategic and tactical decisions. As a result, there will be less environmental unpredictability and less dynamic risk.

The survey results for this research were exclusively completed on Greek SMEs; hence, this is a limitation of the paper. This paper's conclusions should be expanded upon in future research and compared to those of other companies operating in different nations. Semi-structured follow-up interviews with businesses operating in diverse industries are another idea for future studies. In particular, semi-structured interviews give participants a forum for candidly discussing the role that IS strategy played in their achievements. The implementation of the strategy-as-practice can be enhanced by studying the viewpoints of IT executives in order to pinpoint the areas that need work throughout the rollout of IS initiatives. By involving more people in IS strategizing, we can raise the level of flexibility and the degree of alignment between strategy in practice and market requirements.

References

1. Handfield, R., Sroufe, R., Walton, S.: Integrating environmental management and supply chain strategies. Bus. Strateg. Environ. **14**, 1–19 (2005)
2. Blackhurst, J., Craighead, C.W., Elkins, D., Handfield, R.B.: An empirically derived agenda of critical research issues for managing supply-chain disruptions. Int. J. Prod. Res. **43**, 4067–4081 (2005)
3. Cousins, P.D., Giunipero, L., Handfield, R.B., Eltantawy, R.: Supply management's evolution: key skill sets for the supply manager of the future. Int. J. Oper. Prod. Manag. **26**, 822–844 (2006)
4. Lawson, B., Cousins, P.D., Handfield, R.B., Petersen, K.J.: Strategic purchasing, supply management practices and buyer performance improvement: an empirical study of UK manufacturing organisations. Int. J. Prod. Res. **47**, 2649–2667 (2009)
5. Handfield, R.B., Cousins, P.D., Lawson, B., Petersen, K.J.: How can supply management really improve performance? A knowledge-based model of alignment capabilities. J. Supply Chain Manag. **51**, 3–17 (2015)
6. Crook, T.R., Giunipero, L., Reus, T.H., Handfield, R., Williams, S.K.: Antecedents and outcomes of supply chain effectiveness: an exploratory investigation. J. Managerial Issues 161–177 (2008)
7. Newkirk, H.E., Lederer, A.L.: The effectiveness of strategic information systems planning under environmental uncertainty. Inf. Manag. **43**, 481–501 (2006)
8. Shihab, M.R., Rahardian, I.: Comparing the approaches of small, medium, and large organisations in achieving IT and business alignment. Int. J. Bus. Inf. Syst. **24**, 227–241 (2017)
9. Al-Ammary, H., Al-Doseri, J.S., Al-Blushi, Z., ZAl-Blushi, Z.N., Aman, M.: Strategic information systems planning in Kingdom of Bahrain: factors and impact of adoption. Int. J. Bus. Inf. Syst. **30**, 387–410 (2019)
10. Chandak, S., Kumar, N.: E-business processes and its influence on supply chain performance: in the context of Indian automobile industries. Int. J. Recent Technol. Eng. **8**, 862–867 (2019)
11. Chandak, S., Kumar, N.: Impact of e-business processes and information technology tools on supply chain performance of Indian automobile industries. Int. J. Sci. Technol. Res. **8**, 282–285 (2019)
12. Zhang, M., Wang, J., Zhang, J.: Research on construction supply chain management in e-business environment. In: Proceedings of 16th International Conference on Industrial Engineering and Engineering Management, pp. 1565–1568. Hong Kong, China (2009)
13. Jiang, J., Jin, Y., Dong, C.Y.: Research on the e-business logistics service mode based on branch storage and warehouse financing. Int. J. Serv. Technol. Manage. **22**, 203–217 (2016)
14. Lingyun, W., Yingyuan, C.: The evolution of supply chain in e-business environment. In: Proceedings of 12th International Conference on e-Business Engineering, pp. 269–275. Beijing, China (2015)
15. Nguyen, H.O.: Critical factors in e-business adoption: evidence from Australian transport and logistics companies. Int. J. Prod. Econ. **146**, 300–312 (2013)
16. Drechsler, A., Weißschädel, S.: An IT strategy development framework for small and medium enterprises. IseB **16**, 93–124 (2018)
17. Lee, J.J.Y., Randall, T., Hu, P.J.H., Wu, A.: Examining complementary effects of IT investment on firm profitability: are complementarities the missing link? Inf. Syst. Manag. **31**, 340–352 (2014)
18. Ullah, A., Lai, R.: A systematic review of business and information technology alignment. ACM Trans. Manage. Inf. Syst. **4**, 1–30 (2013)

19. Ardito, L., Raby, S., Albino, V., Bertoldi, B.: The duality of digital and environmental orientations in the context of SMEs: implications for innovation performance. J. Bus. Res. **123**, 44–56 (2021)
20. Papadopoulos, T., Baltas, K.N., Balta, M.E.: The use of digital technologies by small and medium enterprises during COVID-19: implications for theory and practice. Int. J. Inf. Manage. **55**, 102192 (2020)
21. Evangelista, P., Sweeney, E.: Technology usage in the supply chain: the case of small 3PLs. Int. J. Logist. Manage. **17**, 55–74 (2006)
22. Matopoulos, A., Vlachopoulou, M., Manthou, V.: Understanding the factors affecting e-business adoption and impact on logistics processes. J. Manuf. Technol. Manag. **20**, 853–865 (2009)
23. Nurmilaakso, J.M.: Adoption of e-business functions and migration from EDI-based to XML-based e-business frameworks in supply chain integration. Int. J. Prod. Econ. **113**, 721–733 (2008)
24. Gunasekaran, A., Ngai, E.W.: Adoption of e-procurement in Hong Kong: an empirical research. Int. J. Prod. Econ. **113**, 159–175 (2008)
25. Becker, W., Schmid, O.: The right digital strategy for your business: an empirical analysis of the design and implementation of digital strategies in SMEs and LSEs. Bus. Res. **13**, 985–1005 (2020)
26. Street, C.T., Gallupe, B., Baker, J.: Strategic alignment in SMEs: strengthening theoretical foundations. Commun. Assoc. Inf. Syst. **40**, 420–442 (2017)
27. Xu, F., Luo, X.R., Zhang, H., Liu, S., Huang, W.W.: Do strategy and timing in IT security investments matter? An empirical investigation of the alignment effect. Inf. Syst. Front. **21**, 1069–1083 (2019)
28. Chen, D.Q., Mocker, M., Preston, D.S., Teubner, A.: Information systems strategy: reconceptualization, measurement, and implications. MIS Q. **34**, 233–259 (2010)
29. Teubner, R.A.: Information systems strategy. Bus. Inf. Syst. Eng. **5**, 243–257 (2013)
30. Galliers, R.D.: Reflections on information systems strategizing. In: Avgerou, C., Ciborra, C., Land, F. (eds.) The Social Study of Information and Communication Technology: Innovation, Actors, and Contexts, pp. 231–262. Oxford University Press, Oxford (2004)
31. Henfridsson, O., Lind, M.: Information systems strategizing, organizational sub-communities, and the emergence of a sustainability strategy. J. Strateg. Inf. Syst. **23**, 11–28 (2014)
32. Karpovsky, A., Galliers, R.D.: Aligning in practice: from current cases to a new agenda. J. Inf. Technol. **30**, 136–160 (2015)
33. Howcroft, D., Newell, S., Wagner, E.: Understanding the contextual influences on enterprise system design, implementation, use and evaluation. J. Strateg. Inf. Syst. **13**, 271–277 (2004)
34. Wagner, E., Howcroft, D., Newell, S.: Special issue part II: understanding the contextual influences on enterprise system design, implementation, use and evaluation. J. Strateg. Inf. Syst. **14**, 91–95 (2005)
35. Newell, S., Huang, J.C., Galliers, R.D., Pan, S.L.: Implementing enterprise resource planning and knowledge management systems in tandem: fostering efficiency and innovation complementarity. Inf. Organ. **13**, 25–52 (2003)
36. Mumford, E.: The story of socio-technical design: reflections on its successes, failures and potential. Inf. Syst. J. **16**, 317–342 (2006)
37. Luftman, J., Ben-Zvi, T.: Key issues for IT executives 2011: cautious optimism in uncertain economic times. MIS Q. Exec. **10**, 203–212 (2011)
38. Luftman, J., Derksen, B.: Key issues for IT executives 2012: doing more with less. MIS Q. Exec. **11**, 207–218 (2012)
39. Luftman, J., Zadeh, H.S., Derksen, B., Santana, M., Rigoni, E.H., Huang, Z.D:. Key information technology and management issues 2012–2013: an international study. J. Inf. Technol. **28**, 354–366 (2013)

40. Earl, M.J.: Experiences in strategic information systems planning. MIS Q. **17**, 1–24 (1993)
41. Chan, Y.E., Reich, B.H.: IT alignment: what have we learned? J. Inf. Technol. **22**, 297–315 (2007)
42. Shollo, A., Galliers, R.D.: Towards an understanding of the role of business intelligence systems in organisational knowing. Inf. Syst. J. **26**, 339–367 (2016)
43. Hirschheim, R., Sabherwal, R.: Detours in the path toward strategic information systems alignment. Calif. Manage. Rev. **44**, 87–108 (2001)
44. Nolan, R.L.: Ubiquitous IT: the case of the Boeing 787 and implications for strategic IT research. J. Strateg. Inf. Syst. **21**, 91–102 (2012)
45. Henfridsson, O., Bygstad, B.: The generative mechanisms of digital infrastructure evolution. MIS Q. **37**, 907–931 (2013)
46. Merali, Y., Papadopoulos, T., Nadkarni, T.: Information systems strategy: past, present, future? J. Strateg. Inf. Syst. **21**, 125–153 (2012)
47. Jarzabkowski, P., Spee, A.P., Smets, M.: Material artifacts: practices for doing strategy with 'stuff.' Eur. Manag. J. **31**, 41–54 (2013)
48. Whittington, R.: Information systems strategy and strategy-as-practice: a joint agenda. J. Strateg. Inf. Syst. **23**, 87–91 (2014)
49. Newell, S., Marabelli, M.: Strategic opportunities (and challenges) of algorithmic decision-making: a call for action on the long-term societal effects of 'datification.' J. Strateg. Inf. Syst. **24**, 3–14 (2015)
50. Spee, A.P., Jarzabkowski, P.: Strategy tools as boundary objects. Strateg. Organ. **7**, 223–232 (2009)
51. Spee, A.P., Jarzabkowski, P.: Strategic planning as communicative process. Organ. Stud. **32**, 1217–1245 (2011)
52. Vaara, E.: Taking the linguistic turn seriously: strategy as a multifaceted and interdiscursive phenomenon. Adv. Strateg. Manag. **27**, 29–50 (2010)
53. Whittington, R.: Learning more from failure: practice and process. Organ. Stud. **27**, 1903–1906 (2006)
54. Chia, R., MacKay, B.: Post-processual challenges for the emerging strategy-as-practice perspective: discovering strategy in the logic of practice. Hum. Relat. **60**, 217–242 (2007)
55. Chia, R.: Strategy-as-practice: reflections on the research agenda. Eur. Manag. Rev. **1**, 29–34 (2004)
56. Chia, R.: From knowledge-creation to the perfecting of action: Tao, Basho and pure experience as the ultimate ground of knowing. Hum. Relat. **56**, 953–981 (2003)
57. Jarzabkowski, P., Paul Spee, A.: Strategy-as-practice: a review and future directions for the field. Int. J. Manag. Rev. **11**, 69–95 (2009)
58. Marabelli, M., Galliers, R.D.: A reflection on information systems strategizing: the role of power and everyday practices. Inf. Syst. J. **27**, 347–366 (2017)
59. Galliers, R.D., Newell, S., Shanks, G., Topi, H.: The challenges and opportunities of 'datification.' J. Strateg. Inf. Syst. **24**, 2–3 (2015)
60. Mirchandani, D.A., Lederer, A.L.: "Less is more:" information systems planning in an uncertain environment. Inf. Syst. Manag. **29**, 13–25 (2014)
61. Newkirk, H.E., Lederer, A.L., Johnson, A.M.: Rapid business and IT change: drivers for strategic information systems planning? Eur. J. Inf. Syst. **17**, 198–218 (2008)
62. Newkirk, H.E., Lederer, A.L., Srinivasan, C.: Strategic information systems planning: too little or too much? J. Strat. Inf. Syst. **12**, 201–228 (2003)
63. Kamariotou, M., Kitsios, F.: An empirical evaluation of strategic information systems planning phases in SMEs: determinants of effectiveness. In: Proceedings of the 6th International Symposium and 28th National Conference on Operational Research, pp. 67–72. Thessaloniki, Greece (2017)

64. Kamariotou, M., Kitsios, F.: Information systems phases and firm performance: a conceptual framework. In: Kavoura, A., Sakas, D., Tomaras, P. (eds.) Strategic Innovative Marketing. Springer Proceedings in Business and Economics, pp. 553–560. Springer (2017)
65. Kitsios, F., Kamariotou, M.: Strategic IT alignment: business performance during financial crisis. In: Tsounis, N.; Vlachvei, A. (eds.) Advances in Applied Economic Research, Springer Proceedings in Business and Economics, pp. 503–525. Springer (2017)
66. Kamariotou, M., Kitsios, F.: Critical factors of strategic information systems planning phases in SMEs. In: Themistocleous, M.; Rupino da Cunha, P. (eds.) Information Systems, EMCIS 2018, Springer LNBIP 341, pp. 503–517. Springer Nature (2019)
67. Kitsios, F., Kamariotou, M.: Strategizing information systems: an empirical analysis of IT alignment and success in SMEs. Computers 8, 74 (2019)
68. Kitsios, F., Kamariotou, M.: Decision support systems and strategic information systems planning for strategy implementation. In: Kavoura, A., Sakas, D., Tomaras, P. (eds.) Strategic Innovative Marketing; Springer Proceedings in Business and Economics, pp. 327–332. Springer, Cham (2017)
69. Kitsios, F., Kamariotou, M.: Information systems strategy and strategy-as-practice: planning evaluation in SMEs. In: Proceedings of Americas Conference on Information Systems (AMCIS2019), pp. 1–10. Cancun, Mexico (2019)
70. Pai, J.C.: An empirical study of the relationship between knowledge sharing and IS/IT strategic planning (ISSP). Manag. Decis. 44, 105–122 (2006)
71. Zubovic, A., Pita, Z., Khan, S.: A framework for investigating the impact of information systems capability on strategic information systems planning outcomes. In: Proceedings of 18th Pacific Asia Conference on Information Systems, pp. 1–12. Chengdu, China (2014)
72. Kitsios, F., Kamariotou, M., Grigoroudis, E.: Digital entrepreneurship services evolution: analysis of quadruple and quintuple helix innovation models for open data ecosystems. Sustainability 13, 12183 (2021)

Managing IS Adoption Challenges in Emerging Technologies: A Longitudinal Case Study of Financial Management Services Automation in a Medium-Sized Enterprise

Henriika Sarilo-Kankaanranta[✉] and Lauri Frank

University of Jyväskylä, Jyväskylä, Finland
{henriika.sarilo,lauri.frank}@jyu.fi

Abstract. Adapting to innovations of the Fourth Industrial Revolution (4IR) has proven to be a challenge for many organizations. This longitudinal and in-depth case study, focusing on the adoption and technology continuation of Robotic Process Automation, encompasses 25 interviews and complementary data collected between 2021 and 2023 in a medium-sized public company. The study's findings underscore that the adoption of technology and its acceptance would greatly benefit from an increased understanding of the diverse challenges that emerge during long-term technological integration, rather than merely relying on initial adoption decisions. Of particular significance is the evolving role and nature of resistance to change over time, as well as the hesitancy in making decisions – both of which have notable implications for the rate of automation adoption. To mitigate resistance toward disruptive innovations, proactive management should, during the early stages, elucidate the reasons for apprehension, communicate the advantages gained by employees, and invest in relatively straightforward implementations that build knowledge and engender trust within the organization. The advent of 4IR innovations necessitates prompt and adaptable resource planning. It is improbable that organizations will achieve success in their adoption journey if they solely rely on outsourced technical competence.

Keywords: Technology adoption · Resistance to Change · Robotic Process Automation · Fourth Industrial Revolution · IT management

1 Introduction

In recent years, several technological and process innovations, such as *cloud software* and *big data*, have disrupted the financial management services business. Adjusting to such new innovations and to the Fourth Industrial Revolution (4IR) has not always been easy for businesses. Such factors as resistance to change and particularly hidden apathy, certain 'technology dithering', have been found to lead to a vicious circle of organizational challenges [1]. While automation technologies, such as Robotic Process Automation (RPA), are yet not fully utilized in Financial Management and Payroll [2], the next disruptive innovation, Artificial Intelligence, is already making its way to accounting services [3 referred in 4].

M. Papadaki et al. (Eds.): EMCIS 2023, LNBIP 502, pp. 183–203, 2024.
https://doi.org/10.1007/978-3-031-56481-9_13

For example, RPA adoption studies concerning the decision making and organizational adoption challenges are relatively scarce [e.g. 5, 6]. While resistance to change itself has been studied, technology acceptance models with resistance to change as a variable have been less examined. It is even more difficult to find longitudinal studies that would examine the relation between the rate of continued adoption and resistance to change.

In the early knowledge phase [7] organizations learn about the innovation and its capabilities and evaluate its suitability for adoption. During the technology evaluation, it is important to understand the specific organizational conditions and characteristics. Otherwise, managing the changes required for adoption could be challenging or even result in failure. For this paper, a longitudinal case study was conducted in an organization that had earlier met challenges with different types of Resistance to Change (RC). With a profound understanding of the case company's RPA adoption journey and its organizational challenges, the path to AI may be easier for the organization to manage. Rather than being retrospective, earlier findings need to be applied also to the initial stages of the adoption process.

The aim of this longitudinal research was to gain in-depth knowledge out of a real-life phenomenon, especially the challenges, obstacles, decisions and successes to find out 1) what factors may have improved the adoption process in time (continuation), 2) how the impact of different factors change over time, and 3) what can be learned from automation of financial management and payroll services to facilitate organizations' progress when adopting 4IR innovations?

A review of literature focusing on RPA adoption and technology acceptance is presented in Sects. 2 and 3. Section 4 presents the data collection and analysis. The key findings from semi-structured interviews and a thematical analysis follow in Sect. 5, and recommendations for future adoption and research within AI in Financial Management Services, are discussed in the sixth and final section.

2 Adopting Robotic Process Automation in Organizations

The worldwide end-user spending on RPA software in 2023 is predicted to be 3,352 million USD with a 17.5% growth from year 2022 and the trend is towards hyperautomation [8]. Western Europe's share of the total spending on RPA was 19% and North America 48,5% in 2022 [8]. Public sector is often perceived to be slower to reform than the private sector. For example, Swedish public sector has been found to be at a "modest level" [9] or in the "starting blocks" [10] of RPA adoption. In public accounting the size of the firm has been found to be the most significant positive factor for RPA utilization and the level of RPA expertise [11]. Many of the challenges related to RPA adoption described below are easier to overcome in environments with higher transaction volumes, investment funds and labour resources.

Accounting and other financial services have been conducive to software robot-based automation because many of their processes are suited for robotization. Use case studies of RPA adoption are carried out mostly in banking, financial services and insurance [12]. Current literature does not cover the adoption in RPA payroll services. Evidence appears to show that employees in public accounting are more positive towards RPA than in other industries [13]. RPA is only one of the tools available in the portfolio of automation technology. Tasks and processes suitable for RPA are mature, highly manual, repetitive and rule-based, high in volume, standardized [14], stable and show low complexity [15]. The chosen implementations of RPA should also be easy to achieve and impactful [14].

Literature acknowledges several challenges for RPA adoption. Limited utilization of the technology and the stakeholders' resistance create a significant hurdle to overcome [12]. These challenges are intertwined with various other factors that can be categorized as issues with technology, managerial or administrative concerns, or interactions and attitudes among individuals. RPA adoption is "primarily driven by lower-level employees" [13] and management needs to be aware of possible user resistance, and attitudes in general because user participation is crucial identifying cases of automation use in their day-to-day work [1].

RPA demands novel and currently unattained skills from financial management professionals [17]. On the other hand, the need for traditional financial management skills will not disappear because new automated processes and tasks need proficient specification, validation, and maintenance. Organizations may fear that their critical competence or basic knowhow is lost when automation replaces human labour [10, 13]. Based on earlier case studies, skilled professionals in accounting firms are not likely to lose their jobs because of automation [e.g. 13, 18]. Instead, job descriptions should evolve over time. RPA implementation projects and the subsequent maintenance phase require substantial work [9]. When senior management is under RPA hype pressure, they may in some cases over-allocate resources, focusing one-sidedly to RPA while 'hunting for use cases', which results other alternatives to be overlooked [5]. Thus, instead of addressing real-world problems, these cases are endorsing technology-push approach to adoption [19 referred in 5].

Earlier studies suggest that to enhance RPA adoption, organizations should invest in governance, standardization, employee awareness, skills; consider collaborative employee-robot relationship; start with stable, simple processes to gain experience; involve stakeholders early, ensure security, audit, governance and IT oversight; consider long-term objectives to avoid issues hindering long-term transformation [12, 15, 18, 20, 21]. When measuring RPA success, organizations should consider qualitative measures such as employee and customer satisfaction along with efficiency and other quantitative measures [18]. Both management and employees in public accounting believe that RPA adoption can increase job satisfaction [13, 16]. Table 1 summarizes the factors affecting the adoption of automation identified in previous literature reviews and case studies.

Table 1. Different factors that affect the adoption of automation.

Issues with technology	Authors
high price of technology and implementation, and other economic reasons	[4, 9, 11]
service capability	[20]
constantly changing technical environment and constant maintenance	[11, 18]
technical difficulties: ease of building robots exaggerated; building robots with multiple-skill-based capabilities are cumbersome; designing an optimum solution using statistical and machine learning techniques difficult	[9–11]
Managerial and administrative concerns	Authors
identification of routine processes for automation	[10, 16]
finding appropriate methods for allocating costs and benefits between clients	[16]
regulatory changes	[17]
hunt for cases	[5]
inadequate handling of customer perspective	[12]
role of administration in managing the requirements, set-up, and implementation of robots underestimated	[9]
insufficient competence in process and IT	[10]
different processes among customers	[16]
Interactions and attitudes among individuals	Authors
resistance to change, lack of trust and willingness to work with robots	[4, 12, 18, 20]
organizational culture	[4]
perceived risk	[20]
reluctance of clients	[16]
collaboration and shared language between IT and other professionals	[10, 17]
tension caused by disconnect between operational experts and RPA automation team	[5]
lack of bottom-up support, dependence on individual enthusiasts and external RPA consultants	[10]

3 Technology Adoption and Resistance to Change

Much of the research on adoption and acceptance of technological innovation in IS is based on the Technology Acceptance Model TAM [22], Theory of Reasoned Action [23], the Theory of Planned Behavior [24], Diffusion of Innovations theory DOI [7] and the Unified Theory of Acceptance and Use of Technology (UTAUT) model [25]. Research has aimed to consolidate these theories in order to provide a more comprehensive explanation of variables and causalities behind user behaviour. These theories generally aim to model user behaviour and the variables relate to the adoption process and acceptance of technology. However, they concentrate on the initial decisions, instead of modelling how the adoption process continues within organizations, and how the user behaviour changes over time. Yet, user behaviour may cause headaches for the IT management throughout the IS lifecycle.

Resistance to change (RC) is behaviour that protects users from the consequences of change [26 referred in 27]. Resistance can be detrimental, cause conflicts and consume time [28], and it has been found negatively to affect adoption of IS and technological innovations in organizations (e.g. big data in healthcare [27, 29]). On the other hand, it can also functionally prevent harmful system implementations and further issues of such as stress, turnover and reduced performance [28]. Van Offenbeek et al. [30] have found that, when resistance to change and how it affects IS implementations are studied, the theories used in it often include variations of Interaction Theory [28], Equity Theory [31] and the more recent Multilevel Theory of Resistance to IT [32]. Literature recognizes the need to combine resistance to change and acceptance or adoption theories and several recent efforts have been made towards that end [e.g. 29, 33–38]. Nevertheless, longitudinal studies of resistance in IS are scarce, and efforts are currently being made to conduct case studies just before or after implementation (a database search of peer reviewed articles was made using keywords: technology adoption OR technology adoption AND resistance to change AND longitudinal study).

One of the key variables of TAM is attitude: a critical determinant affecting users' behavioural intention (BI) to use the technology. RPA is not a technology that financial management and payroll professionals may necessarily use, but these professionals are needed to participate in innovating and implementing and to allow robots to perform their tasks. Hence, user attitudes and behavioural intentions are considered relevant topics for this study. Negative attitudes may cause resistance in users: one type of passive resistance is users' reluctance to participate in change. In a hectic environment and under pressure, the management may overlook this [37].

Although the original TAM model did not encompass resistance to change, more recent versions and modifications have recognized its presence. While not always a key variable, it is considered a covariate influencing behavioural intention [e.g. 4, 27, 36]. Technology acceptance and resistance to change are not necessarily opposites to each other on the same continuum [31]. In mandatory implementations, there may be user resistance to change even when the technology is already in used, and when its adoption is voluntary, potential users may show support towards new technology, and still hesitate to use it [31, 39]. One of the reasons given for user resistance in mandatory implementations is the perceived breach of psychological contract, for example when the system does not meet users' expectations (e.g. technical problems), when implementation causes extra workload or suffers from communication problems [33]. Disconnect of communication between management and employees [33], or between users and people who are responsible for the design and development of the system [40] may cause unnecessary resistance which could be avoided with putting more effort in communication plans and creating common language.

In 4IR technology studies, users' BI and actual use of technology have been found to be affected by variables such as trust, security, and resistance to change [20, 27]; job security and autonomy, uncertainty, ambiguity, and overload [41]; perceived ease of use (PEU) and perceived usefulness (PU) [4, 20].

Demographic factors such as gender and age are sometimes found to be relevant variables [29] while some results explicitly showed that these were not related to the rate of resistance or intention to use IS [36, 38] in work environment.

Since resistance to change, organisational culture and lack of trust have, as stated earlier, been found to be barriers to AI adoption [4], organisations could, when preparing for it, increase willingness to work with robots and mitigate resistance by making an effort to involve employees in such aspects as process planning and decision making [36] and evaluating automation technology capabilities [20]. Users' intentions to participate in mandatory implementation could be improved by increasing personal benefits and thus generating commitment towards a common goal [34].

4   Data Gathering and Analysis

The case firm is a publicly controlled, medium-sized service centre which provides HR and payroll, financial management and ICT services to its owner-clients, Finnish municipalities, municipal corporations, and wellbeing services counties. The boards of directors of these 'in-house firms' are typically led by civil servants and municipal politicians. The case firm's goal is to provide its services as cost-efficiently as possible with a low profit margin.

The data was collected in two points of time between 2021 and 2023. The first round of data gathering was conducted in Q1/2021 (11 interviews of 45–90 min and supporting data) [1] and the follow-up data was gathered Q2/2023 in 14 semi-structured interviews (each approximately 45 min). In both times, the interviews were preceded by pre-discussions with key persons from the ICT and financial teams, where the current situation was briefly covered, and interviewees picked out from current teams. The interviewee-list of data collection point 2 (see Appendix A) included 5 interviewees from the earlier round of 11 interviews conducted in 2021 [1] and 9 new. Some of the key persons had left the organization or changed roles. Some of the financial management (FM) and payroll experts involved in RPA implementation had also changed.

The new interviews of data collection point 2 were thematically analysed by breaking the transcripts down into shorter texts covering one topic and coding them with open codes. The codes were then harmonized and linked to variables (themes) from the earlier case study round (data collection point 1) if there was correlation between the new codes and previous identified themes (Appendix B). Based on the results, one new theme was added.

Three lists were compiled alongside the thematical analysis: 1) the successes and 2) obstacles the interviewees had experienced, and 3) how they described attitudes towards automation in payroll and financial management. These lists were then analysed and thematised in their turn: were they more likely to be informative, were they related to earlier variables, or did they possibly represent a new theme. The themes were also classified by whether their current impact or interpretation was positive or negative. The topics of the success-list were not linked to the variable "success stories" in the results, to avoid over-weighting this theme. Finally, the dependencies between different variables were analysed from the narrative and the results compared to earlier round of interviews.

If interviewees did not bring up resistance to change themselves, the topic was introduced to them at the end of the interview, by asking whether they had sensed any resistance towards automation during the past two years.

The continued adoption or discontinuance of other automation technologies beside RPA have the same need for constant triggers (ideas for new implementations). In this follow-up study, the topic of automation was discussed with the interviewees, not only focusing on RPA implementation, but also taking into account the solutions used by the organization to automate financial and HR service processes.

5   Findings

In Finland, an extensive health and social care reform was carried out on a fast schedule, in which the responsibility for organising health, social and rescue services was transferred from municipalities and health care districts to 'wellbeing services counties' from the beginning of 2023 [42]. The case organization is 'a contracting entity' according to the public procurement and repealing Directive (2014/24/EU) and due to the change of legal status of its clients caused by the reform the senior management deemed it necessary to tender key financial and payroll system contracts. The bidding contest led to a change of the key systems in the beginning of 2023.

When the case organization's senior management was recontacted, they reported that the adoption rate of RPA had improved over the two years and organizational changes had been made to accelerate adoption speed, but they were still not satisfied with the overall situation. While progress has been made during the two years between the two data collections, the adoption rate of RPA was not as high as they desired.

Several improvements of automation were made in the implementation projects, as part of the re-tendering of the systems. Some were available in the newest versions of the systems, some implemented by integration platforms and some with RPA. However, the pace of development of RPA that had started at the time of the previous interview had nearly come to halt because of the reform.

Between the two rounds of interviews (2021–2023) the case organization had also re-tendered its RPA consultant and reorganized the development unit of the ICT services which had suffered from several changes in key personnel. A governance model for RPA development and maintenance had been instituted, but lately several ideas for improvement had been discussed and planned to be executed with the blessing of the senior management. The ICT management told that they were again in the process of hiring a new RPA expert, preferably senior, and that they had been struggling to find new talents for the automation team. The wage levels in the municipal sector are perceived to be lower than in the private sector and the management has seen this as one of the problems of the labour market.

5.1   Fear Fades Over Time, and Resistance Becomes More of a Principle

The organization's RPA adoption journey has already taken seven years and various types of resistance, from active to passive have emerged on this journey. When questioned about prevailing attitudes towards RPA and task automation in general, half of the interviewees reported only positive shifts in attitudes over the past two years. The rest were split: some had observed a mixture of positive and negative sentiments (3) while others had encountered solely negative opinions (4), often manifested as passive resistance. Senior and midlevel management considered attitudes towards automation to be currently positive, while staff members had still encountered negativity among their peers. When the topic of whether those who had observed resistance were perceiving varying attitudes towards RPA as opposed to other automation technologies was raised, their responses indicated either a lack of distinction or uncertainty. Hence, it cannot be inferred from the answers that RPA at this stage of adoption is evoking an abnormal level of resistance to change in the organization.

Interviewees said that employees are no longer afraid of loss of jobs caused by automation and that it is not the cause of resistance at this point of continued adoption. Any hesitation in fully supporting such adoption is more related to earlier disappoint-ments, lack of knowledge or reluctance to hand over tasks to robots. One of the RPA activists explained that the reason could be structural resistance to any change.

In my opinion the way of thinking has changed during past years. Payroll teams explicitly wish for more tools to make their work easier. (Staff, FM&Payroll)

I have experienced that these robots don't do everything. We still have plenty of work. You never know if there is going to be a robot so clever, that it can replace people, but I'm not afraid of it anymore. (Staff, FM&Payroll)

Maybe it is more a sort of passive resistance, than vocal, active resistance. Surely the benefits have been seen as well as that routine work can be given up... The fear and hysteria have gone. People resist for the joy of resistance. (Staff, FM&Payroll)

One cause for resistance and passivity was that some team members worried about their workload being increased due to implementation of new robots, especially if it was unclear to them, how the robots work, and they felt mistrust.

The idea behind is probably the fear that it [the robot] generates extra work in some other tasks. That you must work out the things automation has done. Many people think that when you do something from beginning to end, if you make a mistake, you'll be able to figure it out. (Staff, FM&Payroll)

5.2 Success Stories Support Positive Attitude Towards Automation

Success stories have increased *trust* and perceived *relative advantage* in FM and Payroll teams as knowledge of and skills in automation have accrued in projects. Even though there was still some mistrust, the level of trust in RPA capabilities is now higher in terms of a) reliability in production and b) the view that a robot can be trusted to do what it has been programmed to do. As resistance had decreased and the teams had learned more about the capabilities of automation, new ideas had emerged for the automation backlog. Hence, the earlier problem of getting enough viable ideas from operations due to resistance or lack of knowledge did no longer slow adoption down at this stage.

Employees were advised to post ideas for automation without worrying about how these should be realized. By doing this, innovating was made easier for a larger number of operative personnel. However, interviewees stated that they still needed to put more time and effort into considering what else could be automated in their routine work. Those who had been part of development projects said that automation is not progressing in their organization as would be desirable. Over half of the interviewees stated that either they personally or the FM and payroll teams in general wished for more automation to assist in everyday tasks.

What surprised me positively was that the FM&P teams had on their own iden-tified ideas for automation and evaluated cost-effectiveness and prioritization. (Management, IT)

When we implemented the newer VAT-robot, and since I wasn't part of the earlier implementations, my eyes were opened to how the robot functioned and in a way that was useful to me. (Staff, FM&Payroll)

The key personnel expressed their satisfaction with the recent implementation projects and how the organization had learned to draw up robot specifications. The new robots had mostly worked well, and incidents or errors were caused by source systems rather than defects in robots. Even though IT unit and subcontractors had seen changes key personnel, key experts had remained the same in financial management and payroll teams. It appears that the FM and payroll teams' skills in RPA technology had made more of an impact on the adoption rate of RPA in the organization, than IT experts' skills in FM and payroll processes and knowledge of their jargon. Understanding of the operative processes varies among the subcontractor's consulting team, and FM and payroll experts need to act as mediators not only during the implementation projects but also in the incident and problem management during maintenance phase.

5.3 Adopting Several Automation Technologies Supports Overall Rate of Automation but Lowers Adoption Rate of RPA

The aim is to transition automation in the case firm from internal use only towards customer-oriented service. The existing BI portfolio employs consistent RPA (UiPath and BluePrism have been found to be the most dominant tools in earlier case studies [6]), integration, and data tools, with Microsoft Power Platform recently incorporated. Literature highlights RPA's limitations and impact on adoption and our prior research revealed low utilization rates of RPA due to technology competition. While no formal assessments of automation technology capabilities or portfolio assessments were made in the organization, a portfolio-like approach was in use, and the development team chose a suitable tool for each problem by evaluating the capabilities of the tools available in their portfolio.

The organization had been 'hunting' for ideas from teams but evaluated them by searching for "most suitable solution to the problem at hand". They aimed to increase automation across its services, not just RPA. Thus, if they manage to solve the automation needs, it should not be an issue to choose non-RPA solutions over RPA, considering their goal. However, interviewees expressly noted that the overall rate of adopting automa-tion solutions was not on a satisfactory level. Both personnel and management in FM &

Payroll teams hoped for a future where the level of automation would be higher, especially through better integration. FM & Payroll management prefers to have IS suppliers develop new automation features to systems, instead of developing their own RPA tasks. Building trust towards RPA had taken time, but at the same time, it was unclear whether operational systems had evolved enough to support higher automation either.

> *Well, the more you can build automation, the easier the work gets. After all, all connections from the system have been automated. But we should still be better in development, so that we could better automate [processes]. (Staff, FM&P)*

> *Our customer has started to put pressure that there should be so and so many robots. I don't think that's the purpose itself ... We shouldn't have robots just for fun – we need utility calculations. In principle we should always check first what can be done for automation with [operational] systems. That is the most efficient and reasonable way. (Management, FM&P)*

When discussing RPA as a tool, one of the interviewees from IT described it "boring" to IT experts. It was far more interesting to develop, for example, automation and quality improvements with the integration platform or tools for information and knowledge management. This view of 'nothing new' has already been brought out in earlier interviews and depicted in other case studies as well [10]. Indeed, RPA capabilities are very limited [43] and other tools are needed for adopting more complete processes..

> *In IT teams, perhaps more is currently expected from the arrival of Artificial Intelligence when it comes to working with boring software robotics. AI probably inspires the IT team more, and then connecting it [AI] to automation. (Staff, IT)*

5.4 Dithering with Implementations Caused by Lack of Feasible Business Cases

One of the main challenges brought up the senior management, boils down to the difficulty of building feasible business cases. Despite the initial investments already made in RPA technology and the work time dedicated to RPA services (for gaining knowledge and creating a backlog of ideas, and establishing an organizational framework for automation, development, and operational models including current subcontractor agreements), each implementation not only necessitates a new investment project but also leads to an increase in ongoing operational costs, especially those incurred by additional licenses and software maintenance. Development costs were thought to be high for several reasons: outsourcing is expensive, customers use different FM & payroll systems and other operational systems, and processes and rules varied between customers. Sometimes the volume per customer, and hence the savings in efficiency, were low. To justify the ongoing adoption of RPA, management emphasized not only the quality improvements made possible by robots, but also the potential for increased job satisfaction and work motivation achieved by offloading repetitive and simple tasks to robots, alongside other automation solutions.

> *It is not an obstacle that there would be no tasks where automation can be used. The obstacle is that we have more systems in place and the benefit perspective is not maximal. (Management, FM&Payroll)*

The downside is that the third party doesn't have the same connection to our user base. In practice, it is our expert who is the middleman. It is more difficult to arrange, to be sure to pass on to a third party, what the problem is. We must pay for every hour they spend. That [downside] is the price. (Management, IT)

The [newest] VAT robot was made as an outsourced development work, and it cost so much money that it didn't even occur to me, that it [a similar automation] would be so hard to redo for another to system. The customer gets the investment back in productivity, not us, but we paid for it. ... Good heavens, how much work. (Management, FM&P)

At this point, dithering was suggested to be more of an issue in management team. Interviewees told of a backlog of viable ideas waiting for implementation. Ideas deemed too expensive were left pending and re-evaluated later. If they become more cost-efficient, they will be implemented later. Management also emphasized that their consultants were constantly learning as well, and that estimations changed over time.

5.5 Main Challenges Culminate in a Lack of Resources

The reorganization of IT development was welcomed, since it was believed to clarify responsibilities and speed up the projects. This was collectively seen as a good solution for deficiencies in coordination and management. However, a bigger, and yet to be resolved problem, was the shortage of senior RPA experts. As the interviews turned to good experiences and difficulties, everyone remembered successes, but the discussions were mostly centred on problems and challenges. The main challenges were directly or indirectly related to a lack of resources, both of money and experts. The senior management had already come to the conclusion that RPA adoption would not progress without investing in an internal development team.

We have a quite lean (understaffed) organization. One developer, and if he's sick, then I will help. In case of a problem, we are in contact with the supplier.... One thing that was left unsaid is that compared to many environments, we have very few robots. If there were more robots, then we surely would have too few developers. (Staff, IT)

There were both internal and external reasons behind this lack of resources. External reasons, as mentioned above, related to the organization's challenging history in recruiting RPA developers and coordinators. The new coordinating manager had worked in their position only for month or two. The only full-time developer considered themselves inexperienced and had learned through self-study and with the support of a subcontractor. While IT department had chosen to invest in the learning curve of the junior developer and new managers, no one in the payroll and financial management team was dedicated to automation. The personnel were busy with operative work of their core services and critical transition projects. Due to the limited resources available, RPA implementations competed for the same funds and operative personnel as the other automation technologies, such as new integrations.

We have certainly had the challenge here, that skilled people are very much in demand and difficult to recruit, and therefore we have a consultancy partner for development work. It is also a challenge for us that we had two welfare areas and their transition projects, so our staff was very busy for the last year and a half. (Management, FM&Payroll)

The job situation in the market is bad. Everyone wants these experts, and the competition is tough for sure. Especially when it comes to experienced seniors. The challenge of finding and recruiting [them]. And if the situation is that you don't get them, then you must continue implementations with third parties. There are also some limits to that - it is certainly not possible to launch as much development work if there are not enough coordinators - one's own expertise and work [is needed]. (Management, IT)

Maintenance tasks accounted for 70% of the worktime allocated for IT experts, and not much was left for development. This allocation posed challenges for the experts to advance automation levels. While development work was predominantly outsourced, both IT managers and experts hoped to shift this balance in the opposite direction. It was somewhat unclear to the staff why there was dithering within the organization to take action on this issue.

It's just the question of what the organization intends to allocate resources for. If the management zeroes in on development, resources are made available. The decision lies mostly with them. It's frequently mentioned that progress might not be the same when multitasking with your regular work. (Staff, IT)

The RPA developer introduced citizen development as a new theme to discussion. He believed that automation in Finland would benefit from investing in citizen development. Software providers emphasize that their RPA tools are "no-code" [17] and citizen development has indeed been one approach in the marketing of no-code and low-code tools. When the topic was later discussed with other interviewees, it had a mixed reception. They were cautiously positive about it, but considered the idea problematic, as there was already a shortage of experts on FM and Payroll services in addition to which many of the automation tasks require integration skills. IT management and experts suggested that FM professionals could be suited to some of the maintenance work, but as other research corroborate [17], developing enterprise-level automation is too complex and challenging, for FM professionals to manage alongside their other responsibilities.

6 Conclusion, Recommendations, and Limitations

Disruptive innovations, such as RPA, can be expected to face resistance from users whose jobs it will change. When organizations advance from RPA to hyperautomation and AI adoption, the experience gathered during the RPA journey should be employed to continue the adoption more successfully. In automation, users hand tasks over to technology and learn new skills to replace old tasks. Technologies such as automation and AI persist only if new implementations are constantly introduced, new adoptions triggered, and if professionals whose jobs are undergoing change do not resist to accept their key role in a successful adoption by producing ideas for these new implementations and maintaining the implementations. Users may resist change even when adoption is mandatory. If left unmanaged, it may lead to discontinuance even when resistance is passive, should the negative attitudes restrain potential users from participating in the utilization of technology.

In this longitudinal case study of continued adoption of RPA, passive resistance to change and dithering were found to affect the adoption rate of robots throughout the whole timespan. However, the intensity of resistance fluctuated over time and shifted from active resistance to passive. The users' attitudes towards robots changed towards more favourable, as their fears for losing their jobs was dispelled and they developed trust and knowledge through successful implementations.

The key variables (themes) that have and impact on technology continuance of RPA at two periods of time in adoption continuum and their relations are summarized in Fig. 1. Data Collection Point 1 (DCP1) refers to the interviews and data gathering where initial themes were discovered in 2021 and DCP2 to the follow-up period in 2023.

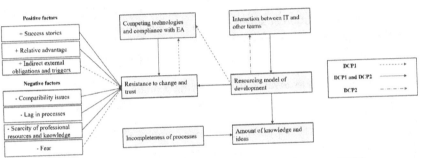

Fig. 1. Key factors impacting the technology continuance of RPA

The findings of this study suggest that organizations should in other 4IR adoptions address resistance throughout the adoption process, starting with potentially active resistance. Their actions should be directed toward dispelling fears by 1) clearly addressing the full potential and limitations of the technology, 2) demonstrating the relative advantage gained by acceptance of the technology, and 3) aiming for early and well scoped implementations. This may require concentrating on organization-specific and context-sensitive capability mappings as well as portfolio-based evaluation of technologies, instead of merely relying on the product descriptions made by the technology supplier. Senior management should allocate sufficient resources for initial implementations at an early stage, thus ensuring that experience is gained rapidly and potential challenges are promptly identified. The first implementation of 4IR technologies should be simple and small enough, suited for the technology under adoption, visible for as many users as possible, and it should solve real problems or fulfil real needs in the organization.

Proactive management of resources is vital throughout the technology's lifecycle. In the context of AI adoption, it is prudent to explore the potential of personnel with skills in Financial Management and Payroll services, and that of emerging junior IT experts. This approach avoids over-reliance on outsourcing or experienced IT personnel alone.

Such caution is necessary due to the tendency of technological hypes to create a shortage of experts in the job market.

Senior management should consider investing in technologies which increase job satisfaction and motivation, when threatened with shortage of skilled professionals. They could consider the possibility of a less ROI based decision making during the first implementations of new automation technologies and AI. The reasoning behind the decision to adopt may include qualitative factors important to the organization and its customers as well as the possible switching costs: what is lost if automation or AI does not proceed at all or the adoption progresses very slowly, or if the qualitative goals are not met, or if knowledge and skills needed in 4IR are not being developed?

This research supports future in-depth, qualitative studies of continued adoption of 4IR technologies. Some of the elements of such theories of technology adoption and acceptance as DOI and TAM variations provide a suitable theoretical starting point for future studies of automation and 4IR, but it is recommended to broaden the scope of studies of technology continuance and longitudinal studies. Studying on organization longitudinally brought to light the variables that cause adoption rates to decrease or increase over time. The study provides for IT management tools for finding the potential actions to be taken in different stages of adoption continuity. In future researches in technology acceptance or adoption and resistance to change, the continuation factor should be taken into account and not focus only on initial decisions, especially when studying such technologies as automation and AI, which need continual triggers. One possible consideration is to advance from studying a single technology to a technology portfolio and so to understand the rate and processes of adoption in organizations. Finally, acceptance of automation technologies cannot be determined from their usage rates among employees, which is what IS acceptance studies traditionally do, because it cannot be measured by usage hours or the number of active users in an organization.

This study has the following limitations: The research aimed to gain in-depth knowledge by using semi-structured interviews in one case organization. Thus, the amount of data is limited, and the research method leaves the data open to interpretation. The reliability of the findings has been tested by comparing them with the findings of case studies and reviews presented in literature and they have been found to carry out same themes, thus the results are context dependent and coherent whole. The research did neither aim to evaluate its case organization's innovativeness compared to others nor how typical resistance to change is in the organization, although the topic was touched upon. Some of the interviewees had also changed between the data collection points, and it is not known whether adoption had anything to do with their decisions to leave the organization. The research does not address the relationship between actual job satisfaction, intentions to stay in the same employment and the adoption of automation technology.

Appendix A DCP2 Interviews

Position/ Role	Employment	Participation in DCP 1
Director, IT services	>2 years	No
Director, Financial management services	>2 years	No, replaced FM service manager
Main user in Financial management systems	<2 years	No
Main user in Financial management systems	>2 years, new position	No
ICT-specialist, RPA	<2 years	No
Director, HR- and payroll services	>2years	Yes
Accountancy specialist	>2years	Yes
Development manager	>2years, new position	No
ICT Designer, in a role supporting the automation team	>2years	No
Project manager	>2years	Yes
Accountancy specialist	>2years	Yes
IT Specialist in knowledge management solutions	>2years	Yes
Specialist in Accounts payable and receivable	>2years	No
Service manager	<2 years	No

Appendix B Restraints, Challenges and Accelerators of Adoption: Summary of Harmonized Open Coding Within Key-Themes in DCP2

Theme (variable)	Restraints and challenges from open coding	Accelerators from open coding
Competing technologies and compliance with EA	- Needs fulfilled by other technologies in the automation portfolio (**Note.** negative in RPA perspective, but positive in the perspective of comprehensive automation) or developments in the information system itself are preferred - Challenges with several customer IS for same processes - >1) need for multiple robots 2) compromises with different customer needs	Good level of integration supports automation positivity
Resourcing model of development	Development with subcontractor: lower ROI, lack of customer knowledge, different financial objectives	Development with subcontractor: good project management, skilled experts
Incompleteness of processes	**Restraints:** - Deficiencies in documentation or operative procedures - Implementation road map either missing or not communicated - Different cultures and processes between provinces	- Lot of work done on process documentation - Suggested improvements on operations model accepted by senior management and to be executed
Interactions between IT and other teams	- Sharing the same level of understanding and lacking common language - Need for developing further co-operation between IT and other operative teams as well as between financial management teams and HR & payroll team	Having IT unit in the same organization

(*continued*)

(*continued*)

Theme (variable)	Restraints and challenges from open coding	Accelerators from open coding
Amount of knowledge and ideas	- Difficulty in prioritizing backlog - Difficulty in efficiency estimations	- Understanding of RPA's capabilities and actions has increased - Goal set to expand RPA knowledge and skills in teams - Skills for specification have improved - Organization has skills to define robot specifications - Teams have identified needs for automation and they are collected to an automation backlog
Resistance to change and trust (excl. list of individual RC and attitude experiences)	**Causes RC and decreases trust**: - Lack of control or visibility in RPA production - Distrust in tech. capabilities	**Reduces RC and creates trust**: - Getting used to technology - Positive attitude
Negative RC factor (-) *Compatibility issues*	- Difficulties in implementing complex rule sets for a robot - Technical challenges - Low flexibility of implementations - Disappointment in tech. capabilities	None
Negative RC factor (-) *Fear*	N/A	- Fear of losing job was gone (No impact any longer)
Negative RC factor (-) *Lag in processes*	- Slow progress with implementations and low adoption rate - Slow incident and problem management - lack of agility	None

(*continued*)

(*continued*)

Theme (variable)	Restraints and challenges from open coding	Accelerators from open coding
Negative RC factor (-) *Scarcity of professional resources and knowledge*	- Not enough automation specialists in-house or outsourced - Personnel changes - Obstacles to recruitment - More urgent projects and excessive operative work - Extra work needed for implementation and maintenance - Misapprehension of objectives	None
Positive RC factor (+) *Indirect external obligations* (Triggers)	None	Variable had positive effect at earlier stages of continued adoption. No effect later in DCP2
Positive RC factor (+) *Relative advantage*	**Relative advantage perceived lower, when:** - RPA is boring to work with - Change does not bring personal benefits - Technology is seen as unnecessary - Unwillingness to hand over tasks to a robot	**Relative advantage perceived higher, when:** - Motivational improvements and job satisfaction is gained - Robots reduce personal workload - A broader, positive effect of automation is observed
Positive RC factor (+) *Success stories* (excl. list of individual successes)	None	- Reliability of implemented robots and integrations - Successful implementation projects (on time and goals met) - General attitude shifted to positive - Measured efficiency improvements - Observed quality improvements
NEW Citizen development	**Obstacles:** - Lack of resources and skills	**Possibilities:** - Maintenance of robots and rule sets

References

1. Sarilo-Kankaanranta, H., Frank, L.: The slow adoption rate of software robotics in accounting and payroll services and the role of resistance to change in innovation-decision process. In: Cuel, R., Ponte, D., Virili, F. (eds.) ItAIS 2021. Lecture Notes in Information Systems and Organisation, vol. 57, pp. 201–216. Springer, Cham (2022). https://doi.org/10.1007/978-3-031-10902-7_14
2. Sarilo-Kankaanranta, H., Frank, L.: The continued innovation-decision process: a case study of continued adoption of robotic process automation. In: Themistocleous, M., Papadaki, M. (eds.) EMCIS 2021. LNBIP, vol. 437, pp. 737–755. Springer, Cham (2022). https://doi.org/10.1007/978-3-030-95947-0_52
3. Demirkan, S., Demirkan, I., McKee, A.: Blockchain technology in the future of business cyber security and accounting. J. Manag. Anal. **2020**(7), 189–208 (2020)
4. Vărzaru, A.A.: Assessing artificial intelligence technology acceptance in managerial accounting. Electron. (Basel) **11**(14), 2256 (2022). https://doi.org/10.3390/electronics11142256
5. Kaniadakis, A., Linturn, L.: Organizational adoption of robotic process automation: managing the performativity of hype. Int. J. Inf. Syst. Proj. Manage. **10**(4), 20–36 (2022). https://doi.org/10.12821/ijispm100402
6. Wewerka, J., Reichert, M.: Robotic process automation - a systematic mapping study and classification framework. Enterp. Inf. Syst. **17**(2) (2023). https://doi.org/10.1080/17517575.2021.1986862
7. Rogers, E.M.: Diffusion of Innovations, 5th edn. Free Press, New York (2003)
8. Gartner Press release. Gartner Says Worldwide RPA Software Spending to Reach $2.9 Billion in 2022. STAMFORD, Conn. (2022)
9. Juell-Skielse, G., Güner, E.O., Han, S.: Adoption of robotic process automation in the public sector: a survey study in Sweden (2022). https://doi.org/10.1007/978-3-031-15086-9_22
10. Lindgren, I., Johansson, B., Söderström, F., Toll, D.: Why is it difficult to implement robotic process automation?: empirical cases from Swedish municipalities (2022). https://doi.org/10.1007/978-3-031-15086-9_23
11. Bakarich, K.M., O'Brien, P.E.: The robots are coming … but aren't here yet: the use of artificial intelligence technologies in the public accounting profession. J. Emerg. Technol. Account. **18**(1), 27–43 (2021). https://doi.org/10.2308/JETA-19-11-20-47
12. Pramod, D.: Robotic process automation for industry: adoption status, benefits, challenges and research agenda. Benchmarking: Int. J. **29**(5), 1562–1586 (2022). https://doi.org/10.1108/BIJ-01-2021-0033
13. Cooper, L.A., Holderness, D.K., Sorensen, T.L., Wood, D.A.: Perceptions of robotic process automation in big 4 public accounting firms: do firm leaders and lower-level employees agree? J. Emerg. Technol. Account. **19**(1), 33–51 (2022). https://doi.org/10.2308/JETA-2020-085
14. Syed, R., et al.: Robotic process automation: Contemporary themes and challenges. Comput. Ind. **115**, 103162 (2020). https://doi.org/10.1016/j.compind.2019.103162
15. Harrast, S.A.: Robotic process automation in accounting systems. J. Corp. Account. Financ. **31**(4), 209–213 (2020). https://doi.org/10.1002/jcaf.22457
16. Cooper, L.A., Holderness, D.K., Sorensen, T.L., Wood, D.A.: Robotic process automation in public accounting. Account. Horiz. **33**(4), 15–35 (2019). https://doi.org/10.2308/acch-52466
17. Kokina, J., Gilleran, R., Blanchette, S., Stoddard, D.: Accountant as digital innovator: roles and competencies in the age of automation. Account. Horiz. **35**(1), 153–184 (2021). https://doi.org/10.2308/HORIZONS-19-145
18. Zhang, C., Issa, H., Rozario, A., Soegaard, J.S.: Robotic process automation (RPA) implementation case studies in accounting: a beginning to end perspective. Account. Horiz. **37**(1), 193–217 (2023). https://doi.org/10.2308/HORIZONS-2021-084

19. Clark, T.D., Jr.: Corporate systems management: an overview and research perspective. Commun. ACM **35**(2), 61–75 (1992)
20. Kim, Y.: Examining the impact of frontline service robots service competence on hotel frontline employees from a collaboration perspective. Sustain. (Basel Switz.) **15**(9), 7563 (2023). https://doi.org/10.3390/su15097563
21. Willcocks, L.P., Lacity, M., Craig, A.: The IT function and robotic process automation. The London School of Economics and Political Science, London, U.K (2015). https://eprints.lse.ac.uk/64519/1/OUWRPS_15_05_published.pdf. Accessed 30 June 2023
22. Davis, F.D.: Perceived usefulness, perceived ease of use, and user acceptance of information technology. MIS Q. **13**(3), 319–339 (1989)
23. Fishbein, M., Ajzen, I.: Predicting and Changing Behavior: The Reasoned Action Approach. Psychology Press, New York (2010)
24. Ajzen I.: The theory of planned behavior. Organizational behavior and human decision processes. **50**(2), 179–211 (1991)
25. Venkatesh, V., Morris, M.G., Gordon, B., Davis, F.D.: User acceptance of information technology: toward a unified view. MIS Q. **27**(3), 425–478 (2003)
26. Zander, A.: Resistance to change—its analysis and prevention. Adv. Manag. J. **1950**(15), 9–11 (1950)
27. Shahbaz, M., Gao, C., Zhai, L., Shahzad, F., Hu, Y.: Investigating the adoption of big data analytics in healthcare: the moderating role of resistance to change. J. Big Data **6**(1), 1–20 (2019). https://doi.org/10.1186/s40537-019-0170-y
28. Markus, M.L.: Power, politics, and MIS implementation. Commun. ACM **26**(6), 430–444 (1983)
29. Shahbaz, M., Gao, C., Zhai, L., Shahzad, F., Arshad, M.R.: Moderating effects of gender and resistance to change on the adoption of big data analytics in healthcare. Complexity (New York N.Y.) 1–13 (2020) https://doi.org/10.1155/2020/2173765
30. Van Offenbeek, M., Boonstra, A., Seo, D.: Towards integrating acceptance and resistance research: evidence from a telecare case study. Eur. J. Inf. Syst. **22**(4), 434–454 (2013)
31. Joshi, K.: A model of users' perspective on change: the case of information technology implementation. MIS Q. **15**(2), 229–240 (1991)
32. Lapointe, L., Rivard, S.: A multilevel model of resistance to information technology implementation. MIS Q. **29**(3), 461–491 (2005)
33. Klaus, T., Blanton, J.E.: User resistance determinants and the psychological contract in enterprise system implementations. Eur. J. Inf. Syst. **19**(6), 625–636 (2010). https://doi.org/10.1057/ejis.2010.39
34. Hwang, Y., Chung, J.-Y., Shin, D.-H., Lee, Y.: An empirical study on the integrative pre-implementation model of technology acceptance in a mandatory environment. Behav. Inf. Technol. **36**(8), 861–874 (2017). https://doi.org/10.1080/0144929X.2017.1306751
35. Kim, H.-W., Kankanhalli, A.: Investigating user resistance to information systems implementation: a status quo bias perspective. MIS Q. **33**(3), 567–582 (2009). https://doi.org/10.2307/20650309
36. Sıcakyüz, Ç., Yüregir, O.H.: Exploring resistance factors on the usage of hospital information systems from the perspective of the markus's model and the technology acceptance model. J. Entrep. Manage. Innovat. **16**(2), 93–131 (2020). https://doi.org/10.7341/20201624
37. Campbell, R.H., Grimshaw, M.: Enochs of the modern workplace: the behaviours by which end users intentionally resist information system implementations. J. Syst. Inf. Technol. **17**(1), 35–53 (2015). https://doi.org/10.1108/JSIT-07-2014-0049
38. Moura, I.V., Brito de Almeida, L., Vieira da Silva, W., Pereira da Veiga, C., Costa, F.: Predictor factors of intention to use technological resources: a multigroup study about the approach of technology acceptance model. SAGE Open **10**(4), 215824402096794 (2020). https://doi.org/10.1177/2158244020967942

39. Molloy, L., Ronnie, L.C.: Mindset shifts for the fourth industrial revolution: insights from the life insurance sector. SA J. Hum. Resour. Manag. **19**(3), e1–e13 (2021). https://doi.org/10.4102/sajhrm.v19i0.1543
40. Long, S., Spurlock, D.G.: Motivation and stakeholder acceptance in technology-driven change management: implications for the engineering manager. Eng. Manag. J. **20**(2), 30–36 (2008)
41. Chung, H., Kim, K.: Service sector response to the fourth industrial revolution: strategies for dissemination and acceptance of new knowledge. Technol. Anal. Strategic Manage. 1–16 (2022). https://doi.org/10.1080/09537325.2022.2110055
42. Finnish institute for health and welfare Homepage: Reform of healthcare, social welfare and rescue services. https://soteuudistus.fi/en/frontpage?p_p_id=fi_yja_language_version_tool_web_portlet_LanguageVersionToolMissingNotificationPortlet&_fi_yja_language_version_tool_web_portlet_LanguageVersionToolMissingNotificationPortlet_missingLanguageVersion=1. Accessed 29 July 2023
43. Gartner Homepage. https://www.gartner.com/doc/reprints?id=1-2B6LCGBU&ct=220921&st=sb. Accessed 10 July 2023

Factors Amplifying or Inhibiting Cyber Threat Intelligence Sharing

Muhammad A. Nainna[✉] [ID], Julian M. Bass[ID], and Lee Speakman

University of Salford, Manchester, UK
a.n.muhammad@edu.salford.ac.uk

Abstract. The increasing frequency of cyberattacks by criminal and state-sponsored actors, has elevated the importance of cyber threat intelligence for organisations. We are interested to understand why practitioner share cyber threat intelligence and the impediments that prevent sharing. This paper addresses practitioners' perceptions of factors that influence cyber threat intelligence sharing. To find out the factors that influence why cyber security practitioners share or don't share. We conducted research interviews with nine cyber security practitioners using a semi-structured, open-ended interview guide which were recorded and transcribed. We also analysed the data using an approach informed by grounded theory. We coded the data, organised the data into themes, and used constant comparison to check our code's consistency and accuracy. Furthermore, we developed memos, from which our theory emerged. Ultimately, our analysis revealed a new phenomenon, which we call Circumstantial Sharing. In circumstantial sharing, practitioners may rigorously discover and mitigate cyber threats and inform top management. However, practitioners my experience pressure from management not to share cyber threat intelligence with external organisations. This is significant because cyber threat intelligence sharing is an important weapon to resist future malicious cyber-attacks. We observed three main impediments to cyber threat intelligence sharing: fear of potential penalties imposed by regulatory authorities, concerns about sharing cyber threat findings with competitors or adversaries and the financial cost of sharing cyber threat intelligence. It is our hypothesis that overcoming the impediments we have observed will facilitate increased cyber threat intelligence sharing and hence help resistance to future cyber-attack.

Keywords: Cyber security · Cyber threat intelligence sharing · Cyber security practitioners · Frameworks · Grounded theory

1 Introduction

The rate of cyber-attacks is increasing, for example in the 1st and 2nd quarter of 2022, SonicWall recorded 2.8 billion malware hits globally, an 11% increase from 2021 [14]. The motivation behind the majority of cyberattacks are financial gains and state sponsor actors, to steal top confidential information [14, 24]. Also, the importance of cyber threat intelligence in an organisation is increasing by the day, with 85% of organisations producing or consuming cyber threat intelligence by 2021 [12]. On the other

M. Papadaki et al. (Eds.): EMCIS 2023, LNBIP 502, pp. 204–214, 2024.
https://doi.org/10.1007/978-3-031-56481-9_14

hand, government and multinational enterprises comes up with cyber threat intelligence sharing frameworks, as a guidance to practitioners on how to share cyber threat intelligence [1, 3, 13, 19]. In this paper, our main research question is: What are practitioner perceptions of factors influencing cyber threat intelligence sharing.? To answer this question, we conducted an analysis in the existing scholarly literature and found that there are cyber security practitioners that share cyber threat intelligence, and there are those that don't share. Further, to find out the factors influencing cyber security practitioners' likelihood to share or not. We conducted research interviews with nine cyber security practitioners using semi-structured, open-ended questions which were recorded and transcribed. We analysed the data using an approach inform by grounded theory. We coded the data, organised the data into themes, and used constant comparison to check our code's consistency and accuracy. Furthermore, we developed memos, from which our theory emerged. After finding enthusiastic, unenthusiastic, and circumstantial cyber threat intelligence sharing, we looked more comprehensively at the circumstantial cyber threat intelligence sharing factors. We found that cyber security practitioners were afraid of potential penalties imposed by regulatory authorities, the financial costs of sharing cyber threat intelligence, and fear of sharing their cyber threat with competitors or adversaries. These are the factors that make cyber security practitioners share in circumstantial instances. This paper is structured as follows: Sect. 2 provides a review of the previous research on cyber threat intelligence sharing, while also offering a brief overview of cyber threat intelligence sharing frameworks and platforms. Moving forward, Sect. 3 outlines the research methodology employed in this study, encompassing details on the selected research sites, data collection procedures, and the adopted data analysis approach. The outcomes of the investigation are presented in Sect. 4, where the study's findings are systematically unveiled and analysed. Section 5 delves into a comprehensive discussion of the results, offering deeper insights and interpretations. Finally, Sect. 6 serves as the conclusion of the study, summarising the three primary research findings and their implications.

2   Related Work

This section introduces cyber threat intelligence sharing and discusses the intersection between cyber threat intelligence sharing frameworks and cyber threat intelligence sharing platforms.

2.1   Overview of Cyber Threat Intelligence Sharing

Sharing cyber threat intelligence is essential for detecting, preventing, and minimising cyberattacks in the modern cybersecurity scenario [22]. Cyber threats are increasingly sophisticated and complicated, requiring a joint strategy to identify and neutralise [35]. A complete security strategy requires cyber threat intelligence exchange to identify new threats and vulnerabilities [6, 29]. Security experts can predict and prevent assaults on vital systems and sensitive data by sharing knowledge about emerging cyber threats [8]. Cyber threat intelligence sharing is becoming increasingly important as the threat landscape evolves, emphasising practitioners' need to prioritise it in their security strategy [27, 28].

The significance of sharing cyber threat intelligence (CTI) has been a subject of exploration within the existing literature [5, 36]. Extensive discourse has revolved around the roles played by CTI frameworks and CTI platforms in facilitating this process [4]. Nonetheless, it is crucial to acknowledge that the mere existence of frameworks does not guarantee optimal outcomes, their effectiveness hinges upon the active engagement and utilisation by cybersecurity practitioners [7]. Considering this, the current study endeavours to delve into the perceptions held by cyber security practitioners concerning the utility and impact of these frameworks. By doing so, we aim to contribute a nuanced understanding of the interplay between CTI frameworks, platforms, and their real-world application from the perspective of those directly involved in the field.

2.2 Cyber Security Frameworks

The advanced persistent threat used by the threat actors, cyber security needs to devise a strategic and collaborative way to stop or minimise cyber threats [9]. To manage possible security hazards, the United Kingdom (UK), the United States (US), and the European Union have all created frameworks to limit and lower such dangers [3, 4, 18]. Establishing these frameworks may be viewed as a response to growing concerns over the security of these nations concerning the prevalence of cyberattacks and the exploitation of vulnerabilities in critical infrastructure [34].

These nations' frameworks indicate their dedication to upgrading security measures and developing effective methods for preventing and reducing security concerns. They are intended to allow successful stakeholder collaboration, information sharing, and best practices [2]. In addition, these frameworks are meant to build a culture of security awareness and preparation, emphasising the necessity for proactive steps to be taken to decrease vulnerabilities and prevent future attacks. The challenges necessitate the application of cutting-edge technology and the creation of innovative methods for detecting and responding to security breaches [13].

Generally, the frameworks produced by the UK, the US, and the European Union constitute a substantial advance in the continuous endeavour to protect against security concerns. They demonstrate a shared commitment to strengthening security and provide a solid basis for continued engagement and cooperation in this vital sector [25, 26].

2.3 Cyber Threat Intelligence Sharing Platforms

The challenges necessitate the application of cutting-edge technology and the creation of innovative methods for detecting and responding to security breaches [13]. Automating the process enables practitioners to share information more efficiently and in real time, leading to more coordinated responses and effective threat mitigation [30]. Cyber threat intelligence sharing platforms provide a secure and efficient way to share cyber threat intelligence and can handle large volumes of data. Additionally, using an open-source platform means the cybersecurity community can customise and improve the software, ensuring it remains relevant and effective [10].

Structured Threat Information eXpression (STIX), Trusted Automated Exchange of Intelligence Information (TAXII), and Malware Information Sharing Platform (MISP) are key technologies for sharing cyber threat intelligence. Incorporating these technologies into security operations improves incident response coordination, promotes efficiency, and enables the automated analysis of vast quantities of data [30, 33]. These tools are crucial for the early detection and mitigation of cyber threats. Developing an automated cyber threat intelligence sharing platform is vital for effective threat mitigation in the current cybersecurity landscape [9].

Derived from our comprehensive analysis of the existing literature, it becomes evident that a substantial portion of research pertaining to the sharing of cyber threat intelligence aligns with the domains of cyber threat intelligence frameworks and platforms. Correspondingly, a noticeable paucity exists in terms of inquiries into the perspectives held by cyber security practitioners concerning these frameworks. This paper, therefore, addresses this particular void in the scholarly discourse.

3   Methodology

As Patton [32] and Creswell [16] advice researchers to choose appropriate methodologies and strategies that align with the research's objectives, as this can significantly impact the persuasiveness and effectiveness of the overall strategy. The research chose the qualitative approach because qualitative methods facilitate the study of issues in depth and detail and are used in a descriptive term to a layman's understanding [16, 20, 32]. Grounded theory data analysis was adopted to in which the research constructs a general, abstract theory of a process, action, or interaction grounded in the participants' perspectives [21]. Nine interviews with cyber security practitioners were conducted as shown in Table 1.

3.1   Research Site

United Kingdom was chosen as the research site for this study. Theoretical sampling applied in selecting the cyber security practitioners from whom the data collected [21, 31]. Glaser et al. [20] state that theoretical sampling is the data collection process for generating theory. Also, the snowballing technique applied to select more cybersecurity practitioners. Most of the cyber security practitioners interviewed has many years of experience in the cyber security industry and work across government and private organisations in the UK.

3.2   Data Collection

The study collect data from research sites through face-to-face and Microsoft teams meetings. 50–60 min open-ended and semi-structured interview questions used to elicit responses from the selected practitioners. This approach allows for flexibility in the interview question structure [15]. The interview questions focus on the perceptions of practitioners on cyber threat intelligence sharing frameworks and assessing the reluctance of practitioners to share threat intelligence and foster collaboration with others. The formulation of the initial interview questions emanated from a meticulous examination of pertinent scholarly literature and drew upon the researcher's expertise in the realm of cyber threat intelligence sharing.

Table 1. Description of participants', academic qualification, years of experience and organisations in the study

Participant	Business Sector	Academic Qualification	Years Of Experience	Organisation Size
Web Developer	Private	Postgraduate	20 Years	Medium enterprise
Cyber security instructor 1	Public	Postgraduate	17 Years	Large enterprise
Law enforcement officer	Public	Postgraduate	30 Years	Large enterprise
Cloud security specialist	Private	Graduate	40 Years	Large enterprise
IT risk assurance analyst	Private	Postgraduate	2 years	Large enterprise
Cyber security expert	Private	Postgraduate	4 years	Medium enterprise
Cyber awareness specialist	Private	Graduate	5 years	Medium enterprise
Cyber security analyst	Private	Postgraduate	5 years	Large enterprise
Cyber security instructor 2	Public	Postgraduate	15 years	Large enterprise

3.3 Data Analysis

An approach informed by grounded theory data analysis was adopted to provide a systematic and flexible approach to analysing data without preconceived notions. This allows for the emergence of patterns and themes directly from the data itself. The methodological choice facilitates a deeper understanding of the phenomena under investigation, as it encourages an inductive and open-ended exploration of the data [20]. Employing grounded theory data analysis, the study aimed to find new insights and theories that are based on the empirical reality of the data [20].

After each interview, the researcher listened to the recordings multiple times to ensure accurate transcription to prevent any potential distortion in the interpretation of their responses [21]. The transcript is used for coding to process and identify a passage in the text or other data items, for searching and identifying concepts, and for finding relations between them [23]. The researcher coded the data, as Glaser et al. (1968) recommended that if an analyst wishes to convert qualitative data into an accurately quantifiable form so that the research can provisionally create a hypothesis, the researcher code the data first and then analyse the data.

The researcher organises the data into themes by interpreting the data set to identify, analyse, and report repeated patterns [32]. Constant comparison is used to check our code's consistency and accuracy and ensure that the passages coded the same way are similar [15, 21]. Also, the researcher documents his thinking and ideas by developing memos about codes and the interconnections among them [11]. Because it is through memoing that similarities and differences are identified, relationships are explored, and hypotheses are inspired [11]. Finally, the researcher takes note of theoretical saturation at a point where no new dimension or relationship emerges during data analysis [17].

4 Findings

Cyber threat intelligence sharing is essential for cyber security practitioners looking to improve their cybersecurity readiness, reduce risk, help other practitioners and regulatory compliance requirements. Nonetheless, certain factors impede or restrict practitioners from engaging in the dissemination of cyber threat intelligence. During this research, a distinct pattern emerged among the participating cyber security practitioners. The patterns are going to be discussed in the section below.

This section delves into three memos concerning cyber threat intelligence sharing. The first memo portrays practitioners who display a keen enthusiasm for sharing cyber threat intelligence. In the second memo, we encounter practitioners who readily share cyber threat intelligence within their organisation but encounter difficulties when attempting to share it beyond organisational boundaries. Lastly, the third memo outlines practitioners who abstain entirely from sharing cyber threat intelligence.

4.1 Factors for Enthusiastic Cyber Threat Intelligence Sharing

The sharing of cyber threat intelligence holds paramount significance in the realm of cybersecurity. Numerous facilitating factors come into play, fostering a conducive environment that encourages cyber security practitioners to exchange valuable cyber threat intelligence amongst themselves.

In the research interview conducted, some interviewees showed enthusiasm to share cyber threat intelligence with other organisations unimpeded. According to practitioner (cyber security instructor 1).

"If an institution faces certain attack, they probably send it or communicate to the government relevant departments... it is disseminated to all relevant organisations" *(Cyber security instructor 1).* A practitioner said, *"There's very clearly defined end processes based around the national intelligence model"* (law enforcement). Another practitioner is of the opinion *"I would say that organisations are productive in sharing cyber threat intelligence where it applies to them"* (IT risk assurance analyst).

Also, the practitioners take General Data Protection Regulation (GDPR) seriously. Which makes them aware of the importance of keeping personal data safe while sharing threat intelligence. A practitioners said, *"My manager was taking [the GDPR] seriously"* (Web developer).

The sharing of cyber threat intelligence is of paramount significance in the field of cybersecurity. Several facilitating factors contribute to a conducive environment that encourages practitioners to share cyber threat intelligence. Notably, government initiatives play a crucial role in promoting cyber threat intelligence sharing among cybersecurity practitioners. Additionally, practitioners' adherence to government regulations, such as the (GDPR), also contributes to the facilitation of sharing cyber threat intelligence in this domain.

4.2 Factors for Circumstantial Cyber Threat Intelligence Sharing

Whilst the sharing of cyber threat intelligence holds paramount importance, numerous practitioners encounter notable challenges in effectively disseminating such crucial information, particularly when dealing with external organisations.

In certain instances, practitioners diligently identify and mitigate cyber threats, promptly notifying top management of their findings with the intention of sharing cyber threat intelligence with external organisations. However, it is possible that the higher-ranking management decides not to disseminate the cyber threat intelligence to other organisations, withholding their reasons for doing so. According to practitioner (Web developer).

"We went through a great deal of investigation as to whether the information commissioner's office need to be informed,... it went to the university higher ranking, what dent it I don't know".

Another practitioner said, *"From within policing, the sharing of intelligence is very and strictly managed"* (law enforcement). Another challenge organisations faces when sharing cyber threat intelligence that may hinder them sharing with external organisations is there is no single and accepted language in sharing that everyone agrees on. According to a practitioner (Cloud security specialist). *"Having a language, a way of describing things that everyone agrees on".* Also, (Cyber security analyst) said *"The financial cost of sharing and receiving cyber threat intelligence hinder some organisation from sharing and receiving cyber threat intelligence."*

The effective dissemination of vital cyber threat intelligence poses considerable challenges for numerous practitioners, especially when dealing with external organisations. In many instances it's the top management that are responsible to execute sharing cyber threat intelligence with other organisations, not the cyber security practitioners. The top management may decide not to share, and the reasons for such non-disclosure may remain undisclosed. Moreover, a significant obstacle hindering the sharing of cyber threat intelligence with external organisations lies in the absence of a universally accepted language for sharing, financial cost of sharing and receiving cyber threat intelligence.

4.3 Factors Preventing Cyber Threat Intelligence Sharing

While acknowledging the significance of cyber threat intelligence sharing, it remains evident that a considerable number of practitioners exhibit limited enthusiasm or hesitation in divulging such cyber threat intelligence sharing to external organisations.

The cybersecurity industry's capitalist nature fosters a lack of trust among practitioners, which hinders their willingness to share cyber threat intelligence with one another. According to practitioner (Cyber security analyst) *"There is lack of trust among cyber security practitioners."* another practitioner is of the opinion that *"If you want to see capitalism or the free market in action, go and look at the cyber security market"* (Cloud security specialist). Another practitioner (Cyber security instructor 2) said, *"I'm not sure if I would really trust private organisations."*

As previously mentioned, a subset of cyber security practitioners exhibits hesitancy or diminished enthusiasm when it comes to sharing cyber threats with external organisations. This reluctance stems from inherent trust concerns within the cyber threat intelligence sharing entities. Practitioner (Cloud security specialist) is of the opinion that,

"When you sort of look at who is collecting this information, it's tempting to think that it's all some you know the army or the state, but in fact it isn't. The majority of this information is being collected by commercial organisations".

The practitioners also, emphasised that no organisation wants to share their cyber threat intelligence findings with their competitors or adversaries. A practitioner (Cyber security expert) said, *"you don't want to share [cyber threat intelligence] with the adversary."* A practitioner (Cyber awareness specialist) also said *"you probably don't want to share [cyber threat intelligence] with your competitor"* furthermore another practitioner said.

"The last thing you want to do as a defending organisation is to pass on to the adversaries the knowledge of what you have about their tools, techniques, practices, and everything else" (Cloud security specialist).

In the realm of cyber threat intelligence sharing, it is apparent that a significant portion of cyber security practitioners demonstrates a reluctance or hesitancy in sharing cyber threat intelligence to external entities. This apprehension can be attributed, in part, to the capitalist nature of the cybersecurity industry, which fosters an atmosphere of mistrust among practitioners, impeding their willingness to collaborate and share insights with each other. As a result, these trust issues extend to interactions with other organisations, leading to a reluctance to share valuable cyber threat intelligence. Consequently, there is a danger that such information could fall into the hands of potential adversaries.

Enthusiastic cyber threat intelligence sharing denotes a proactive exchange of relevant information among cybersecurity practitioners. Stakeholders willingly contribute and disseminate CTI with a sense of urgency because they share a commitment to strengthen group security measures. Such enthusiasm often manifests in the prompt disclosure of threat indicators, vulnerabilities, and strategic insights, reflecting cooperation within the cybersecurity community. Enthusiastic CTI sharing is characterized by a commitment to mutual benefit and a recognition of the collective imperative to fortify cyber defences.

Unenthusiastic CTI sharing reflects a reluctant engagement in disseminating cyber threat information. This stance may arise from various factors, including concerns about the potential ramifications of sharing sensitive data, a lack of trust among stakeholders, or institutional barriers that inhibit the free flow of information. Unenthusiastic CTI sharing is marked by a cautious approach, with practitioners exhibiting hesitancy in divulging critical intelligence due to perceived risks or uncertainties.

Circumstantial CTI sharing refers to situations in which the sharing of cyber threat intelligence depends on certain outside factors. Contrary to enthusiastic or unenthusiastic sharing, circumstantial sharing is sporadic and dependent on situational factors like the fear of potential penalties imposed by regulatory authorities, organizational priorities, and legal and regulatory considerations. Practitioners engaged in circumstantial CTI sharing evaluate the relevance and urgency of the circumstances before deciding to disclose intelligence. This approach acknowledges the dynamic nature of cybersecurity challenges and emphasizes the need for a judicious and context-aware strategy for sharing threat intelligence.

5 Discussion

One of the objectives of the cyber threat intelligence sharing frameworks created by NCSC, NIST, and ENISA, among many others, is to encourage cyber threat intelligence sharing among cyber security practitioners [4, 5, 17]. This study responds to the call for further research on factors amplifying or inhibiting cyber threat intelligence sharing [15, 16]. Thus, at the end of our study, we found three categories of practitioner cyber threat intelligence sharing behaviour. The practitioners who are enthusiastic about sharing cyber threat intelligence, the practitioners that share cyber threat intelligence in circumstantial instances, and the practitioners that are unenthusiastic about sharing cyber threat intelligence. The intriguing aspect of our findings is that:

1. The enthusiastic cyber security practitioners that share cyber threat intelligence see the significance of sharing cyber threat intelligence with other practitioners. Evidently, practitioners who ardently engage in the exchange of cyber threat intelligence hold the perspective that their productivity is enhanced through the reciprocal sharing and reception of such intelligence insights.
2. It is apparent that a significant portion of cyber security practitioners demonstrate a reluctance or hesitancy to share cyber threat intelligence with external entities. This apprehension can be attributed, in part, to the capitalist nature of the cybersecurity industry, which fosters an atmosphere of mistrust among practitioners, impeding their willingness to collaborate and share insights with each other.
3. The circumstantial cyber security practitioners share cyber threat intelligence within their organisation, but don't want to share with external organisations. Because of the cost of sharing cyber threat intelligence, fearing potential penalties imposed by regulatory authorities, and don't want to share their cyber threat findings with competitors or adversaries.

After discovering the enthusiastic, unenthusiastic, and circumstantial cyber threat intelligence sharing. We are more interested on why cyber security practitioners share cyber threat information in a circumstantial instance.

Circumstantial cyber threat intelligence sharing refers to the selective dissemination of cyber threat information by practitioners, occurring infrequently rather than consistently. The frequency of sharing is influenced by various factors. Numerous organizations have established policies governing the internal and external sharing of CTI, with a notable emphasis on personal trust as a pivotal determinant in these practices.

6 Conclusion

This study explores practitioner perceptions of factors influencing cyber threat intelligence sharing. We performed recorded and transcribed research interviews with nine cyber security practitioners. We used a semi-structured, open-ended interview guide which were analysed using an approach informed by grounded theory. Data was coded, organised into themes, and subjected to constant comparison, to check for coding consistency and accuracy. Furthermore, we used the data to develop memos, from which our descriptive theory was derived.

We discovered three categories of cyber threat intelligence sharing practitioners. The enthusiastic cyber threat intelligence sharing practitioners, the circumstantial cyber threat intelligence sharing practitioners, and the unenthusiastic cyber threat intelligence sharing practitioners. We contribute an in-depth analysis of the factors influencing circumstantial cyber threat intelligence sharing.

The main impediments to sharing we found are that: practitioner don't want to share cyber threat findings with competitors or adversaries, they fear potential penalties imposed by regulatory authorities, and they are concerned about the financial cost of sharing cyber threat intelligence. This is significant because advocates of cyber threat intelligence sharing will need to overcome these impediments to increase practitioners' propensity to share. Using these findings, we can make efforts to eliminate these impediments.

As a continuation of our research agenda, we intend to develop a more comprehensive taxonomy describing the multifaceted factors that hinder the exchange of cyber threat intelligence among cybersecurity practitioners. The extended taxonomy will serve as a valuable resource for both academia and industry.

References

1. Center for internet security. https://www.cisecurity.org/about-us
2. Cyber-threat intelligence information sharing. https://www.gov.uk/government/publications/cyber-threat-intelligence-information-sharing/cyber-threat-intelligence-information-sharing-guide
3. National cyber security centre. https://www.ncsc.gov.uk/section/about-ncsc/what-we-do
4. National institute of standards and technology. https://www.nist.gov/about-nist
5. Abu, M., Selamat, S., Ariffin, A., Yusof, R.: Cyber threat intelligence–issue and challenges. Indones. J. Electr. Eng. Comput. Sci. **10**(1), 371–379
6. Alahmari, S., Renaud, K., Omoronyia, I.: A model for describing and maximising security knowledge sharing to enhance security awareness. In: Information Systems: 16th European, Mediterranean, and Middle Eastern Conference, EMCIS 2019, vol. Proceedings 16, Dubai, United Arab Emirates
7. Ardo, A.A., Bass, J.M., Gaber, T.: An empirical investigation of agile information systems development for cybersecurity. In: Themistocleous, M., Papadaki, M. (eds.) EMCIS 2021. LNBIP, vol. 437, pp. 567–581. Springer, Cham (2022). https://doi.org/10.1007/978-3-030-95947-0_40
8. Badsha, S., Vakilinia, I., Sengupta, S.: Privacy preserving cyber threat information sharing and learning for cyber defense. In: 2019 IEEE 9th Annual Computing and Communication Workshop and Conference (CCWC)
9. Balson, D., Dixon, W.: World economic forum cyber information sharing reports 2020. https://www3.weforum.org/docs/WEF_Cyber_Information_Sharing_2020.pdf
10. Bauer, S., Fischer, D., Sauerwein, C., Latzel, S., Stelzer, D., Breu, R.: Towards an evaluation framework for threat intelligence sharing platforms. In: Proceedings of the 53rd Hawaii International Conference on System Sciences. https://hdl.handle.net/10125/63978
11. Birks, M., Chapman, Y., Francis, K.: Memoing in qualitative research. J. Res. Nurs. **13**(1), 68–75. https://doi.org/10.1177/1744987107081254
12. Brown, R., Lee, R.: SANS cyber threat survey report 2021. https://www.sans.org/white-papers/40080/
13. Chris, J., Lee, B., David, W., Julie, S., Clem, S.: NIST guide to cyber threat information sharing. https://nvlpubs.nist.gov/nistpubs/specialpublications/nist.sp.800-150.pdf
14. Conner, B.: SonicWall cyber threat report 2022. https://www.sonicwall.com/2022-cyber-threat-report/
15. Corbin, J., Strauss, A.: Basics of Qualitative Research: Techniques and Procedures for Developing Grounded Theory. SAGE Publications

16. Creswell, J.: Research Design Qualitative, Quantitative and Mixed Methods Approaches. SAGE
17. Creswell, J., Creswell, J.: Research Design: Qualitative, Quantitative, and Mixed Methods Approaches. Sage Publications
18. European Union Agency Cybersecurity: About ENISA - The European Union Agency for cybersecurity. https://www.enisa.europa.eu/about-enisa
19. Cyber threat intelligence CTI overview, European Union Agency for Cybersecurity, (ENISA). https://www.enisa.europa.eu/topics/publications/year-in-review
20. Glaser, B., Strauss, A.: The Discovery of Grounded Theory: Strategies for Qualitative Research
21. Glaser, B., Strauss, A., Strutzel, E.: The discovery of grounded theory; strategies for qualitative research. Nurs. Res. 17(4), 364
22. Guarascio, M., Cassavia, N., Pisani, F., Manco, G.: Boosting cyber-threat intelligence via collaborative intrusion detection. Future Gener. Comput. Syst. 135, 30–43
23. Hoda, R., Noble, J., Marshall, S.: Developing a grounded theory to explain the practices of self-organizing agile teams. Empir. Softw. Eng. 17(6), 609–639. https://doi.org/10.1007/s10 664-011-9161-0
24. Jabbour, K., Devendorf, E.: Cyber threat characterization. Cyber Defense Rev. 2(3), 79–94. http://www.jstor.org.salford.idm.oclc.org/stable/26267387
25. Jasper, S.: Us cyber threat intelligence sharing frameworks. Int. J. Intell. Counter Intell. 30(1), 53–65
26. Kwon, R., Ashley, T., Castleberry, J., Mckenzie, P., Gourisetti, S.: Cyber threat dictionary using MITRE ATT&CK matrix and NIST cybersecurity framework mapping
27. Nozomi Network Lab, Enhancing Threat Intelligence with the MITRE ATT&CK Framework. https://www.nozominetworks.com/blog/enhancing-threat-intelligence-with-the-mitre-attck-framework
28. Ma, D., Zhou, J., Zuo, M.: Inter-agency information sharing for Chinese e-government development: a comparison between vertical and horizontal dimensions. Inf. Technol. Dev. 28(2), 297–318. https://doi.org/10.1080/02681102.2020.1801566
29. Moyo, M., Sadeck, O., Tunjera, N., Chigona, A.: Investigating cyber security awareness among preservice teachers during the COVID-19 pandemic. In: Themistocleous, M., Papadaki, M. (eds.) EMCIS 2021. LNBIP, vol. 437, pp. 527–550. Springer, Cham (2022). https://doi.org/10.1007/978-3-030-95947-0_38
30. OASIS Group: Information Management. https://www.oasisgroup.com/
31. Oktay, J.: Grounded Theory. Oxford University Press
32. Patton, M.: Qualitative Research Evaluation Methods: Integrating Theory and Practice. Sage Publications
33. MIS Platform: About malware information sharing platform (MISP). https://www.misp-standard.org/about/
34. Srinivas, J., Das, A., Kumar, N.: Government regulations in cyber security: framework, standards and recommendations. Future Gener. Comput. Syst. 92, 178–188
35. Tounsi, W., Rais, H.: A survey on technical threat intelligence in the age of sophisticated cyber attacks. Comput. Secur. 72, 212–233. https://doi.org/10.1016/j.cose.2017.09.001
36. Zibak, A., Simpson, A.: Cyber threat information sharing: perceived benefits and barriers. In: Proceedings of the 14th International Conference on Availability, Reliability and Security

Empirical Investigation of Practitioners' Perceptions of Agile Testing Coordination in Medical Information Systems Development

Yazidu B. Salihu(✉) ⓘ, Julian M. Bass ⓘ, and Gloria E. Iyawa ⓘ

Department of Computer Science and Software Engineering, University of Salford, Manchester, UK

y.b.salihu@edu.salford.ac.uk

Abstract. Despite adopting agile practices during large-scale agile software development testing, practitioners developing medical information systems face critical challenges coordinating agile software testing. The lack of collaboration and coordination among practitioners developing medical software testing may result in delays in preventing software bugs early and throughout the development process, the cost of rework, and the risk of software failure. To enhance the quality of medical software products, there is a need to prioritise agile software testing in the medical software process.

The study explores various agile testing practices practitioners employ to develop medical information systems. The empirical data about software testing was collected through semi-structured interviews involving ten selected practitioners with experience developing standalone and embedded agile medical software in the United Kingdom (UK) and India. A snowballing approach was used to identify these practitioners. An approach to data analysis informed by grounded theory, including open coding, constant comparison, memoing and theoretical saturation, was used to analyse the data.

We discovered heterogeneous ongoing test automation practices involving unit testing, integration testing, functional regression testing, and clinical testing conducted in the automated or manual testing bay. We classified the practice as roles, ceremonies, and artefacts. The findings also revealed the lack of ceremonies that allow practitioners to interact and discuss with one another and further coordinate, collaborate, and effectively communicate testing strategies in agile medical information systems development. This lack of ceremonies forms the main contribution of this research.

Keywords: Agile Software Testing · Medical Information Systems Development · Practitioners · Tailored Agile · Automation Scripts · Agile Concepts

1 Introduction

Medical information system practitioners have been intensifying efforts to provide quality medical products by using agile methods in their software development. The Agile Manifesto encourages practitioners to collaborate effectively during software testing [1].

© The Author(s), under exclusive license to Springer Nature Switzerland AG 2024
M. Papadaki et al. (Eds.): EMCIS 2023, LNBIP 502, pp. 215–229, 2024.
https://doi.org/10.1007/978-3-031-56481-9_15

Agile software testing is crucial to the software process due to the increased demand for quality and to prevent software bugs early and throughout the development software process [1, 2], even in the medical software development process [3]. Test practices can be automated or manual, which involves various testing practices [4]. In software processes, agile methodologies such as Scrum, eXtreme Programming (XP), and Lean are widely used [5]; undoubtedly, Scrum became the prevailing methodology in the medical information systems development [6, 7]. Agile scrum concepts consist of Roles, ceremonies, and artefacts to deliver quality software to the customer [8]. However, organising effective collaboration among testing practitioners to ensure the quality of medical systems development remains a critical challenge that requires attention [1, 2].

Numerous studies [1, 2, 9, 10] have looked at the issue of software testing among the software development teams in a generic way. However, none of these studies investigates how practitioners coordinate to perform agile testing in medical information systems development. Therefore, this study will empirically bridge this gap by exploring practitioners' agile testing practices in developing medical information systems. The study will be achieved by answering the following research questions.

RQ1: What are practitioners' perceptions about agile testing practice in medical software development?

RQ2: How do practitioners coordinate to enhance quality testing in developing agile medical information systems?

To address these research questions, empirical data was collected through semi-structured interviews with ten highly experienced practitioners and analysed using a data analysis approach informed by grounded theory. The research contribution of this paper is to reveal a lack of ceremonies allowing practitioners to collaborate and coordinate testing strategies in medical information systems development. Ceremonies are collaborative mechanisms that provide transparency and enable visibility in the agile software process. The research highlights various ongoing test case automation involving different roles and artefacts.

The remaining sections of this paper are structured as follows: firstly, the review of the related literature on agile practice and software testing in medical information systems development, including coordination and communication theory—secondly, a presentation of the methodology adopted covering research sites, data collection including demography roles, and experience of study participants and data analysis. Thirdly, findings are organised based on prevailing memos and categorised on the agile concepts. Fourthly, the discussion is organised based on the findings and aligned with research questions—Finally, the conclusion is provided with a discussion of the limitations and future work.

2 Related Work

This section highlights and discusses the existing studies on agile practices and software testing in medical information systems development. The literature has been carefully and thoroughly synthesised to justify the need for the research.

2.1 Overview of Agile Practice

The agile approach tolerates new requirements in any development phase [5]. It is a mindset based on fundamental principles and behaviours and the effort of 17 software practitioners and consultants in 2001. The agile application in software development is guided by core values and principles, known as the manifesto [11]. The core values emphasised "Individuals and interactions over processes and tools", "Working software over comprehensive documentation", "Customer collaboration over contract negotiation", and "Responding to change over following a plan" [11]. The agile manifesto encourages practitioners to coordinate and collaborate in the software development process [12]. Agile scrum concepts include roles, ceremonies, and artefacts crucial in mitigating software failure [13]. Existing studies show that agile has been adopted to address issues in the traditional approach studies [5, 14]. These issues include inadequate coordination in the collaboration [15], inflexibility, high delivery costs, software quality issues, and excessive documentation [5]. The common agile ceremonies are sprint planning, daily standup, sprint review, and retrospectives, which form the broader agile practice adopted in the software development process, including large-scale organisations. For practitioners to build quality software artefacts, coordination among testing practitioners is highly required [1]. It helps guide practitioners' collaboration and clear communication in the agile software development [8].

Additionally, agile focuses on delivering quality software through customer collaboration [16] in developing information systems [17]. The agile methods "practice" includes Scrum, XP, and lean; Scrum and XP methods are widely adopted in software development industries [18]. Similarly, some existing studies [13, 19, 20] suggest that large-scale software practitioners use a hybrid approach known as agile tailoring in their software process. Agile tailoring is the practice of combining agile with the planned-based method. The practice was envisaged even in a regulated medical environment focusing on developing regulated medical products [21], where evidence of agility remains a lingering issue.

2.2 Software Testing in Medical Information Systems Development

There is a paradigm shift in today's computing world due to the increased demand for quality software products. Software testing has seemed crucial since computing systems were introduced in medical care, and software systems were categorised as safety-critical systems [22]. Software testing is essential in developing various artefacts in non-critical and safely critical industries, including healthcare, communication, automobiles, military, entertainment, electricity, and household appliances [23]. However, one of the most devastating software incidents in the history of medical information systems can be traced back to the Therac-25 overdose [22]. It was an incident involving the software of a computerised radiation treatment matching known as Terac-25, which claimed the lives of six patients and left many patients critically injured [22]. A similar medical information systems failure known as London Ambulance Service computer-aided dispatch (LASCAD) happened 1992. The software failure claimed the lives of over twenty people within fifteen hours [24]. The information systems development or software process consists of requirements, design, implementation, testing, and deployment phases [10,

25]. Software automation practice is broadly classified as automated and manual test scripts [4] involving unit tests, integration testing, systems tests, and user acceptance tests [25, 26].

The application of agile practice was a deliberate effort to address the shortcomings in traditional methodologies [5, 27]. It was observed that some practitioners directly incorporate agile practices into their software development processes. In contrast, others use traditional and agile approaches, known as agile tailoring, in developing medical information systems [28]. Numerous research studies have significantly contributed to agile tailoring in large-scale organisations [13, 19, 20, 29, 30]. A researcher conducted an in-depth empirical study with 46 practitioners to mitigate software failure risk [13]. The researcher identified 25 artefacts and mapped each to a particular event. Bass [13] emphasised the need to have corresponding ceremonies as a basis for the quality assurance of software products. In the context of a medical software product development project, the research claimed that using agile methodology will result in cost savings and decrease the need for reworking [31]. Even though most safety-critical systems are into agile tailoring, they have yet to reach the level of agility in the development process.

In addition, medical software products are subject to rigorous testing to safeguard the well-being of patients, ensure compliance with regulatory standards and legal obligations [32], and prevent issues of software recall [3]. The research conducted by [2] attributed the software problem to a lack of coordination among software developers and testers. Another study attributed it to a lack of collaboration efforts from the testing practitioners in adhering to multidimensional requirements in the software process [1]. Therefore, coordination and effective collaboration among testers, developers, scrum masters, and customers is necessary to achieve software product quality. Agile software testing encourages such effort to be performed early and continues throughout the product lifecycle; unlike planned-driven practice, testing is carried out at the end of product development.

Several studies discuss agile software coordination and collaboration in the software development [8, 33, 34]. A researcher, [8], conducted a case study and interviewed 33 software practitioners to understand inter-team coordination mechanisms in large-scale agile software development. The research identified 27 inter-team coordination mechanisms and characterised them, including coordination meetings, coordination roles, and coordination artefacts. He claimed that technical, organisational, physical, and social characteristics are essential to inter-team coordination [8]. This research is built upon their argument that test coordination strategies are necessary for dispersed practitioners to succeed in the large-scale agile software development [15]. [35] conducted empirical studies involving 23 software practitioners to improve the agile software development process. The study identified 26 security practices and mapped them to roles and a lack of collaborative ceremonies. The researcher [12] investigated 24 software practitioners in twelve co-located teams using a case study involving 175 individuals over four years. The research findings show that customer engagement, software architecture, and inter-team coordination are the key challenges and need refinement. Equally, effective collaboration among software testing practitioners is necessary to develop secure information systems [2, 35]. Practitioners were constrained to deal with the issues of security, privacy, safety, and reliability early in the medical software process [36]. We observed studies about collaboration and coordination among testing practitioners of medical information systems are not given much attention. The consequences may amount to a rework [37], a software failure [13] or even a loss of life [22]. Therefore, our study will bridge this gap by investigating the mechanisms and contributing to the existing body of knowledge.

3 Method

This study adopted a qualitative multimethod approach in response to the research questions postulated earlier [38]. We used a data analysis approach informed by grounded theory. This is relevant in software engineering research because it could adjourn predetermined perceptions and examine novel concepts that emerged from the data [39]. This implies that we are not using grounded theory as a methodology. The data collection in this study involved semi-structured, open-ended interviews conducted in cyclical form using snowballing, the subset of purposive sampling and a network of professionals. This is relevant, primarily when recruiting participants is difficult [40]. The participants of this study practitioners in this study include software developers, testers, scrum masters, and project managers in the selected research site developing standalone and embedded medical devices. In the initial step, an exploratory pilot study was undertaken to facilitate the refinement of questions and any revisions and to familiarise oneself with this research methodology. During the second phase, a deductive synthesis was conducted on each interview to enable analysis. This is relevant, especially in the study involving a multimethod approach [38].

3.1 Research Sites

We identified the appropriate research site, as shown in Table 1. We collected data from ten selected UK and Indian practitioners developing standalone and embedded medical devices. The Indian study site coded IMDC is a globally recognised multi-national medical device industry focusing on developing high-level medical devices and software, which include Computer Tomography, Electrocardiography, Magnetic Resonance Imaging and Molecular Imaging devices. The company software headquarters is located in Bangalore, India, and has thousands of staff members across the globe. Similarly, at the UK research site coded UKITCon, four experienced practitioners from medical-related companies were interviewed. The practitioners were involved in various medical software development projects, including developing medical-embedded sensors. The practitioners were recruited based on a snowballing sampling process, academic contact connections, and a network of professionals. All the practitioners are using agile methods in their software development. The diversity and experience of the practitioners provide credibility to the data.

3.2 Data Collection

Data was collected through interviews with ten UK and Indian software practitioners developing standalone and embedded agile medical software. Table 1 comprehensively depicts the geographical locations, roles, and corresponding responsibilities. The data was gathered through semi-structured, open-ended questions in English, virtual and face-to-face, after securing written approval from each participant. We also obtained the consent of the practitioners before recording all interviews; each interview lasted 45–60 min. Subsequently, the transcription of the interviews was done manually before typing in Microsoft Word version 16.76. The manual approach guarantees accurate transcription and reminds the interviewer of the social and emotional dynamics throughout the interview [41]. According to research findings, manual recording and transcribing data from interviews have been the most efficient approach for optimising the obtained data [42]. The interviews were performed using a guide consisting of open-ended questions, allowing the participants to address any topic that arose, regardless of whether it was explicitly covered.

Table 1. Description of participants & demography in the study

Code	Role	Exp (years)	Types of Business	Location
IMSD_SSE	Senior Software Engr	7	Medical Devices Development	India
IMSD_TL1	TEST Lead	7	Medical Device Development	India
IMSD_SA1	Software Architect	9	Medical Device Development	India
IMSD_SA2	Software Architect	12	Medical Device Development	India
IMSD_TL2	TEST Lead	8	Medical Device Development	India
IMSD_PM1	Project Manager	16	Medical Device Development	India
UKITCon_PM2	Project Manager	6	Healthcare	UK
UKITCon_SD1	Software Dev/Analyst	20	Healthcare	UK
UKITCon_SD2	Software Developer	2	Medical sensor development	UK
UKITCon_SDL1	Software Dev Lead	5	Heath app Development project	UK

3.3 Data Analysis

An approach to data analysis informed by grounded theory, including open coding, constant comparison, memoing and theoretical saturation, was used to analyse the data [39]. It allows the discovery of novel knowledge through a high level of abstraction [43]. The researcher meticulously checked each recorded interview to ensure correct transcription and minimise distortion of meaning. Transcribed data was imported into NVivo version 12 for qualitative data analysis.

The transcribed interviews were subjected to line-by-line open coding, and 76 codes were discovered. Line-by-line coding identifies categories, emergent concepts, and behaviour patterns [42]. The constant comparison technique helped identify categorised concepts and compare them to other categories without bias. This led to the emergence of new categories after numerous refinements.

During the memo-writing phase, the researcher captured the conceptual links of each category and revised it to be more formal. The researcher believed the approach would ensure memos are well-written Field [42] and understood correctly. Thus, constant comparison was employed to analyse categories and compare participant responses from the two sites. The written memos cited interview excerpts as evidence.

The theoretical saturation procedure ended the investigation when no new categories appeared, or data conceptualisation was needed to provide meaningful insight by adding more interviews [44].

4 Findings

The findings of this research are organised based on a constant comparison process and presented in four main memos. The memos describe the informed grounded theory analysis aspect that emerged from the data. These memos include tailoring agile testing, unit testing, integration testing, and coordination among testing practitioners in medical information systems development and are classified into roles, ceremonies, and artefacts.

4.1 Tailoring Agile Testing in Medical Information Systems Development

Software testing is considered a significant practice for ensuring the quality of software programs in the development process. Agile software development emphasises early testing and throughout the software development lifecycle, unlike traditional software development, in which testing is implemented towards the end of the project. According to the practitioners (IMSD_SSE, IMSD_TL1, IMSD_SA1, IMSD_TL2, and UKITCon_SDL1), testing medical software is associated with complexity and stringent processes that require careful attention. Practitioners were highly experienced in large-scale medical software development. The practitioners used agile and planned-driven methods, commonly known as tailored agile, as highlighted by one of the senior software developers:

"Generally, maybe we are not using the pure Agile, ...we have altered the process here and there. But ... whatever you have started, you must finish first, then only take up the subsequent work" (IMSD_SSE).

In medical software testing, the practitioners used testing automation practice involving automated and manual test scripts. The testing roles, like test lead, clinical tester, developer, scrum master, and architect, were engaged in developing various test artefacts, like unit testing, integration tests, test scripts, acceptance criteria tests, and test results, to automate the application feature in the development process. The test lead responsible for software verification and working with two scrum masters highlighted: *"Basically... two scripts will be there; one is manual, and the other is automation. So, I will play basic rules in the automation scripts, ...analyse the automation result, and...raise new issues, or see if some new issues have been already moved into testing"* (IMSD_TL1, Test Lead).

A senior software engineer working on a cardiac function project, a scanning software relating to computer tomography (CT), stated:

"Earlier, there was a clear distinction between testing and developing, like, developers do the development job, and a testing person will do the testing job. But in the last one or two versions, we have rolled out, like, everyone does everything" (IMSD_SSE, Software Engineer).

The senior software engineer further stressed that he generates artefacts like software design documents (SDD) through automated tests, writes user cases, and updates use case documents. The roles, which include scrum master, architect, clinical tester, and test lead, will be reviewed and certified based on the feature requirements. Another practitioner working as a software release test lead within the scrum highlighted:

"Earlier, we followed ... separate verification team... the guy who does the development he was capable enough to do the testing as well" (IMSD_TL2, Test Lead).

According to the software architect working on a post-processing application for cardiac function therapy, "developers are the team's heart". They were highly engaged in coding, feature development, unit testing, and putting the feature into the build system. However, clinical use cases are tested by specialised clinical testers in the manual bay: *"... ensured the clinical use cases are tested well, so the manual testers take care of that because they are the ones with the clinical experts ...come from clinical background ...radiologists ...are part of it"* (IMSD_SA1, Software Architect).

The practitioner further highlighted: *"But extensive clinical testing is done by that clinical tester, not by us"* (IMSD_SA1, Software Architect).

However, test leads explained that they compromise in certain features in multiple features released for testing, leading to: *"... if there are some 10, 15 features it is always difficult even for the test lead to do all the testing ...10 tasks in a day, one or other test we will compromise"* (IMSD_TL1, Test Lead) *"... some piece of codes which you have tested within the developer machine ...but I have never got it from the integration team because it was developed in the last week of the previous sprint. We did not have enough time to turn back to the tester"* (IMSD_TL2, Test Lead). The result shows that some issues will be found during the testing but cannot be raised as defects to reprioritise in the backlog.

4.2 Unit Testing in Medical Information Systems Development

Unit testing enhances software code quality in the early development process. The software development process involves meticulously examining an application's most minor tested feature code, a "unit", to ensure its appropriate functionality. The software architect working on molecular imaging (IM) and responsible for non-functional requirements, like maintainability, performance, and any quality issues, revealed:

"We run all the unit tests at the code level; it gives us a bit of confidence... things are working as they are supposed to. Also, we did not break anything that was working with the new implementation" (IMSD_SA2, Software Architect).

A senior software engineer suggests that the developer can do the unit testing because they write the code and are familiar with the feature. *"I know the feature well, so I need to write unit testing ...a small part of that has to be tested"* (IMSD_SSE, Software Engineer). Practitioners who took part in the study further stated that coverage is defined from user acceptance and without defects for the safety-critical feature to be tested agilely (automated test): *"If it is a safety, then I have to cover it 100%, all the classes write an extensive unit test for that... is called the hazard key"*, and for non-safety critical: ... *"I cover only 70% of the user cases, like 70%/80% is agreeable... If defects increases/decreases that also the Scrum Master updates...."* (IMSD_SSE, Software Developer).

The test lead 1, whose work is basically around automation script, and he is responsible for every analysis, development, and software verification and working with two scrum masters highlighted: *"Basically here we will use a DWH tool for the reporting. Once the execution is done the results will be pushed to the centralised server. Once these results are available in the centralised server"* (IMSD_TL1, Test Lead).

4.3 Integration Testing in Medical Information Systems Development

Integration testing is commonly recognised as the subsequent stage in the software testing procedure, which occurs after the completion of unit testing. The test lead informs that a role is responsible for feature integration, both forward and reverse integration tests in a team foundation server environment: *"We have specific people who build these forward integration and reverse integration...he does the forward integration and ... see whether...feature was broke because of this forward integration"* (IMSD_TL1, Test Lead). He further highlighted: *"After that, we have to work our things, and by the end of the day, we also have to do reverse integration so that our changes should be in the common baseline"*. It is a software testing technique that examines the functionality and compatibility of various components or units inside a software project. It aims to identify any flaws or issues arising when these components are combined, ensuring they effectively collaborate.

As attributed by a senior software engineer working on a cardiac function project, a scanning software relating to computer tomography (CT): *"Normally, we have a four-eye review. It means one person is reviewing; one is the author, and one is reviewing ...If it is a safety code, there is an inspection review by two people...After check-in, one person will validate the review, and the other will validate. After him, another person will validate, and we cover 100%"* (IMSD_SSE1, Software Engineer).

During the process of forward integration and reverse integration, commonly known as FIRI, the software architect highlighted: *"We do the process of forward integration and reverse integration. Every day, we have some FI done from the CT integration branch to our branch so that other teams' changes come into us"* (IMSD_SA1, Software Architect).

4.4 Coordination Among Testing Practitioners

Test coordination is a simple but complicated mechanism envisaged in medical software development. Test coordination activities require interdependent roles to collaborate and interact effectively throughout the processes. The test leads highlighted: *"... tester interaction with other teams would be-is not a common scenario when it comes within a scrum. ...there are no planned interaction meetings"* (IMSD_TL2, Test Lead).

It is noted that multiple parallel scrums are running at various business units, which include Molecular Imagine (MI), Computerized Tomography (CT) and Magnetic Resonance (MR). The test lead further explained: *"I am sure that whatever happens in MI, ... not have any visibility to what is happening in CT or MR in other business units ... from a tester point of view..."* Most practitioners have highlighted that their communication is mostly via email and over-communicator calls.

The software engineer, who is a senior software developer: *"Irregularly, we talk to them. Sometimes, ...we will talk to them by email – we will send an email, and then we will talk to them over Communicator. We will have a Communicator call also"* (IMSD_SSE1, Software Developer).

The issue of rework sometimes arises due to effective test coordination. A senior test lead said he would sometimes write 100 test cases for release and share them with the product owner to review; the product owner would reject and write a new requirement. A practitioner stated that they experience lagging issues because of a lack of shared understanding and transparency between developers and testers: *"A developer ...mindset will be toward ...the development of user cases, whereas the tester ...like to reproduce a bug"* (IMSD_TL1, Test Lead).

The senior software engineer stated: *"For a common feature, we interact with them, but it is not regular"* (IMSD_SSE1, Software Engineer).

Another practitioner, a Health IT consultant, and a senior project manager clearly stated: *"Involvement of ...different teams and having a team that is interconnected ... try to make them feel as if they are one team, we also have resistance"* (ITConsul_PM2 Project Manager). *Customers* are using it. They may find some issues, and they send us a ticket to the German counterpart, and from there we do get to analyse, and they will send it to us" (IMSD_SSE1, Software Engineer).

Effective communication fosters trust, comprehension, and collaboration among software testing practitioners. Testers can facilitate the smooth running of the software testing process by establishing efficient channels with the software development team and stakeholders. Another software consultant and developer/analyst revealed: *"The development team is there ...For a common feature, we will interact with them, but it is not very regular"* (UKITCon_SDL1). The practitioners reveal that sometimes an issue ticket will be raised. The only way to inform the developer is by sending an email: *"Irregularly, we talk to them. Sometimes, ...we will talk to them by email – we will send an email, and then we will talk to them over Communicator. We will have a Communicator call also"* (IMSD_SSE1, Software Developer). All the practitioners say they are using Scrum and Kanban but only at the verification level, which gives them flexibility and transparency to track the status of each testable feature. One software test lead highlighted: *"Instead of talking to the people if I go with the Kanban, I can then talk with the Kanban board"* (IMSD_TL1).

5 Discussion

This study aimed to investigate practitioners' perceptions of agile testing practices in medical software development and investigate the coordination among agile practitioners to enhance quality testing in agile software processes. To the best of the researchers' knowledge, this is the first time such a study has been conducted in the context of medical information systems development. The study unearths novel challenges from a lack of agile ceremonies to facilitate software coordination and collaboration among software testing roles.

5.1 Practitioners' Perceptions of Agile Testing Practice in Medical Software Development (RQ1)

To respond to our research question, "What are the practitioners' perceptions about agile testing practice in medical software development?" The study revealed that practitioners employed test automation scripts like manual and automated tests involving unit tests, integration tests, functional regression tests, acceptance criteria, and clinical testing scripts using various containers as part of quality assurance. Each of these scripts uses a distinct and exclusive procedure for testing the functionality of the program feature and remains an essential component of medical information systems development. The findings suggest numerous software testing activities implicate automated and manual test scripts [45]. When comparing our results with the literature, testing practices are crucial for software quality [1], even though most of the literature discusses agile testing in a generic way, not specifically for medical software processes. Therefore, the medical software process is subjected to clinical testing to prevent the risk of failure [46]. There was consensus among testing practitioners regarding testing a software feature, especially unit "code" tests. Any software development team member (software engineers, software analysts, and software architects) could run the unit testing during the early development process. This gives confidence to the practitioners at the onset of the development process. In either reverse, forward, or continued integration, the practice examines the functionality and compatibility of features. This aligns with the existing literature, which identified the practices as artefacts for improving quality standards [13]. However, testers compromised testing in some situations, especially in multiple features. Features lined up for testing tend to ignore issues without raising defects to reprioritise the backlog. We found dependencies issues where if the testers raised a defect, the development team took a long time to rectify. The practitioners claim they are doing actual agile; based on the study's findings, they are into agile tailoring or hybrid approaches by combining agile scrum, Kanban, and other methods. There is a need for further studies to understand the risk of compromising testing. We discover many corresponding roles and artefacts within the medical software testing process adopted by the practitioners, but where are the ceremonies? This issue led to our second research question.

5.2 Coordination Among Practitioners to Enhance Quality Testing in Agile Software Processes (RQ2)

Our second research question is, "How do practitioners coordinate to enhance quality testing in developing agile medical information systems?". Agile software testing is a collaborative effort involving everyone, including customers, particularly among developers and testers [1]. Our findings reveal that the practitioners have a consensus on utilising the Kanban board to ensure quality in the software testing verification process. The physical Kanban board gives flexibility and transparency in the software development process [16]. Using the Kanban board is a good practice; however, coordination mechanisms, including stand-up meetings, sprint reviews, and retrospectives, are not given priority. This is an exciting finding as it has not previously been reported in the literature. Existing studies [8] stated that using a Kanban board would help facilitate tests and experiments and support the top management in visualising progress; however, it is not being used effectively.

In addition, a lack of effective test coordination between the tester and developers will also result in software quality issues [2]. Agile encourages practitioners to coordinate, collaborate, and have more meetings throughout the agile software testing process [8]. This helped build trust and respect among teams, though it remained a severe issue for testing practitioners developing medical information systems. Previous studies reveal that effective communication between agile software testers and developers will help build quality software [2]. The mechanisms for coordination and communication theories emphasise effective interaction and discussion amongst roles to collaborate and coordinate agile software testing activities with defined agile ceremonies [2, 8]. Therefore, ceremonies are needed to organise agile medical information systems development testing to prevent software risk and improve software quality assurance.

6 Conclusion

Agile software testing aims to discover and prevent bugs early and throughout the software process. This paper focuses on coordinating agile testing to enhance the quality of medical information systems development.

The study used an approach informed by grounded theory to analyse practitioners' interview transcript data. The interviews were conducted with selected UK and Indian software practitioners developing standalone and embedded agile medical software. We performed meticulous line-by-line data coding, memoing, and constant comparison to obtain our findings.

We uncovered heterogeneous ongoing software testing practices by analysing practitioners' interview transcripts: automated and manual test scripts involving unit testing, integration testing, functional regression testing, and clinical testing conducted by the practitioners. We discovered that practitioners are doing tailored agile involving more roles and artefacts.

The significant contribution of this research is that no ceremonies are dedicated to discussing the practitioners' testing practices in the context of this study. Consequently, our recommendation is that agile testing would be more effective if ceremonies were dedicated to discussing testing practices as part of quality enhancement in medical software development.

We plan to conduct more empirical research and ceremonies around test case development and test result analysis of agile medical testing in the software process.

References

1. Srivastava, A., Mehrotra, D., Kapur, P.K., Aggarwal, A.G.: A literature review of critical success factors in agile testing method of software development. In: Singh, P.K., Singh, Y., Kolekar, M.H., Kar, A.K., Chhabra, J.K., Sen, A. (eds.) ICRIC 2020. LNEE, vol. 701, pp. 859–870. Springer, Singapore (2021). https://doi.org/10.1007/978-981-15-8297-4_69
2. Cruzes, D.S., Moe, N.B., Dybå, T.: Communication between developers and testers in distributed continuous agile testing. In: 2016 IEEE 11th International Conference on Global Software Engineering (ICGSE) (2016)
3. Majikes, J.J., Pandita, R., Xie, T.: Literature review of testing techniques for medical device software. In: Proceedings of the 4th Medical Cyber-Physical Systems Workshop (MCPS 2013), Philadelphia, USA (2013)
4. Sharma, R.: Quantitative analysis of automation and manual testing. Int. J. Eng. Innov. Technol. **4**(1) (2014)
5. Abrahamsson, P., et al.: Agile software development methods: review and analysis (2017)
6. McCaffery, F., Trektere, K., Ozcan-Top, O.: Agile – is it suitable for medical device software development? In: Clarke, P., O'Connor, R., Rout, T., Dorling, A. (eds.) SPICE 2016. CCIS, vol. 609, pp. 417–422. Springer, Cham (2016). https://doi.org/10.1007/978-3-319-38980-6_30
7. Demissie, S., Keenan, F., McCaffery, F.: Investigating the suitability of using agile for medical embedded software development. In: Clarke, P., O'Connor, R., Rout, T., Dorling, A. (eds.) SPICE 2016. CCIS, vol. 609, pp. 409–416. Springer, Cham (2016). https://doi.org/10.1007/978-3-319-38980-6_29
8. Berntzen, M., et al.: A taxonomy of inter-team coordination mechanisms in large-scale agile. IEEE Trans. Softw. Eng. **49**(2), 699–718 (2023)
9. Sharma, S., Kumar, D.: Towards a shift in regression test suite development approach in agile. Recent Adv. Comput. Sci. Commun. **15**(5), 668–675 (2022)
10. Najihi, S., et al.: Software testing from an agile and traditional view. Procedia Comput. Sci. **203**, 775–782 (2022)
11. Beck, K., et al.: Manifesto for agile software development (2001). http://www.agilemanifes to.org
12. Dingsøyr, T., et al.: Exploring software development at the very large-scale: a revelatory case study and research agenda for agile method adaptation. Empir. Softw. Eng. **23**(1), 490–520 (2018)
13. Bass, J.M.: Artefacts and agile method tailoring in large-scale offshore software development programmes. Inf. Softw. Technol. **75**, 1–16 (2016)
14. Cohen, D., Lindvall, M., Costa, P.: An introduction to agile methods. Adv. Comput. **62**(03), 1–66 (2004)
15. Berntzen, M., Stray, V., Moe, N.B.: Coordination strategies: managing inter-team coordination challenges in large-scale agile. In: Gregory, P., Lassenius, C., Wang, X., Kruchten, P. (eds.) XP 2021. LNBIP, vol. 419, pp. 140–156. Springer, Cham (2021). https://doi.org/10.1007/978-3-030-78098-2_9
16. Rahy, S., Bass, J.: Information flows at inter-team boundaries in agile information systems development. In: Themistocleous, M., da Cunha, P.R. (eds.) EMCIS 2018. LNBIP, vol. 341, pp. 489–502. Springer, Cham (2019). https://doi.org/10.1007/978-3-030-11395-7_38
17. Ardo, A.A., Bass, J.M., Gaber, T.: An empirical investigation of agile information systems development for cybersecurity. In: Themistocleous, M., Papadaki, M. (eds.) EMCIS 2021. LNBIP, vol. 437, pp. 567–581. Springer, Cham (2021). https://doi.org/10.1007/978-3-030-95947-0_40

18. Pikkarainen, M., et al.: The impact of agile practices on communication in software development. Empir. Softw. Eng. **13**, 303–337 (2008)
19. Bass, J.M., Haxby, A.: Tailoring product ownership in large-scale agile projects: managing scale, distance, and governance. IEEE Softw. **36**(2), 58–63 (2019)
20. Garcia, L.A., OliveiraJr, E., Morandini, M.: Tailoring the Scrum framework for software development: literature mapping and feature-based support. Inf. Softw. Technol. **146**, 106814 (2022)
21. McHugh, M., McCaffery, F., Casey, V.: Barriers to adopting agile practices when developing medical device software. In: Mas, A., Mesquida, A., Rout, T., O'Connor, R.V., Dorling, A. (eds.) SPICE 2012. CCIS, vol. 290, pp. 141–147. Springer, Cham (2012). https://doi.org/10.1007/978-3-642-30439-2_13
22. Leveson, N.G., Turner, C.S.: An investigation of the Therac-25 accidents. Computer **26**(7), 18–41 (1993)
23. Pressman, R.S., Maxim, B.R.: Software Engineering: A Practitioner's Approach. 8th edn. McGraw-Hill Education, New York (2015)
24. Fitzgerald, G., Russo, N.L.: The turnaround of the London ambulance service computer-aided despatch system (LASCAD). Eur. J. Inf. Syst. **14**(3), 244–257 (2005)
25. Bass, J.M.: Agile Software Engineering Skills. Springer, Cham (2023). https://doi.org/10.1007/978-3-031-05469-3
26. Pan, J.: Software testing. Dependable Embed. Syst. **1999**(5), 1 (2006)
27. Boehm, B., Turner, R.: Management challenges to implementing agile processes in traditional development organizations. IEEE Softw. **22**(5), 30–39 (2005)
28. Rahy, S., Bass, J.M.: Managing non-functional requirements in agile software development. IET Softw. **16**(1), 60–72 (2022)
29. Conboy, K., Fitzgerald, B.: Method and developer characteristics for effective agile method tailoring: a study of XP expert opinion. ACM Trans. Softw. Eng. Methodol. **20**(1), Article 2 (2010)
30. Campanelli, A.S., Parreiras, F.S.: Agile methods tailoring – a systematic literature review. J. Syst. Softw. **110**, 85–100 (2015)
31. McHugh, M., McCaffery, F., Coady, G.: An agile implementation within a medical device software organisation. In: Mitasiunas, A., Rout, T., O'Connor, R.V., Dorling, A. (eds.) SPICE 2014. CCIS, vol. 477, pp. 190–201. Springer, Cham (2014). https://doi.org/10.1007/978-3-319-13036-1_17
32. Gordon, W.J., Stern, A.D.: Challenges and opportunities in software-driven medical devices. Nat. Biomed. Eng. **3**(7), 493–497 (2019)
33. Strode, D.E.: A dependency taxonomy for agile software development projects. Inf. Syst. Front. **18**(1), 23–46 (2016)
34. Strode, D., Dingsøyr, T., Lindsjorn, Y.: A teamwork effectiveness model for agile software development. Empir. Softw. Eng. **27**(2), 1–50 (2022)
35. Ardo, A.A., Bass, J.M., Gaber, T.: Towards secure agile software development process: a practice-based model. In: 2022 48th Euromicro Conference on Software Engineering and Advanced Applications (SEAA) (2022)
36. Kostić, M.: Challenges of agile practices implementation in the medical device software development methodologies. Eur. Proj. Manage. J. **7**(2), 36–44 (2017)
37. Sherif, E., Helmy, W., Galal-Eedeen, G.H.: Managing non-functional requirements in agile software development. In: Gervasi, O., Murgante, B., Hendrix, E.M.T., Taniar, D., Apduhan, B.O. (eds.) ICCSA 2022. LNCS, vol. 13376, pp. 205–216. Springer, Cham (2022). https://doi.org/10.1007/978-3-031-10450-3_16
38. Morse, J.M.: Principles of mixed methods and multimethod research design. Handb. Mixed Methods Soc. Behav. Res. **1**, 189–208 (2003)

39. Hoda, R., Noble, J., Marshall, S.: Using grounded theory to study the human aspects of software engineering, in Human Aspects of Software Engineering, p. Article 5. Association for Computing Machinery, Reno (2010)
40. Patton, M.Q.: Qualitative Research & Evaluation Methods: Integrating Theory and Practice. Sage Publications (2014)
41. Vaivio, J.: Interviews – learning the craft of qualitative research interviewing. Eur. Account. Rev. 21(1), 186–189 (2012)
42. Adolph, S., Hall, W., Kruchten, P.: Using grounded theory to study the experience of software development. Empir. Softw. Eng. 16, 487–513 (2011)
43. Glaser, B., Strauss, A.: Grounded theory: the discovery of grounded theory. Sociol. J. Br. Sociol. Assoc. 12(1), 27–49 (1967)
44. Corbin, J.M., Strauss, A.: Grounded theory research: procedures, canons, and evaluative criteria. Qual. Sociol. 13(1), 3–21 (1990)
45. Crispin, L., Gregory, J.: Agile Testing: A Practical Guide for Testers and Agile Teams. Pearson Education (2009)
46. Kannan, V., et al.: Agile co-development for clinical adoption and adaptation of innovative technologies. In: 2017 IEEE Healthcare Innovations and Point of Care Technologies, HI-POCT 2017. Institute of Electrical and Electronics Engineers Inc. (2017)

Assessing Main Factors Adopted within the EU Harmonised Online Public Engagement (EU HOPE)

Aliano Abbasi[✉] and Muhammad Mustafa Kamal

School of Strategy and Leadership, Coventry University, Coventry, UK
Alianoabbasi30@gmail.com, ad2802@coventry.ac.uk

Abstract. EU Cohesion Policy is crucial leading policy for the public and state's interaction and cooperation activities not only within regional but also national and cross-border to improve a sustainable and harmonised regional growth across the EU countries. To achieve such sustainable development, a citizen engagement within their local government's decision making is vital. A scattered citizen engagement practices and policies across the local governments requires an integrated framework to achieve sustainability within the EU Cohesion Policy. This study is assessing a set of factors which highlight the initiatives of EU Harmonised Online Public Engagement (EU HOPE) and their relationship with Cohesion Policy. Citizen engagement factors within some of the EU cross-border pilot studies have been investigated against relative factors adopted within EU Harmonised Online Public Engagement (EU HOPE). A relation between the common citizen engagement factors within the pilot studies and found factors within EU HOPE have also been illustrated in the end of the research.

Keywords: Cohesion Policy · Citizen engagement · EU Cross-Border · Intersubjectivity Knowledge · EU Interoperability Framework · Digital Participatory Platform

1 Introduction

Local development has always been one of the key elements in EU Cohesion Policy, aiming to increase the policy impact on the local territories potential development, particularly within sustainable urban development resilience and the necessity of its integrated strategies. One of the five objectives of Cohesion policy is 'a Europe closer to the citizens: sustainable and integrated development through local initiatives to foster growth and socio-economic local development of urban, rural, and coastal areas' [16]. In cohesion policy 2014-20, The European Commission identified some positive impacts on the involvement of local administration in the policy making. The involvement empowered local governments to establish an engagement not only with neighboring urban authorities, but within national and European networking levels too [4].

To establish a citizen engagement approach within public service design and delivery, the public engagement policy needs to respect the bottom-up public management in

M. Papadaki et al. (Eds.): EMCIS 2023, LNBIP 502, pp. 230–247, 2024.
https://doi.org/10.1007/978-3-031-56481-9_16

decision making. Smart city is a vision of adopting digital technologies within public services and public management. However, the idea of a "smart city" runs the risk of becoming an urban labelling phenomenon [22] that is utilized more as a means of municipal branding than as a genuine advancement in the use of technology to empower inhabitants. Although top-down and bottom-up initiatives are not mutually exclusive forces; on the contrary, they can enhance each other and boost a city's capacity for innovation [5]. The author claims the coexistence of a "smart city" (top-down) and "smart citizens" (bottom-up) approach can foster innovation, for instance local governments can proactively design programs to integrate residents' voices in their smart project development (Fig. 1).

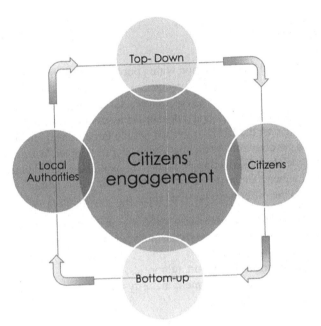

Fig. 1. Circular Process of Citizens' Engagement [24]

Even though the value of bottom-up efforts is underlined within the study, still, citizens frequently lack the means to act without governmental assistance [5]. Such smart city services require very expensive infrastructures and legislative adjustments, and as is typically the case with disruptive technology, it can be challenging for the citizens to comprehend the potential applications [5.] Therefore, by integrating the citizen engagement as the theory of bottom-up within a smart city ecosystem, the digital engagement strategy will result in regional development sustainability which are the Cohesion Policy initiatives within the EU. [29] confirm the regional sustainability of citizen engagement within the EU's new Green Deal strategy, which is also called the EU's Moon- shot mission and its global smart specialisation strategy, includes sustainability, regional integration, and social inclusion. The authors explain the differences between previous

Smart Specialisation agenda S3 and the latest S4+, claiming although smart specialisation strategies have created the groundwork, it is now necessary to combine bottom-up leadership with the new directionality of public inclusivity and sustainability.

The purposes of the current research are three folded.

- *Firstly*, it attempts to empirically examine the relationship between citizens engagement and Cohesion Policy, highlighting the importance of bottom-up public administration and public digital factors adoption.
- *Secondly*, it investigates several citizen engagement involvements within the EU cross-border pilot studies across different EU local governments and cities,
- *Thirdly*, evaluate the factors which are stressed in the theoretical studies against those practices.

2 Methodology

The necessary of digital factor adoption within public engagement services, the absence of the EU Harmonised Online Public Engagement (EU HOPE) framework amongst the EU local governments and the evident of a successful EU single Digital Gateway within cross-border practices, create a crucial demand to assess the same digital adoption factors within SDG against the EU Harmonised Online Public Engagement factors, which improve the local governments' perceptions of the digital adopted factors. Maturity models have been crucial in e-Government studies in explaining the planned, anticipated, and actual developments in the digital transformation of public administration. [15] and [32] have outlined the differences between three types of maturity models:

- Governmental models created to assist authorities in determining and enhancing their degree of maturity by governments, consultancies, and academics (often making use of pre-set toolkits and models).
- Holistic models intended to support authorities,
- Evolutionary models of eGovernment maturity that rely on successive evolutionary stages,

Neither evolutionary eGovernment and nor holistic maturity models, which emphasize project execution and organizational capacities inside a single organization [32, 34, 41], are suitable for the goals of this study. Nevertheless, according to [31], eGovernment maturity models frequently in possibility of combining positive (i.e. a model highlighting relationships between variables and factors), normative (i.e. a prescriptive model of the situation and an example to follow) and evaluatory elements (e.g. a benchmark for comparisons). According to [27], their strategy provides a framework for governments to adopt and is an illustration of a normative model (i.e. a prescriptive model of what impacted factors are in place and a model to match [9, 10]. The maturity model adopted for this research is also adopted by [21] research, which is an exemplary positive model (i.e. a model which indicates the interaction between variables and factors in place [10, 30]. Thus, the subject of shared innovation adoption determinants within the EU Harmonized Online Public Engagement is being investigated in this study. Initially, a review literature is conducted on the significance of citizen participation in the EU Cohesion Policy and how it relates to the bottom-up engagement strategy. Second, the study adheres to the widely used EIF framework, which ensure an alignment with other digital public participation methods. The layers of the EIF: 1) Semantic, Legal, Organizational, and Technical. The European Commission created the EIF to provide detailed instructions on how to establish digital public services that are compatible and to raise the standard of public services throughout Europe [13].

The EIF criteria and a mapping against them, allows us to explore the topic comprehensively. Thirdly, the identified factors would be reviewed within various Cross border Cooperation pilot studies across the EU. Lastly, the research compares the factors from EIF and other participatory methods with the existing approaches in cross-border pilot studies and provide an evaluation for the importance of the factors within practical approaches. At the end the study provides an overview of the findings and highlight the importance of adopting EU HOPE philosophy within EU local governments citizen digital engagement, along with the recommendations based on the descriptions, analysis and evaluation in the respective parts of this paper.

This study investigates the EIF criteria and the mapping academic against them. Thirdly, the elements that were identified will be examined in several EU-wide pilot studies related to cross-border cooperation. Finally, the study evaluates the significance of the factors within workable ways by contrasting the elements from EIF and other participatory techniques with the current approaches in cross-border pilot projects. In conclusion, the study offers a summary of the results and emphasizes the significance of implementing the EU HOPE philosophy in EU local governments' citizen digital engagement. Additionally, recommendations are provided based on the descriptions, analysis, and evaluation presented in this study.

3   Europe Closer to Citizens

The report in [36] states that to improve cohesion policy within the EU, several elements need to be respected:

- Public judgement for better policy outcome, rather than rely on public opinion.
- Improve genuineness decision making to avoid deadlock.
- Improve citizens' trust and empower the public by providing citizens an effective role to impact public decision making.
- Improve public inclusiveness to a much more diverse communities and group of citizens,
- Improve integrity and prevent disinformation, corruption, and polarisation.

Thirty percentage of the European Union's budget is dedicated to Cohesion Policy to support a harmonious but consistent development across the union. [40] states that the policy is to promote partnership among all actors including public, private and state administration. However, citizens as the end users, should play a crucial role in the investment and shaping the decision making but also hold the authorities accountable and transparent in their policies. [40] claims citizen engagement is a priority across the Europe development and is considered as the key player in Horizon Europe, which builds a unique opportunity to study, assess, and improve mechanisms for such public engagement. The report indicates "Effective citizen engagement involves three stages of intervention: communication and awareness raising; co-design and co-creation; and co-implementation". Those three stages are the outcome of a bottom-up decision-making approach which empower citizens in their interaction with the public administration.

3.1 Bottom-Up Approach (Intersubjective Management)

Intersubjectivity science is related to an orientation which takes an appropriate action within a management operation or decision making. The science is moving from classical linear management, which is mainly top-to- bottom approach, to postnonclasical within a cybernetic scientific rationality which is based on communication and control in living organisms, [39]. The author explained the concept as each member of the public can contribute to enhance their cultural heritage within a bottom-up community's discipline promotion. [35] believes public contribution will lead to improving the culture's potential in the whole society, including contribution to the ethics of participation in decision making. The intersubjectivity is not only a theory of management but also socio-techno science.

4   The Necessity of Harmonised Online Public Engagement (EU HOPE)

Facilitating and developing digital platforms to construct the public engagement governance between local administrators and public citizens are challenges that local governments need to overcome. [16] highlights the gap in practising a standard policy in implementing digital engagement within the government-citizen collaboration and communication. Such a gap indicates that local governments are not fully aware of the crucial factors of adopting digital innovation within their digital citizen-engagement and every state or municipality applies their own understanding and version of public-engagement which is scattered within their smart city development program.

These various engagement practices in a scale of European local governments, are providing different experiences to citizens once they are travelling from one to another city or even country within the EU. Although culture, education and language are different across the EU region, a harmonised online public engagement platform, protocols or standard can be provided to the EU local governments as a non-mandatory consultation and implementation. Any EU resident could interact with their local municipality's platform and experience the same in other cities throughout the union thanks to the harmonised and integrated digital factors and values which have been embedded within online public engagement approaches across the European local governments. The harmonised digital adaptation factors could impact the European values, social sustainability, governing quality, digital engagement culture and skills amongst civil servants and citizens, whilst helping Cohesion policies to be delivered more appropriately with least friction. [33] quoted that supranational organisation as the European Union (EU) have planned by "Europe for Citizens Programme (2018)" to highlight the necessity of encouraging the democratic and civic participation of citizens at the EU level, by improve the necessity of citizen empower across EU public administration and stimulating interests with the EU policy making process. However, the key success of single citizen's digital-engagement requires a digital synchronisation and integrated platform between the states' public digital-engagement systems.

4.1 Digital Innovation Factors Within the EU HOPE

[13] designed the New European Interoperability Framework (EIF) to be adopted by public authorities and practitioners to follow as the guidance within digital public services, which as the result improves the quality of the public services interoperability within the EU. The framework consists of factors including Legal, Organisational, Technical and Semantic interoperability [13, 25] (Table 1).

Table 1. New European Interoperability Framework [13]

European Interoperability Framework (EIF) Principles	
Organisational	aims to integrating and documenting the business process and within exchanged relevant information.
Legal	Aims to establishing interconnected public administrations, able to cooperate under different legal framework, strategies, and policies.
Semantic	focuses on data and confirm the data format exchanging, the information is understood and preserved across the data exchange among various parties.
Technical	ensure the interaction and communication is established across various linked technical resources, infrastructures, systems, and services.

4.2 Digital Participatory Platforms (DPP)

This study adopts similar approaches which have been practiced in academic theories such as Intersubjective Science Management and Digital Participatory Platforms (DPP) to validate the factors adopted within the New EIF Framework. [16] went through an extensive literature review, to provide a clear categorization of Digital Participatory Platforms' (DPP) challenges. The authors believe that their study provided a clear categorization of challenges as a starting point for a better understanding of the factors hampering local government and citizens collaboration based on digital participatory platforms (DPPs). Set of main challenging factors introduced for Digital Participatory Platform by [16] are as listed in table below (Table 2).

Table 2. Digital Participatory Platform (DPP) [16]

Digital Participatory Platform (DPP)	
Contextual or in other word Semantic factors	internet accessibility, digital illiteracy, and the digital divide; Institutional framework
Technological factors	technological advancements and data management
Organisational factors	process-related challenges; intra-organizational culture; availability of human resources

4.3 Intersubjective Agreements of Actors

Heterogeneous actors within Intersubjective agreement must decide on the principles of decision-making and some "rules of the games" shared by everyone. The framework for the actor's agreement within the intersubjective knowledge includes semantic, experimental, logical, operational, and normative [23]. Such knowledge is also provided with the aid of ontologies by [38] who listed them as activity, corporate culture, facts, decision-making, legal and normative ontology (Table 3).

Table 3. Intersubjective Agreement Ontologies Framework [39]

Organisational	Culture	Technological	Decision-Making	Legal
Availability Of Diverse Infrastructure	Attracting Youth and Talented Residents	Avoiding Infrastructure Duplication	Citizen Participation	Transparency In Decision Making.
Unified Initiatives and Master Plans	Improved The Quality of Life's Standard.	Marketing And Advertisement		
A Strong Economy	Promote the Sense of Belonging			
Access to European Funds and Related Budget				
A Strong Territorial Policy Plan				
Strong Political Commitment				

5 Single Digital Gateway, A Harmonised Cross-Border Solution

Cross border digital governments are the picture of implementing those factors within the established integrated system between two or several countries (Krimmer et al. (2021). Such a system improves the trust among various stakeholders by simplifying the identification through a framework in an interoperable recognition of national identification in a cross-border system's settings [11]. Therefore, the Single Digital Gateway (SDG) has been adopted in 2018 by the EU to address the data system exchange and integration across the union [11]. The SDG regulation aims to build one single point in which the citizens and business would be able to receive information, give feedback and to access digital public procedures. The SDG will be implemented into the EU portal "Your Europe" to act as a single access point to existing national portals [2].

The Once-Only Principle (OOP) is the guiding principle of the Single Digital Gateway (SDG) Regulation [26], which mandates that European Member States develop and link to a single European portal and infrastructure to enable citizens, businesses, and public administrations interacting with public services across borders. However, the OOP seems to be understood differently in different countries. In some, it implies that data is stored only once and linked to a single source, while in others, it means that businesses and citizens only need to supply personal information once, but a copy of the data can be stored in different places [26]. In order to achieve the principle of interoperability-by-default, ministers of the EU Member States specifically agreed to work together to implement the OOP for the important public services and to abide by the EIF for the cross-border digital public services [25, 42].

6 Cross-Borders Pilot Studies

6.1 Cross-Border Data Flow in The Digital Single Market: Study on Data Location Restrictions

The study's objectives included a) pinpointing and assessing obstacles, both legal and otherwise, to the free movement of data within the EU, b) determine how these obstacles affect both private and public sector's users and cloud computing services providers. The context of this study aims to improve the Digital Single Market Strategy (DSM Strategy), which was approved by the Commission in May 2015 [18], to eliminate obstacles within the free flow of data throughout the EU (DSM). The free data flow process is relevant to legal, semantic, technological and organisational factors as it coordinates with other states and public organisations and contains

[18] three steps that are taken to develop a harmonised coordination data process amongst the Member States. When data is covered by the free movement of data policy, a Member State are required to:

- Monitor the public department's laws to find out whether any new regulations have been added which will restrict the free flow of data.
- if so, then simplify the process by translating them into a pattern which is required within DSM's principals,
- and if the results indicate any technical requirement for the data flow, then coordinate with other state members to harmonise the level of technical or operational requirements.

recommended three policies to be approached in various ways when the data flow is shared across the border. These methods to organise this process are:

- To firmly establish the free flow of data by strictly enforcing the Treaties, the Services Directive, and the e-Commerce Directive. Whilst no new legal framework is implied, it is necessary to remind Member States of how the current legal system affects their needs for where to store their data.
- Developing a new horizontal legal framework, that establishes the principle of data free movement and specifies the conditions and procedures under which Member States may adopt exceptions.

- Having a non-regulatory agreement and policy recommendation with other member states to establish the free movement of data, although the agreement between states might be too weak to have an appropriate and quick progress and may not deliver the same impact and result as the principle of free movement of data by Single Digital Market.

The study proves the necessity of legal and regulatory adjustment between member states to have a harmonised understanding and implementation across the borders. The same necessity is required within EUHOPE across the local governments to have a similar level of understanding of digital engagement principals from a legal perspective within their public digital engagement practices. Such legal harmonisation can help EU local governments to respect the integrated legal factors within their public engagement platforms when it comes to citizens' data collection, processing, analysis, and data processing algorithms.

6.2 European Experience of Citizens' Participation in Cross-Border Governance

The study [24] was aiming to have a general understanding of the methods used to manage transfrontier activities and plans. The pilot study assessed 9 cross-border cooperation across Europe and another 20 cross-border cooperation projects with built-in public participatory activities. The aim of the study was to have an overview of cross-border initiatives from a citizen participation perspective, but also describe the main features of those plans, and the impact of participation on decision making within cross border establishment. Eventually, identify the legal concerns through European legal instruments. The study [24] claims the impact of sustainability within the Cross-Border Cooperation CBC process, heavily relies on citizen's engagement and involvement to witness an improvement in the CBC strategies and action. Such sustainability impact within CBC is:

- Promote the socio-economic development of the border area.
- Develop economical scale to provide better services.
- Widen cultural perspectives.

A border region serves as a hub for cultural output, offering new perspectives on social life, whilst the communities have the power to modify and redefine the idea of a border. The study claims trust has a direct relation with political accountability. In that regard, one of the organisational principles which political accountability within public institution has been described within CBC notion by [24] as:

- Citizens need to understand the concept of CBC and its operation.
- Poor performance will lose the trust among citizens, and they will begin to believe it as a waste of public time and funds.
- An efficient communication and a culture of trust need to be adopted by local governments to avoid misunderstanding in their engagement with citizens and stakeholders.
- The type of decision-making process also impacts the level of transparency and confidence in integrity among citizens.

- Considering not only the local governments residents' interest but also the interest of the citizens of all the neighbouring and partner local governments.

Eventually, the study [24] indicates main factors which play a role in engaging citizens within decision making process at cross border cooperation as:

- residents' understanding and awareness of cross-border governance management systems and actions.
- pre-existing social capital network and public organisation function links across the border.
- citizens' involvement in the early decision-making process through seminars and events.
- resources availability including funding and,
- Finally, transparency of the Cross Border Cooperation process by considering democratic and technical accountabilities.

The study highlights the importance of data transparency, culture of trust, organisational, technological and resource, and funding availability within implementing citizen engagement decision making. Citizen interest and good understanding in public engagement concepts across a networking region would increase the credibility and importance of citizen engagement across the EU.

6.3 Cross-Border Cooperation (CBC) in Southern Europe—A Case Study. The Eurocity Elvas-Badajoz

The research is assessing the impact of Cross-Border Cooperation (CBC) between cities of Elvas and Badajoz to analyse a protocol which was established in 2013 between those cities to build a concept of Eurocity Elvas-Badajoz. The aim of this study [3] is to identify and analyse crucial factors which impact territorial success in CBC and its importance to be emphasised in local and regional sustainable development. The study summarised a list of influencing factors to achieve territorial success within CBC areas. The factors are aligned with the Digital Participatory Platform (DPP) [16] and are listed in the table below (Table 4).

6.4 Engaging Citizens in Cohesion Policy, DG REGIO and OECD Pilot Project Final Report

The study conducted a survey among various stakeholders, to assess the citizen engagement approaches within Cohesion Policy projects across five European cities. The result [36] demonstrates that almost none of the Managing Authorities (MA) and/or Intermediate Body (IB) directly engage individual public citizens in their participation activities. The study concludes that the ignoring citizen involvement is the main reason why citizens are being underrepresented and have not been heard, although the authorities have realised the gap to be addressed. To tackle the gap of how to engage with citizens, the OECD [36] recommended several steps of participation, including:

- Information: which government provides information to the citizens and stakeholders which is the initial step for the engagement.

Table 4. Factors set impact territorial success in CBC within Eurocity Elvas-Badajoz [3]

Organisational	Culture	Technological	Decision-Making	Legal
Availability Of Diverse Infrastructure	Attracting Youth and Talented Residents	Avoiding Infrastructure Duplication	Citizen Participation	Transparency In Decision Making.
Unified Initiatives and Master Plans	Improved The Quality of Life's Standard.	Marketing And Advertisement		
A Strong Economy	Promote the Sense of Belonging			
Access to European Funds and Related Budget				
A Strong Territorial Policy Plan				
Strong Political Commitment				

- Consultation: establishing two ways communication feedback based, regarding a particular concern, to the government and vice-versa.
- Engagement: it's a dialogue where the citizens and stakeholders are provided with the information, data, and digital tools, by the public authorities to engage in establishing service design and policymaking. It includes proposing projects or policies, setting the agenda for studying and project processing, to the level of decision making.

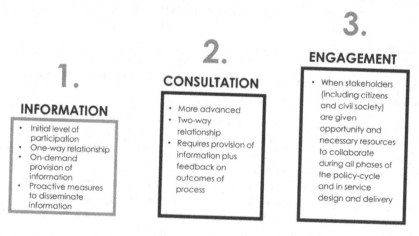

Fig. 2. Three Pillars of Stakeholder Participation [36]

The survey [36] also indicated all five cities which have participated in this study, are only familiar with the Information and Consultation, apart from Regione Emilia-Romagna which has experience in co-creation activities directing to the Engagement level. Therefore, the study listed a set of recommendation for establishing the necessary actions over citizen engagement with Cohesion Policy:

- Occasional mandatory public consultation is not sufficiently enough. The study suggests the good example of the citizen engagement practice is at the early stage where citizens are getting involved in the design of a public project which impacts directly on their daily life.
- Learning how to involve citizens in various aspects and layers of Cohesion Policy, it requires to be embedded within the regional policies. For instance, the cross-border projects need to be included fully in facilitation, multilingualism, etc. Also encourage public authorities to engage stakeholders and citizens in their project design, implementation, monitoring and supporting projects.
- Providing legitimate data and facts, resources and political support improves innovative citizen engagement and citizens empowerment. The engagement requires time, staff, fund as well as improving participation culture in the designed engagement.

Building the three pillar engagement is one of the crucial points in having an integrated and equally understanding of public engagement across various local governments. Harmonising the level of public understanding in such a citizen engagement concept across the EU states borders is vital to approach an inconsistent culture, trust and quality of public engagement.

6.5 Flanders – The Netherlands Citizens' Panel on Addressing Cross-Border Issues and Stimulating Cross- Border Contacts

The study [37] is based on a project to involve individuals and civil society in the implementation of Cohesion Policy in which the OECD and the European Commission were partnered to deliver the report. Five regions across Europe were selected to explore innovative ways of citizens and civic society engagement when it comes to the importance of strategic engagement-based decision-making within the Cohesion Policy. The aim of the study [37] is to practise citizens' experience and knowledge within public participation and analyse how to overcome citizens' most pressing needs in such engagement. The most challenging factor within the cross-border area is having disparate legal and organisational public engagement arrangements on either side of the border which impacts the local citizens within the regions. The capacity of various organisations to communicate with the public in different ways results in a lack of mutual understanding of each other's citizen-engagement tactics on both sides [37]. To overcome such complication, the study first highlights some of the common problems in citizen engagement:

- Citizens have different priorities, therefore there is a wide range of engagement prospective among them.
- Citizens usually don't have access to free time and available resources on daily bases to make judgement on policy making and its impact on their life.
- Those citizens who engage voluntarily with online platforms, won't represent the wider community.

However, the study proposed a new form of citizen engagement, called the Citizen Panel to address the unavailability of citizens within the public engagement. In Citizens' Panel process [37], A deliberative exercise is established with a group of chosen representatives from mixed communities, and they are asked to take part in listening to the discussions, analysing the opinions of the experts, exercising critical thought, and not being biassed in the course of an open discussion and defensible decision-making. Deliberative exercise is an innovative thinking which is made possible by open, honest discussion (factual), that is based on thorough, "open book" information and careful consideration of opposing ideas [20]. Advocacy and argument are completely dissimilar from this. The study [37] claims that deliberative processes would give better sense of the shared priorities of a vast representative group of citizens living in the area. The study carries on claiming that the Citizens Panel would improve cross-border policy by bringing people from both sides to collaborate in their common social, economic, and environmental matters which is at the heart of Cohesion Policy.

Such an approach can include a representative of non-local residents or migrant's community of the region within public engagement service deliveries, who are new in the local speaking language, culture, and public services' processing system, therefore, their level of engagement wouldn't be at a limited level. Every citizen needs to have the chance and ability to assume their proper roles as participants in this process [20]. Such an approach can add value to an established EU Harmonised Online Public Engagement, creating an equalised level of engagement across the EU, regardless citizens background, language, digital skills, age, and ability.

6.6 The 23 Citizen-Based Topics for Future EU Research - Project Names: Fostering Equal Opportunities in the Digital Era

The study was conducted by Citizen and Multi-Actor Consultation (CIMULACT), funded by the European Commission through Horizon 2020, to engage citizens specifically in health evaluation, demographic change, and wellbeing. The purpose of this study [8] is to examine the problems that European individuals face due to the various levels of healthcare accessible in different locations. The study believes that technologies playing a vital part to deliver the engagement. Several improvements are being indicated within technological and cultural factors to impact the citizen accessibility to the digital public services, including:

- EU citizens to have accessibility to internet equally, freely, whenever, and everywhere,
- Digital empowerment becomes a key factor in closing generational, economic, and social disparities.
- Promoting digital learning culture, which specifically is related to the way of data collection and used,
- citizens have access to free educational materials and courses.

The paper also found out that the EU citizens do not receive the same health care standard everywhere across the union, it [8] therefore highlights the necessity of promoting a harmonised healthcare and delivering the equal resources distribution of knowledge in the healthcare system in different EU countries. The study suggest:

- Identify and categorise various indicators which impact analysing good or bad practices in various countries health care systems in the EU. This could result in the sharing of information and data with open access and the adoption of standards by all parties involved to build a harmonised European health network and standardise medical treatment.
- Recognizing and developing local expertise in healthcare and establishing person-centre which aim to promote health awareness by educating citizens and healthcare experts.

Linking the local health care with European health care standards can improve the quality and values of healthcare received by citizens which has a direct social and economic impact [8]. The study is perfecting targeting the importance of building a harmonised citizen digital engagement framework to deliver the initiatives. In this regard, organisation, technological, cultural and semantic factors have been addressed by stressing the public education, citizen-centric, resource accessibility, and standardising the health service framework to achieve the harmonised and resource efficiency within public health care across the EU.

7 Discussion

states that within the framework of a European Grouping of Territorial Cooperation, their informal cooperation with neighbouring municipalities, known as intermunicipal associations, which is composed of municipalities from other EU Member States, might be institutionalised and made more systematic. Nevertheless, the common needs and challenges, particularly in the request for the delivery of key public services despite increasingly tight budgets, are shared by most municipalities throughout Europe [19]. Therefore, to assess a harmonised framework across intermunicipal public engagement, the case study typology could be categorised as hypothesis-generating case studies, according to [28], after the approach was applied to *"examine one or more cases for the purpose of developing more general theoretical propositions, which can then be tested through other methods"*.

The table below illustrates the common factors between intersubjectivity management science factors (science of bottom-up), Digital Participator Platform DPP, the EIF 2017 and their relationship with various citizen engagement across the EU. [5] believes everyone can extend the citizen engagement framework by accepting an interconnection agreement that ensure the principals are upheld. Therefore, the table below is indicating that most theoretical and managerial factors have been respected in various engagement practices which highlight the importance of those elements within the concept of the EU Harmonised Online Public Engagement to gain the vital impact of the integrated citizen digital engagement strategies within the EU Cohesion Policy.

The Table 5, is a list of EU Harmonised Online Public Engagement (EU HOPE) Factors which includes Semantic, Decision Making, Technological, Organisational, Legal and Facts, are combination of common factors amongst the Intersubjective Management Science, EIF and DPP. The table indicates that Decision making factor has only been mentioned within Intersubjective Management Science, although as it has been

Table 5. A Comparison Between EU Harmonised Online Public Engagement (EU HOPE) Factors and the Intersubjective Management Science, EIF And DPP Factors

The EU Harmonised Online Public Engagement (EU HOPE) Factors	Semantic Including Culture, Language	Decision Making Including Concept-Judgement-Conclusion	Technical Operation	Legal Principal	Organisational Principal	Fact and Trust
Intersubjectivity Management Science	x	x	x	x	x	x
The European Commission EIF (2017)	x	-	x	x	x	x
Digital Participatory Platforms (DPP)	x	-	x	x	x	-
Number Of Matching Factors	3 out of 3	1 out of 3	3 out of 3	3 out of 3	3 out of 3	2 out of 3

Table 6. The EU Harmonised Online Public Engagement (EU HOPE) Factors Comparison with Cohesion Policy Cross-Border Pilot Approaches' Factors

Cohesion Policy Cross-Border Pilot Studies	Semantic Including Culture, Language	Decision Making Including Concept-Judgement-Conclusion	Technical Operation	Legal Principal	Organisational Principal	Fact and Trust
Cross-Border Data Flow in The Digital Single Market: Study on Data Location Restrictions	x	-	x	x	x	x
European Experience of Citizens' Participation in Cross-Border Governance	x	x	x	x	x	x
Cross-Border Cooperation (CBC) in Southern Europe—A Case Study. The Eurocity Elvas-Badajoz	x	x	x	x	x	x
Flanders – The Netherlands Citizens' Panel on Addressing Cross-Border Issues and Stimulating Cross-Border Contacts	x	x	-	x	x	x
Engaging Citizens in Cohesion Policy, Dg Regio and OECD Pilot Project Final Report	x	x	x	-	x	x
23 Citizen-Based Topics for Future EU Research - Project Names: Fostering Equal Opportunities in The Digital Era	x	x	x	-	x	-
Number Of Matching Factors	6 out of 6	5 out of 6	5 out of 6	4 out of 6	6 out of 6	5 out of 6

stated earlier in this research, it contributes to the bottom-up principle of public engagement within public decision-making. Also, Fact and Trust has been excluded from the DPP set of challenging factors within citizen engagement. facilitate a broader consensus and shared solutions would increase public trust towards the government. Facilitating a wider consensus and achieving common solutions would boost public trust in the administration [1, 17] which has not been discussed within the DPP.

The Table 6, A comparison Over the EU Harmonised Online Public Engagement (EU HOPE) factors and Cohesion Policy cross-border Pilot approaches' factors in relation to citizen engagement. Almost all the EU HOPE's factors have been discussed and covered within the pilot studies apart from the Legal aspect of cross- border public collaboration, which has been absent from two pilot practices across all 6 studies. Such links highlight the vital and close relation between the common public engagement factors in two different concepts but same initiatives within citizens and public engagement within local and cross border scale. However, an interpretive qualitative approach is suggested to discover the insights of citizen engagement experience and practices within European

Online Public Engagement across the local European governments, by testing the study in the form of qualitative structure interviews.

8 Conclusion

The article discusses the importance of local development in EU Cohesion Policy, particularly in sustainable urban development resilience. It emphasizes the need for citizen engagement in public service design and delivery, and the potential of digital technologies in public services and management. The concept of a "smart city" is explored, highlighting the risk of it becoming more of a branding tool than a genuine advancement in technology use. It was suggested that a combination of top-down (smart city) and bottom-up (smart citizens) approaches can foster innovation. This study evaluated the innovation adoption factors from Cross-border cooperation (CBC) pilot studies that were integrated into citizen engagement. The CBC is a great example of building neighbourly relationships between local citizens/stakeholders and authorities across the EU and the goal is to foster the harmonious development of border communities.

It also examined the objectives and suggestions for improving harmonization services throughout Europe, aiming to highlight a unified approach to digital public engagement across the EU. This would firstly involve harmonized initiatives that all EU local governments could adopt within their citizen digital engagement implementation. Secondly, an integrated and harmonised citizen digital engagement framework to adopt within a smart city ecosystem would result in regional development sustainability in the EU. In other word, on one hand citizens would experience the same understanding of engagement's values, information, and level of data analysis across the union, and on other hand local authorities would have a common understanding of the digital engagement concept and long-term commitment, whilst they pursue their own business operation and other priorities within their citizen digital engagement.

Although several factors have been investigated within the set of pilot studies, the economic factor has been missed from the digital innovation adoption set of factors in this research. It should be stated that the cost and economical factor is related to having a stable economy in the region [3]. Therefore, cost and financial factors need to be investigated within the harmonised online public engagement. it also notes that citizens often lack the means to act without governmental assistance, especially in the context of expensive infrastructures and legislative adjustments required for smart city services. As the result, further research is required to study how to improve citizens digital skills and the create either new or amendment legislations within the harmonised framework across the inside and outside of the involved EU public organisations.

References

1. Agger, A.: Towards tailor-made participation: how to involve different types of citizens in participatory governance. The Town Planning Review, 29-45. (2012).
2. Akkaya, C. and Krcmar, H.: Towards the implementation of the EU-wide "once-only principle": perceptions of citizens in the DACH-region. In International Conference on Electronic Government (pp. 155-166). Springer, Cham. (2018)

3. Alexandre Castanho, R., Loures, L., Cabezas, J., Fernandez-Pozo, L.: Cross-border coopera-tion (CBC) in Southern Europe-An Iberian case study. Eurocity Elvas-Badajoz. Sustain. **9**(3), 360 (2017)

4. Berkowitz, P., Hardy, S., Muravska, T. Bachtler, J.: EU Cohesion Policy: reassessing performance and direction. London, New York: Routledge 10, p.9781315401867 (2016)

5. Capdevila, I., and Zarlenga, M.I.: Smart city or smart citizens? The Barcelona case. J. Strategy Manag. **8** (3). (2015).

6. Castanho, R., Loures, L., Fernández, J. Pozo, L.: Identifying critical factors for success in Cross Border Cooperation (CBC) development projects. Habitat International 72 (2018)

7. Castanho, R.A., Loures, L., Cabezas, J. Fernández-Pozo, L.: Cross-Border Cooperation (CBC) in southern europe—An iberian case study. Eurocity Elvas-Badajoz. Sustain. **9** (3) (2017)

8. Citizen and Multi-Actor Consultation (CIMULACT).: 23 citizen-based topics for future EU research - Fostering equal opportunities in the digital era. (2020) http://www.cimulact.eu/wp-content/uploads/2017/02/Access-to-equal-holistic-health-services.pdf

9. Cordes, J.J.: Reconciling normative and positive theories of government. Am. Econ. Rev. JSTOR **87**(2), 169–172 (1997)

10. Desmarais-Tremblay, M.: Normative and positive theories of public finance: contrasting Musgrave and Buchanan. J. Econ. Methodol. **21**(3), 273–289 (2014)

11. Dumortier, J.: Regulation (EU) No 910/2014 on electronic identification and trust services for electronic transactions in the internal market (eIDAS Regulation). In: EU Regulation of E-Commerce. Edward Elgar Publishing (2017)

12. European Commission. Directorate General for Communications Networks, C. and Technol-ogy, Spark Legal Network, Tech4i2, Time.lex.: Cross-border data flow in the digital single market: study on data location restrictions: executive summary (2017)

13. European Union. New European Interoperability Framework. Publications Office of the European Union. (2017). https://ec.europa.eu/isa2/sites/default/files/eif_brochure_final.pdf

14. European Union Passed Regulation (EU) 2018/1807: framework for the free flow of non-personal data intheEuropeanUnion.(2018).https://eur-lex.europa.eu/legal-content/EN/TXT/PDF/?uri=CELEX:32018R1807

15. Fath-Allah, A., Cheikhi, L., Al-Qutaish, R.E., Idri, A.: EGovernment maturity models: a comparative study. Int. J. Softw. Eng. Appl. **5**(3), 72–91 (2014)

16. Falco, E., Kleinhans, R.: Beyond technology: Identifying local government challenges for using digital platforms for citizen engagement. Int. J. Inf. Manag. **40** (2018)

17. Forester, J.: The deliberative practitioner: encouraging participatory planning processes. Mit Press (1999)

18. Graux, H., Ypma, P., Foley, P.: Cross-border data flow in the digital single market: study on data location restrictions: executive summary. A study prepared for the European Commission DG Communications Networks, Content & Technology (2017)

19. Guderjan, M. and Verhelst, T.: Local government in the European Union: completing the integration cycle. Springer Nature (2021)

20. Hartz-Karp, J.: Harmonising divergent voices: sharing the challenge of decision making. Public Adm. Today **2**, 14–19 (2004)

21. Hiller, J.S., Belanger, F.: Privacy strategies for electronic government. E-Government **200**, 162–198 (2001)

22. Hollands, R.G.: Will the real smart city please stand up? City: Analysis of Urban Trend, Culture, Theory. Policy, Action **12**(3) (2008)

23. Huebner K.: The truth of the myth. (in Spanish). Respublika, Moscow, section 1.6 (1996).

24. Institute of International Sociology.: European Experience of Citizens' Participation in Cross-Border Governance (2015). https://rm.coe.int/1680686b1b

25. Kamal, M.M., Ziaee Bigdeli, A., Themistocleous, M., Morabito, V.: Investigating factors influencing local government decision makers while adopting integration technologies. Inf. Manag. **52**(2), 135–150 (2015)
26. Krimmer, R., Kalvet, T., Olesk, M., & Cepilovs, A.: The Once Only Principal Project, Position Paper on Definition of OOP and Situation in Europe (updated version), **2**(34198.86089) (2017). https://doi.org/10.13140/RG
27. Layne, K., Lee, J.: Developing fully functional E-government: a four stage model. Govern. Inf. Quart. **18**(2), 122–136 (2001)
28. Levy, J.S.: Case studies: types, designs, and logics of inference. Conflict Manag. Peace Sci. **25**(1), 1–18 (2008)
29. McCann, P., Soete, L.: Place-Based Innovation for Sustainability. Publications Office of the European Union, Luxemburg (2020)
30. Musgrave, R.A.: Theory of Public Finance; a Study in Public Economy. McGraw-Hill, New York (1959)
31. Nielsen, M.M.: The Demise of eGovernment Maturity Models: Framework and Case Studies. Tallinn University of Technology Press, Tallinn (2020)
32. Persson, A. and Goldkuhl, G.: Stage-models for public e-services -Investigating conceptual foundations. In: 2nd Scandinavian Workshop on E-Government. Copenhagen: Scandinavian Workshop on E- Government, pp. 1–20 (2005)
33. Richard, E., David, L.F.: The future of citizen engagement in cities—The council of citizen engagement in sustainable urban strategies (ConCensus). Futures **101**, 80–91 (2018)
34. Ross, J.W., Weill, P., Robertson, D.: Enterprise Architecture as Strategy: Creating a Foundation for Business Execution. Harvard Business Press, Cambridge (2006)
35. SKobelev, P.O., Borovik, S.Y.: 'On the way from industry 4.0 to industry 5.0: from digital manufacturing to digital society'. Int. Sci. J. "Ind. 4.0" **311**(6) (2017)
36. The Organization for Economic Cooperation and Development (OECD).: ENGAGING CITIZENS IN COHESION POLICY, DG REGIO and OECD Pilot Project Final Report. (2022)
37. The Organization for Economic Cooperation and Development (OECD).: Flanders-The Netherlands Citizens' Panel on Addressing Cross-Border Issues and Stimulating Cross-Border Contacts. (2021). https://www.oecd.org/gov/open-government/fl-nl-cross-border-citizens-panel.pdf
38. Vittikh, V.A., Ignatyev, M.V., Smirnov, S.V.: Ontology in the intersubjective theories (in Russian). Mekhatronika Avtomatizatsia Upravlenie **5**, 69–70 (2012)
39. Vittikh, V.A.: Introduction to the theory of intersubjective management. Group Decis Negot. **24**, 67–95 (2015). https://doi.org/10.1007/s10726-014-9380-z
40. Warin, C., & Delaney, N.: Citizen science and citizen engagement. Achievements in Horizon 2020 and recommendations on the way forward. Directorate-General for Research and Innovation Science with and for society. (2020). https://op.europa.eu/en/publication-detail/-/publication/c30ddc24-cbc6-11ea-adf7-01aa75ed71a1
41. Weill, P.: Don't just lead, govern: how top-performing firms govern IT. MIS Quart. Exec. **3**(1), 1–17 (2004)
42. Krimmer, R., Dedovic, S., Schmidt, C. and Corici, A.A.: Developing cross-border e-Governance: exploring interoperability and cross-border integration. In: International Conference on Electronic Participation (pp. 107-124). Springer, Cham. (2021)

How Can Innovation Systems be Sustainable? An Approach for Organizations in Times of Crises, Alongside Economic and Information Technology Issues

Rafael Antunes Fidelis⬩, Antonio Carlos dos Santos⬩,
Paulo Henrique de Souza Bermejo(✉)⬩,
Diogo Bernardino de Oliveira Lima Bezerra⬩, and Rafael Barreiros Porto⬩

Universidade de Brasília, Brasília, Brazil
{paulobermejo,rafaelporto}@unb.br

Abstract. Although information technology is widely considered in innovation processes, environmental, social, and political issues are commonly neglected in times of crisis. This is especially unwelcome since these issues require scarce time and resources due to the instability and vulnerability of organizations in these times. When treated in terms of Innovation Systems (IS), these issues culminate in dedicated IS (DIS) to sustainability. This article presents the theoretical foundations of DIS through a Systematic Literature Review (RSL) based on the Templier and Paré protocol, which analyzed 44 studies. This study examines how DIS behaves to meet environmental, social, political, and economic aspects in crisis times. The results reveal a predominance of governmental institutions as protagonists in promoting IS in times of crisis. We also observed recurrent cooperation between different actors, demonstrating the importance of partnerships. Our study fosters discussion on the behavior of IS dedicated to sustainability in times of crisis and highlights the main elements of their functioning. Finally, this paper provides a research agenda for future studies.

Keywords: Innovation System (IS) · Sustainability · Times of crisis · Systematic Literature Review (SLR) · Organizational vulnerabilities · Cooperation

1 Introduction

The literature highlights the necessity to analyze the functioning of dedicated Innovation Systems (DIS) in times of crisis [1]. This research aims to examine how DIS behaves in times of crisis to fill this gap. A better understanding of how these systems work is crucial for the sustainable generation of new business opportunities. This theme is also essential to mitigate the harmful impacts inherent in times of crisis, strengthening the achievement of the Sustainable Development Goals (SDGs) [2], and contributing to the search for a sustainable way of life for present and future generations through economically, socially, and ecologically appropriate ways [3].

During times of crisis, organizational vulnerability is magnified. The crisis–which can be conceptualized as an event with a low probability of occurrence and elevated risk of damaging impact [4]–stimulates organizational innovations in a fast way [4–5]. Such innovations can occur through Innovation Systems (IS), which can be defined as instruments that involve the creation, diffusion, and use of knowledge to promote learning, innovation, and the construction of competencies among organizations [6, 7].

IS are oriented primarily by technological and economic interests [8, 9]. On the other hand, sustainability aspects are becoming relevant today, requiring fundamental changes in technologies, industries, organizations, consumption patterns, and lifestyles [9, 10]. Thus, IS has begun to adopt a paradigm dedicated to the aspects related to sustainability in its broadest sense-environmental, social, political, and economic [2, 9, 11, 12].

In the face of the natural sense of urgency in moments of crisis, the participative character and the complex development process inherent to the DIS present themselves as possible obstacles to organizational innovation. If, on the one hand, it is desirable to contemplate environmental, social, and political aspects in the process of sustainable innovation, on the negative side, efforts to involve such themes require time and resources, generally scarce in times of crisis [6, 9, 13]. Crises, such as, for example, the war in Ukraine, the pandemic of COVID-19, the humanitarian crisis of Syrian refugees, or the dam bursts in Mariana and Brumadinho in Brazil, are threats even to the implementation of sustainable development, typically neglected in periods of instability [14, 15].

The United Nations (UN), through its Agenda of Sustainable Development, highlights the relevance that the theme has for the current and future generations of society on a global level. Measures encouraging adherence to the agenda by institutions and individuals are desirable if the agenda is effective and the objectives are to be achieved [16]. Several researchers have addressed the theme of sustainability in their articles, demonstrating its importance, including for the academy [3, 17–24].

The corporation's success today is linked to sustainability aspects in its strategy, competitive vision, and innovation policies. By contemplating these aspects, organizations contribute to increasing their competitiveness in global markets meeting international requirements and regulations [2]. Sustainability has become a competitive advantage within organizations because it provides social and environmental value.

A systemic understanding [1, 25, 26] of DIS behavior in times of crisis is relevant for guiding innovation to extra-economic benefits, with practical implications for organizations, consumers, practitioners, and policymakers [2, 27]. Clarity about this behavior can favor the accurate formulation of public policies and guide consumers and professionals to adopt sustainable practices [1, 28].

This paper is organized into five sections. After this introduction, we present the research method. The third section offers a thematic analysis of the recurring issues related to the functioning of DIS in times of crisis and discusses the theoretical aspects of the topic. The fourth section addresses the research agenda, indicating future paths. Finally, the fifth section brings the conclusions of this study.

2 Research Method

The research was done using the systematic literature review (SLR) method. The SLR was chosen because it is an objective, reliable, thorough research process [29] and structured in multiple stages [30]. With this method, we intend to contribute to the academic world by suggesting new paths for future research and the professional world by providing access to a reliable knowledge base [31, 32]. Thus, we conducted the following steps:

Step 1. Formulation of Research Questions and Terms. We performed a preliminary investigation with five sets of keywords in six databases to formulate the search terms. The fact that variations of sustainability-focused IS, such as DIS, are relatively new has narrowed the search results, resulting in fewer articles for an SLR. Thus, we opted for the following strategy: use broader search terms [("innovation system*") AND ("crisis")]; next, during Step 3, we analyzed the abstracts to obtain a group of articles adhering to the differential aspects of DIS, therefore eligible for investigation. This way, we arrived at a group of articles that address its relevant aspects in times of crisis, even though they do not expressly bring the DIS term.

Step 2. Database Search. As in Step 1, we conducted a preliminary search to delimit the databases. In this experiment, we looked at six widely known databases in management research. Next, we delimited our search to three databases presenting the most articles. Thus, Web of Science (WoS), Elsevier's Scopus, and Science Direct were chosen. We have refined the search for articles in the English language and academic journals. The search was not limited to a specific period. At this stage, 356 articles comprised the sample. The articles were organized using the Mendeley © reference software.

Step 3. Download and Selection of Articles. The 356 articles in Step 2 were checked to eliminate duplicates, thus removing 41 repeated articles using Mendeley ©. Then, we identify the impact factor of the journals based on the Journal Citation Reports (JCR) of 2020. This action is justified because of the growing number of academic productions of lower quality [33]. Thus, adopting the JCR-based cutoff criterion ensures a minimum quality standard for the sources researched [34]. The 64 articles were removed because their journals did not have JCR. The final articles were selected manually by reading the abstracts, titles, and keywords. Thus, 208 articles were excluded for not addressing the research questions. Therefore, the complete search resulted in a final list with 44 articles.

Step 4. Data Selection. The forty-four articles were read and analyzed in total. Based on Templier and Parré [35], a standardized data extraction process was performed for all articles using this protocol.

Step 5. Thematic Analysis. During the content analysis of the articles, we identified and investigated the recurring themes [35]. We performed content analysis at this stage according to the Bardin protocol [90]. The excerpts related to the functioning of DIS in times of crisis were reviewed, addressing the most recurrent elements of the DIS function, including (i) management and governance; (ii) multi-stakeholder interactions; (iii) investment, promotion, and credit; and (iv) information and communication technology. The results were consolidated in tables, synthesizing the collected evidence [36]. Finally, we suggest selected topics aiming to support future IS research. Figure 1 shows the steps of the methodology used in SLR.

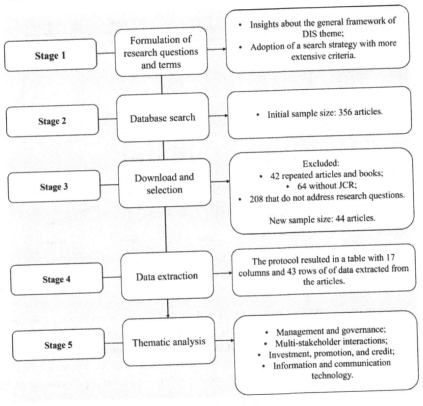

Fig. 1. Methodological steps.

3 Results

A thematic analysis achieved the results of this article. The main functions of DIS in times of crisis were organized into four functional categories. They are (a) management and governance; (b) multi-stakeholder interactions; (c) investment, promotion, and credit; and (d) information and communication technology. We identified four aspects of DIS, that are: environmental, social, political, and economic aspects. Figure 2 presents the relationships between the function of the DIS and these aspects, thus representing the elements of its functioning.

Therefore, we define the functioning of DIS based on the behavior of their functions in environmental, social, political, and economic aspects. For example, suppose that the El Nino climate phenomenon caused a severe drought in Colombia and impacted coffee producers. In this scenario, a DIS could be a cooperative action between coffee producers to implement a production model based on the circular economy. From this perspective, the function of DIS would be cooperation between producers, which are related to environmental aspects (generating cleaner production), social aspects (reducing humanitarian damage caused by the drought crisis), and economic aspects (increasing production affected by the crisis).

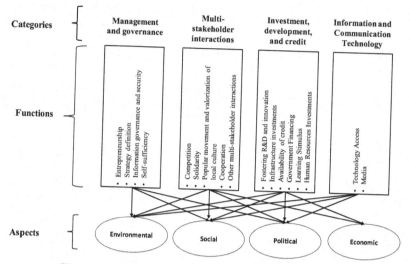

Fig. 2. Overview of the functioning of DIS in times of crisis

As seen in Fig. 2, the category of multi-stakeholder interactions and the category of investment, promotion, and credit directly relate to the four aspects of DIS. The management and governance category and the information and communication technology category only relate to environmental, social, and political aspects. More details about the items in each category that underlie the relationship with these aspects are described in the following subsections.

3.1 Management and Governance

Strategic management, governance, and entrepreneurship were identified as relevant elements in the function of DIS in times of crisis. Entrepreneurship, above all, is crucial to transform the potential of new knowledge, networks, and markets into concrete actions to generate and seize new business opportunities. In addition, the management and mobilization of financial, human, and physical resources are essential inputs required for all activities in the innovation system [37].

Innovation support policies should consider potential diversification strategies. Different industries are affected differently by the same environmental conditions [38]. In strategy definition, it is necessary to observe the complementarity between DIS factors [39]. An effective strategic vision is fundamental for innovation policies to better reflect each sector's specific dynamics [38]. The strategic definition uses governmental and political tools to face crises [40].

In implementing the strategy, one should look for the readiness and credibility of public authorities to design and implement sound strategies, the willingness of companies involved in strategic planning, and the availability of private investment funds [41]. In addition, the strategy should promote systemic innovation and faster adaptation of organizations to the challenges and opportunities of the crisis [40, 41]. An example of this type of strategy is presented in [40] study. These authors demonstrate a strategy to increase investment in local innovation in the Guangdong province of China, through private sector investment combined with government financial funds in ten large innovation projects over three years in the 2010s.

In times of crisis, there may be a departure from business as usual due to challenges and opportunities [42]. In many cases, the actors act self-sufficient and versatilely, seeking to take advantage of these opportunities during the crisis. In this context, the do-it-yourself concept stands out, evidencing the drive of the DIS to subsistence needs and the reuse of pre-crisis dynamics and resources to produce specific solutions for the crisis period [1, 43]. In addition, the use of agile governance and adaptive governance concepts can also contribute to the rapid response and effective adaptability of DIS actors, especially governmental ones [44].

The strategies presented in the DIS can be of innovation, stability, or discontinuity. Innovative character refers to significant changes. Stability changes are for the permanence of the status quo during the crisis. On the other hand, discontinuity changes correspond to change actions that are reversed from the pre-crisis scenario [43]. The operation described in this section is structured in Table 1.

Table 1. Operation of the DIS regarding management and governance

Aspect	Function	Actors	References
Environmental; political	Entrepreneurship	Public sector; Private sector	[37, 42]
Environmental	Strategy definition	Public sector; Private sector	[37–39, 41, 44–46] [37–39, 41, 44–46] [37–39, 41, 44–46]
Political	Information governance and security	Public sector	[47, 48]
Environmental; Social	Self-sufficiency	Population; Organization	[1, 43] [1, 43] [1, 43]

3.2 Multi-stakeholder Interactions

A crisis can cause interconnected problems in several systems. Thus, DIS applied to solve such sustainable problems is ineffective when it emphasizes indicators of only one nature (environmental, for example) and ignores other relevant aspects (such as economic, social, and political). Therefore, the multidisciplinary group of researchers assigned to face the problem is desirable because the lack of adequate and systemic relations among different actors in various sectors can result in conflicts and parallelism of responsibilities [49, 50]. In this sense, DIS in times of crisis includes not only government planning but also infrastructure components and collaboration between institutions and actors, such as universities, industry, consumers, countries, technology parks, and incubators [1, 51–56].

On the other hand, specific crises can be faced by cooperating with a group of actors in the same sector and implementing innovative and collaborative actions for shared growth [57]. Thus, the importance of community-centered strategies is highlighted by cooperative business models that transcend isolated narratives of individualized success through a shared model of productivity and profit. DIS should, therefore, value the role of all actors, including local communities, in building knowledge for innovation [1, 52, 54, 55]. Cooperation should be encouraged, even to increase the resilience of the DIS [56]. However, despite the importance of multi-stakeholder involvement, DIS in times of crisis is typically a public service responsibility of governments [52].

In facing crises, the government can use the public-private partnership (PPP) model, which can contribute to the cooperation among all players in developing the DIS and promote the speed of the process [46, 58]. In addition, the model brings transparency and clear and present control of the public sector authorities [46]. However, there are situations where governments are silent. Even in this scenario, research organizations must develop new activities and operation mechanisms that can transform their role and affect their relationship with the economy, interacting positively with organizational managers, new entrepreneurs, associations, and societies [51, 59].

In addition to cooperation, competition, whether between organizations or countries, can also influence the function of DIS. The competitive landscape does not scare away DIS. On the contrary, it stimulates innovation [60]. A competitive environment can catalyze the development of technologies, which, with appropriate regulation, contribute to sustainable innovation in times of crisis [37, 53]. However, it is worth noting that the fruits of competition strongly depend on the maturity levels of technological diffusion among actors of the same DIS. The imbalance of such diffusion is observed between emerging countries and more stable economies [53, 60]. For example, the [60] study focused on innovation for better energy consumption in China, India, the United States, and Europe. It demonstrated that economic maturity and specific policies adopted in each country or region had a different impact.

The function of the DIS can also be marked by the movement of popular participation in innovation processes, regional resources, local knowledge, and strategies to combat the crisis. Thus, participatory innovation and community integration are critical elements of these DIS. Commonly, these movements are treated as alternative models, but they can present significant impacts on innovation in times of instability. The literature highlights some important common elements among the DIS coming from popular movements: a) collaborative character, b) protagonism of the population, and c) governmental, private, and third-sector support. Besides the main focus on valuing popular and local culture, these DIS also highlight the importance of auxiliary actions by the government and other actors in the process [52, 54, 61].

The crisis presents an unprecedented window of opportunity to promote innovation. But the social vision and mission in times of instability should not be forgotten, as these periods expose and reinforce the deep inequalities of marginalized populations. DIS should incorporate acts of solidarity, evident in actions to support people affected during a crisis or to help reduce the immediate impacts [1, 55]. In addition, legitimization and awareness are relevant elements of the DIS [56]. A certain level of legitimacy is required for actors to commit to and invest in innovation [37, 62]. The operation described in this section is structured in Table 2.

Table 2. DIS operation regarding multi-actor interactions.

Aspect	Function	Actor	Reference
Environmental	Competition	Countries	[53]
Social; political	Solidarity	Population; Organization	[1, 45, 55, 56] [1, 45, 55, 56] [1, 45, 55, 56]
Social	Popular movement and valorization of local culture	Population	[52, 54, 61] [52, 54, 61] [52, 54, 61]
Social; political	Cooperation	Private sector	[40, 52, 53, 57]
Economic; environmental; social; political	Other multi-stakeholder interactions	Public sector; Private sector; Countries; Educational Institutions; Organization	[39, 40, 50–54, 58, 59, 63–69]

3.3 Investment, Promotion, and Credit

Companies do not always have the financial resources for innovation, especially in times of crisis. Therefore, the inflow of capital, whether through credit availability or government funding, is a relevant factor in the DIS in periods of crisis [37, 62, 70, 71]. The lack of a structured financial environment reduces innovation performance. On the other side, its existence increases the availability of financial instruments and the access entrepreneurs have to them, contributing to innovation [72, 73].

In several studies, it has been observed that government support, through the efficient and strategic distribution of funds, has promoted stability, taking the crisis as an opportunity for sustainable transformation of the economy towards an environment based on innovation and knowledge [40, 43, 74]. However, in some cases, government support through subsidies can be frowned upon, as it tends to create dependence on the companies. The artificial growth of organizations can end up being incompatible with the demand capacity of the market or with the level of infrastructure in the region [68, 75]. That is why it is essential that the infrastructure also be the target of government investment [68].

R&D investments, whether public or private, are recurrent in DIS. These investments are determinant in the search for new fundamental knowledge to achieve economic and social benefits, especially in regions most impacted by the crisis [76, 77]. DIS must involve actions before the crisis period. R&D investment in strategically selected companies by the government can stimulate innovation, reduce the damaging impacts of crises, and contribute to economic recovery. These previous investments generate knowledge and learning for the companies: vital elements for innovation and facing crises [78, 79]. However, it is not enough that there is an investment in R&D. It is necessary to transform such investment into commercially exploitable outputs. The formation of a market depends on government spending on Science, Technology, and Innovation, the valorization of entrepreneurial activities, and the encouragement of new companies' formation [80].

It is worth noting that legislating and regulating companies' governance, structure, and innovation performance can prepare them for times of crisis. This allows companies to innovate autonomously without government funding [70, 71]. Regardless of the catalyzing actors of the investment actions (companies or government), the qualification of human resources proved paramount in mitigating the effects of innovation deceleration in times of crisis. Thus, people management is a determining factor in the ability of institutions to react to the transformations resulting from these periods [40, 81–83].

Human resources training represents a fundamental element in all phases of the DIS. For example, in developing Systems of Technological Innovations, sustainability transitions depend fundamentally on knowledge and learning mechanisms and interactive processes between all the actors involved and the local context where the technologies are embedded [37]. This empowerment can occur through flexible modalities, culminating in more resilient innovation systems [45]. To illustrate, the [1] study analyzed 707 innovation projects during the COVID-19 pandemic and found that human needs must be addressed in every innovation system, especially during crises. The human needs perceived include affection, understanding, participation, and creation.

Social rights to universal access to education may suffer during a crisis when the need and search for income promote the prioritization of work over education. Thus, DIS must innovate educational models and methods with the union between productive work and education in times of crisis. The literature highlights the importance of looking at successful models from other countries and adapting the link between learning and productive work to provide a valuable stimulus to educational innovation [65]. The operation described in this section is structured in Table 3.

3.4 Information and Communication Technology

Technological innovations are increasingly emerging as viable solutions in times of crisis. But sometimes, they are not enough to meet social demands, and it is necessary to increase the population's access to technologies [56, 62]. This access guarantees social rights and meets minimum human needs [86, 87]. To do this, DIS is needed to overcome environmental, economic, technological, political, and sociocultural barriers. Such solutions must consider in their operation the following elements: a) economic-be of low cost for acquisition, operation, and maintenance; b) environmental-use renewable resources; c) technical-be easy to introduce, operate, maintain, and manage; and d) sociocultural-consider the significant socio-cultural barriers regarding the introduction and use [87].

In the observed studies, as the main catalyst of DIS, the government adopts complex technologies, such as Artificial Intelligence (AI), to technologically simple resources,

Table 3. Operation of the DIS to investment, promotion, and credit

Aspect	Function	Actor	Reference
Economic; environmental; social; political Economic; environmental; social; political	Fostering R&D and innovation	Public sector; Private sector	[37, 43, 48, 60, 63, 65, 72, 76, 77, 80, 84, 85] [37, 43, 48, 60, 63, 65, 72, 76, 77, 80, 84, 85]
Economic; Environmental	Infrastructure investments	Public sector	[37, 68, 72]
Economic; Environmental	Availability of credit	Public sector; Private sector	[37, 41, 61, 70] [37, 41, 61, 70]
Economic; Political	Government Financing	Public sector	[40, 42, 45, 48, 71, 74, 75] [40, 42, 45, 48, 71, 74, 75] [40, 42, 45, 48, 71, 74, 75]
Environmental; social; political	Learning Stimulus	Public sector; Private sector	[37, 40, 45, 48, 65, 74] [37, 40, 45, 48, 65, 74] [37, 40, 45, 48, 65, 74]
Economic; environmental; social	Human Resources Investments	Small and medium enterprises; Countries; Public sector; Private sector	[37, 40, 45, 48, 81, 82] [37, 40, 45, 48, 81, 82] [37, 40, 45, 48, 81, 82]

such as mobile device applications and social media, for interaction with citizens [58]. These channels can propel an isolated project into a wide-ranging expansion, bringing new opportunities for innovation [47, 57]. An example of the broad expansion of technologies and their innovation impacts are the new socioeconomic contexts created in the public sector of Canada by the widespread adoption of AI and its far-reaching consequences [47]. The operation described in this section is structured in Table 4.

4 Limitations and Suggestions for Future Research

Many articles address social, political, and environmental aspects of innovation systems. However, they do not directly mention the term "dedicated innovation systems". Although the search and selection criteria were carefully worked on, there is a possibility that this search was not exhaustive.

Table 4. Operation of the DIS as to Information and Communication Technology

Aspect	Function	Actor	Reference
Environmental; social; political	Technology Access	Population; Public sector Population; Public sector	[42, 48, 62, 86, 87] [42, 48, 62, 86, 87] [42, 48, 62, 86, 87]
Social	Media	Private sector	[57, 58]

The items that form each category were organized following the Bardin protocol [90] and resulted in a thematic classification. However, other forms of classification can be helpful to understand the functioning of DIS in crises, such as ethnically and geographically, essentially in times of crises generated by armed conflicts.

We still found some unresolved challenges in the literature or promising aspects to be studied, such as:

1. Given the observed relevance of communication in DIS [47, 57], the relationship between different communication strategies and the success or failure of IS could be analyzed as the focus of future research.
2. The results of DIS may gain different relevance and application due to changing scenarios. In the post-crisis moment, an innovative solution developed for the crisis may become more relevant, irrelevant, or useful for other purposes or even a challenge to be met [47]. We suggest that future research examine the repercussions of the innovations after the crisis period in which they emerged.
3. The transversality and integration of actions presented themselves as a relevant and recurrent factor in the DIS. On the other hand, the coordination of multiple and distinct actions in a dedicated way is, in several studies, a challenge to be faced [80]. We recommend that empirical studies develop proposed frameworks to facilitate this coordination.
4. The literature notes that a country's level of development and economic stability can influence the DIS's functioning [60]. Further research can explore, in greater detail, this relationship.
5. Recognizing the value of local knowledge represents a challenge for experts [52, 54, 61]. Thus, future research can analyze mechanisms to catalyze communication and knowledge sharing between local agents and other IS actors.
6. Another aspect to be studied in further research is the challenges of implementing policies to reduce inequalities and stimulate popular participation in the scope of DIS [55].
7. The credibility of the authorities and the awareness of the population were identified as determining factors for the actors of a DIS to commit to the goals of innovation and invest in it. Likewise, the level of legitimacy of the strategies adopted is an essential element for the functioning of the IS. [37, 41, 56, 62]. We recommend further studies to analyze the influence of the credibility of the authorities, the legitimacy of the strategies, and the population's awareness of the IS's functioning.
8. DIS involves the action of multiple actors. However, in times of crisis, this relationship is typically the responsibility of governments [52]. In a scenario of governmental

omission, the other actors must assume the position of protagonists [1, 51–56, 59]. Making up for the absence of governmental actions in the DIS could challenge future research.

9. The competitive landscape influences the behavior of DIS [37, 53, 60]. Thus, future studies can examine how a competitive environment relates to innovation in times of crisis. In addition, the compatibility between organizations' growth, infrastructure, and market demand was also addressed in the observed studies [68, 75]. This issue can be addressed in more detail in further studies.

10. Pre-crisis actions, whether by government or private institutions, can prepare the actors of a DIS to face a future crisis [70, 71]. Some of the studies addressed here bring as examples of these actions the investment in R&D, qualification of human resources, and regulation of governance and innovation in organizations [40, 78, 79, 81–83]. The relationship between these prior actions and the success of a DIS in times of crisis may be the subject of further studies.

5 Concluding

Environmental, social, and political necessities are commonly neglected in the innovation process in times of crisis, as they require time and resources that are usually scarce in such periods. These issues pose significant challenges for today's societies to achieve more sustainable modes of production. In this context, dedicated Innovation Systems (DIS) stands out for addressing these extra-economic necessities. The objective of this research was achieved by examining the functioning of DIS in times of crisis through a Systematic Literature Review (SLR).

This study shows a predominance of governmental institutions as protagonists in the functioning of the DIS in times of crisis. We can also highlight private and academic institutions' coordinated and recurrent action. Partnerships with public-private and third-sector organizations are recurrent. Another relevant element present in DIS is the cooperation among agents to implement and ensure the effectiveness of sustainable results in environmental, economic, political, and social aspects.

Our study discusses DIS behavior in times of crisis, bringing the critical factors for its functioning. Some of these factors are learning; HR qualification; public and private investment; credit availability; experience of local culture; motivation, diffusion, communication, and awareness of the population; adaptation; networking; cooperation; competition; regulation; sustainability; social inclusion; credibility; political flexibility; access to information and communication technology; stability; interdisciplinary systemic application; and solidarity.

The present study presents theoretical and practical implications. Regarding the theoretical contributions of this study, this research demonstrates the importance of observing social, political, and environmental aspects in IS, in addition to economic factors. The analysis of these extra-economic factors in IS generates the concept of DIS, which is especially relevant in times of crisis. Observing current literature, we identified seventeen DIS functions in times of crisis, which can be organized into four categories: (a) management and governance; (b) interactions between multiple stakeholders; (c) investment, development, and credit; and (d) information and communication technology. In empirical terms, the results are helpful in formulating public and private policies and strategies, as it is possible to observe factors that influence the proper functioning of DIS. The research also contributes to organizations seeking to achieve the Sustainable Development Goals (SDGs) through sustainable IS practices.

Declarations of Interest: None. This research received no specific grant from funding agencies in the public, commercial, or not-for-profit sectors.

References

1. Dahlke, J., Bogner, K., Becker, M., Schlaile, M.P., Pyka, A., Ebersberger, B.: Crisis-driven innovation and fundamental human needs: a typological framework of rapid-response COVID-19 innovations. Technol Forecast Soc Change **169**, 120799 (2021). https://doi.org/10.1016/j.techfore.2021.120799
2. Zartha Sossa, J.W., Lopez Montoya, O.H., Acosta Prado, J.C.: Determinants of a sustainable innovation system. Bus Strategy Environ. **30**(2), 1345–1356 (2021). https://doi.org/10.1002/bse.2689
3. Zhu, B., Nguyen, M., Siri, N.S., Malik, A.: Towards a transformative model of circular economy for SMEs. J. Bus. Res. **144**, 545–555 (2022). https://doi.org/10.1016/j.jbusres.2022.01.093
4. Pearson, C.M., Clair, J.A.: Reframing crisis management. Acad. Manag. Rev. **23**(1), 59–76 (1998). https://doi.org/10.5465/AMR.1998.192960
5. Oborn, E., Pilosof, N.P., Hinings, B., Zimlichman, E.: Institutional logics and innovation in times of crisis: telemedicine as digital 'PPE.' Inf. Organ. **31**(1), 100340 (2021). https://doi.org/10.1016/j.infoandorg.2021.100340
6. Carlsson, B., Jacobsson, S., Holmen, M., Rickne, A.: Innovation systems: analytical and methodological issues. Res. Policy **31**(2), 233–245 (2002). https://doi.org/10.1016/S0048-7333(01)00138-X
7. Cassiolato, J.E., Pessoa de Matos, M.G., Lastres, H.M.M.: Innovation systems and development. Int. Develop. 566–581 (2014). https://doi.org/10.1093/acprof:oso/9780199671656.003.0034
8. Carlsson, B., Jacobsson, S.: In search of useful public policies—key lessons and issues for policy makers, 299–315 (1997). https://doi.org/10.1007/978-1-4615-6133-0_11
9. Pyka, A.: Dedicated innovation systems to support the transformation towards sustainability: creating income opportunities and employment in the knowledge-based digital bioeconomy. J. Open Innov.: Technol. Market Complexity **3**(4) (2017). https://doi.org/10.1186/s40852-017-0079-7
10. Markard, J.: The life cycle of technological innovation systems. Technol. Forecast Soc. Change 153(July 2018), 119407 (2020). https://doi.org/10.1016/j.techfore.2018.07.045
11. Ghazinoory, S., Nasri, S., Ameri, F., Montazer, G.A., Shayan, A.: Why do we need 'Problem-oriented Innovation System (PIS)' for solving macro-level societal problems?," Technol. Forecast Soc. Change **150**(October 2018), 119749 (2020). https://doi.org/10.1016/j.techfore.2019.119749
12. Hekkert, M.P., Suurs, R.A.A., Negro, S.O., Kuhlmann, S., Smits, R.E.H.M.: Functions of innovation systems: a new approach for analysing technological change. Technol. Forecast Soc. Change **74**(4), 413–432 (2007). https://doi.org/10.1016/j.techfore.2006.03.002
13. Williams, T.A., Gruber, D.A., Sutcliffe, K.M., Shepherd, D.A., Zhao, E.Y.: Organizational response to adversity: fusing crisis management and resilience research streams. Acad. Manag. Ann. **11**(2), 733–769 (2017). https://doi.org/10.5465/annals.2015.0134

14. Filho, W.L., Brandli, L.L., Salvia, A.L., Rayman-Bacchus, L., Platje, J.: COVID-19 and the UN sustainable development goals: threat to solidarity or an opportunity? Sustain. (Switz.) **12**(13), 1–14 (2020). https://doi.org/10.3390/su12135343

15. Ranjbari, M., et al.: Three pillars of sustainability in the wake of COVID-19: a systematic review and future research agenda for sustainable development. J. Clean. Prod. **297**, 126660 (2021). https://doi.org/10.1016/j.jclepro.2021.126660

16. ONU, "United nations general assembly. Int. J. Marine Coastal Law **25**(2), 271–287 (2015). https://doi.org/10.1163/157180910X12665776638740

17. Hundscheid, L., Wurzinger, M., Gühnemann, A., Melcher, A.H., Stern, T.: Rethinking meat consumption – How institutional shifts affect the sustainable protein transition. Sustain. Prod. Consum. **31**, 301–312 (2022). https://doi.org/10.1016/j.spc.2022.02.016

18. Ho, J.Y., Yoon, S.: Ambiguous roles of intermediaries in social entrepreneurship: the case of social innovation system in South Korea. Technol. Forecast Soc. Change **175** (2022). https://doi.org/10.1016/j.techfore.2021.121324

19. Klerkx, L., Begemann, S.: Supporting food systems transformation: the what, why, who, where and how of mission-oriented agricultural innovation systems. Agric. Syst. **184**, Sep (2020). https://doi.org/10.1016/j.agsy.2020.102901

20. Nair, R., Viswanathan, P.K., Bastian, B.L.: Reprioritising sustainable development goals in the post-covid-19 global context: will a mandatory corporatesocial responsibility regime help? Adm. Sci. 11(4) (2021). https://doi.org/10.3390/admsci11040150

21. Yin, X., Chen, J., Li, J.: Rural innovation system: revitalize the countryside for a sustainable development. J. Rural. Stud. **93**, 471–478 (2019). https://doi.org/10.1016/j.jrurstud.2019.10.014

22. Hernandez, R.R., Jordaan, S.M., Kaldunski, B., Kumar, N.: Aligning climate change and sustainable development goals with an innovation systems roadmap for renewable power. Front. Sustain. **1**, 1–13 (2020). https://doi.org/10.3389/frsus.2020.583090

23. Kilkiş, Ş: Sustainability-oriented innovation system analyses of Brazil, Russia, India, China, South Africa, Turkey and Singapore. J. Clean. Prod. **130**, 235–247 (2016). https://doi.org/10.1016/j.jclepro.2016.03.138

24. Davis, K.: How will extension contribute to the sustainable development goals? A global strategy and operational plan. J. Int. Agric. and Exten. Educ. **23**(1), 7–13 (2016). https://doi.org/10.5191/jiaee.2016.23101

25. Forrester, J.W.: Counterintuitive behavior of social systems. Technol. Forecast Soc. Change 3(C), 1–22 (1971). https://doi.org/10.1016/S0040-1625(71)80001-X

26. Schlaile, M.P., Urmetzer, S., Ehrenberger, M.B., Brewer, J.: Systems entrepreneurship: a conceptual substantiation of a novel entrepreneurial 'species.' Sustain. Sci. **16**(3), 781–794 (2021). https://doi.org/10.1007/s11625-020-00850-6

27. Dahlke, J., Bogner, K., Becker, M., Schlaile, M.P., Pyka, A., Ebersberger, B.: Crisis-driven innovation and fundamental human needs: a typological framework of rapid-response COVID-19 innovations. Technol. Forecast Soc. Change **169** (2021)

28. Pan, S.L., Zhang, S.: From fighting COVID-19 pandemic to tackling sustainable development goals: an opportunity for responsible information systems research. Int. J. Inf. Manage. **55**(July), 102196 (2020). https://doi.org/10.1016/j.ijinfomgt.2020.102196

29. Mangas-Vega, A., Dantas, T., Sánchez-Jara, J.M., Gómez-Díaz, R.: Systematic literature reviews in social sciences and humanities a case study. J. Inf. Technol. Res. **11**(1), 1–17 (2018). https://doi.org/10.4018/JITR.2018010101

30. Denyer, D., Neely, A.: Introduction to special issue: innovation and productivity performance in the UK. Int. J. Manag. Rev. **5–6**(3–4), 131–135 (2004). https://doi.org/10.1111/j.1460-8545.2004.00100.x

31. Briner, R.B., Denyer, D.: Systematic review and evidence synthesis as a practice and scholarship tool. The Oxford Handbook of Evidence-Based Management, no. November 2015 (2012). https://doi.org/10.1093/oxfordhb/9780199763986.013.0007

32. Compagnucci, L., Spigarelli, F.: The Third Mission of the university: a systematic literature review on potentials and constraints. Technol. Forecast Soc. Change **161**(March), 120284 (2020). https://doi.org/10.1016/j.techfore.2020.120284

33. Pinto, A.C., de Andrade, J.B.: Fator de impacto de revistas científicas: qual o significado deste parâmetro? Quim. Nova **22**(3), 448–453 (1999). https://doi.org/10.1590/s0100-404219 99000300026

34. Garfield, E.: The meaning of the Impact Factor. Int. J. Clin. Health Psychol. **3**(2), 363–369 (2003)

35. Dixon-Woods, M., Agarwal, S., Jones, D., Young, B., Sutton, A.: Synthesising qualitative and quantitative evidence: a review of possible methods. J. Health Serv. Res. Policy **10**(1), 45–53 (2005). https://doi.org/10.1258/1355819052801804

36. Denyer, D., Tranfield, D.: Producing a Systematic Review. The SAGE Handbook of Organizational Research Methods, pp. 671–689 (2009)

37. Wesche, J.P., Negro, S.O., Dütschke, E., Raven, R.P.J.M., Hekkert, M.P.: Configurational innovation systems – Explaining the slow German heat transition. Energy Res. Soc. Sci. **52**, 99–113 (2019). https://doi.org/10.1016/j.erss.2018.12.015

38. Zabala-Iturriagagoitia, J.M., Porto Gómez, I., Aguirre Larracoechea, U.: Technological diversification: a matter of related or unrelated varieties?. Technol. Forecast Soc. Change **155** (2020). https://doi.org/10.1016/j.techfore.2020.119997

39. Garcés-Ayerbe, C., Cañón-de-Francia, J.: The relevance of complementarities in the study of the economic consequences of environmental proactivity: analysis of the moderating effect of innovation efforts. Ecol. Econ. **142**, 21–30 (2017). https://doi.org/10.1016/J.ECOLECON. 2017.06.022

40. Kroll, H., Tagscherer, U.: Chinese regional innovation systems in times of crisis: the case of Guangdong. Asian J. Technol. Innov. **17**(2), 101–128 (2009). https://doi.org/10.1080/197 61597.2009.9668675

41. Komninos, N., Musyck, B., Reid, A.I.: Smart specialisation strategies in south Europe during crisis. Eur. J. Innov. Manag. **17**(4), 448–471 (2014). https://doi.org/10.1108/EJIM-11-2013-0118

42. Sampat, B.N., Shadlen, K.C.: The covid-19 innovation system. Health Aff. **40**(3), 400–409 (2021). https://doi.org/10.1377/hlthaff.2020.02097

43. Anderson, C.R., McLachlan, S.M.: Exiting, enduring and innovating: farm household adaptation to global zoonotic disease. Glob. Environ. Chang. **22**(1), 82–93 (2012). https://doi.org/ 10.1016/J.GLOENVCHA.2011.11.008

44. Janssen, M., van der Voort, H.: Agile and adaptive governance in crisis response: lessons from the COVID-19 pandemic. Int. J. Inf. Manage. **55** (2020). https://doi.org/10.1016/J.IJI NFOMGT.2020.102180

45. Lennox, J., Reuge, N., Benavides, F.: UNICEF's lessons learned from the education response to the COVID-19 crisis and reflections on the implications for education policy. Int. J. Educ. Dev. **85** (2021). https://doi.org/10.1016/J.IJEDUDEV.2021.102429

46. Park, J., Chung, E.: Learning from past pandemic governance: early response and Public-Private Partnerships in testing of COVID-19 in South Korea. World Dev. **137** (2021). https:// doi.org/10.1016/J.WORLDDEV.2020.105198

47. Kuziemski, M., Misuraca, G.: AI governance in the public sector: three tales from the frontiers of automated decision-making in democratic settings. Telecomm. Policy **44**(6) (2020). https:// doi.org/10.1016/J.TELPOL.2020.101976

48. Labunska, S., Gavkalova, N., Pylypenko, A., Prokopishyna, O.: Cognitive instruments of public management accountability for development of national innovation system|Viešojo valdymo atskaitomybės pažintiniai instrumentai kuriant nacionalinę inovacijų sistemą. Public Policy Adm. **19**(3), 114–124 (2019). https://doi.org/10.5755/j01.ppaa.18.3.24727
49. Ghazinoory, S., Khosravi, M., Nasri, S.: A systems-based approach to analyze environmental issues: problem-oriented innovation system for water scarcity problem in Iran. J. Environ. Dev. **30**(3), 291–316 (2021). https://doi.org/10.1177/10704965211019084
50. Huff, T.: Malaysia's multimedia super corridor and its first crisis of confidence. Asian J Soc Sci **30**(2), 248–270 (2002). https://doi.org/10.1163/156853102320405843
51. Papon, P.: Research institutions in France: between the Republic of science and the nation-state in crisis. Res. Policy **27**(8), 771–780 (1998). https://doi.org/10.1016/S0048-7333(98)000 89-4
52. Ballantyne, P.: Accessing, sharing and communicating agricultural information for development: emerging trends and issues. Inf. Dev. **25**(4), 260–271 (2009). https://doi.org/10.1177/ 0266666909351634
53. Bauner, D.: International private and public reinforcing dependencies for the innovation of automotive emission control systems in Japan and USA. Transp. Res. Part A Policy Pract. **45**(5), 375–388 (Jun.2011). https://doi.org/10.1016/J.TRA.2010.12.008
54. Brooks, S., Loevinsohn, M.: Shaping agricultural innovation systems responsive to food insecurity and climate change. Nat. Resour. Forum. **35**(3), 185–200 (2011). https://doi.org/ 10.1111/j.1477-8947.2011.01396.x
55. O'Hara, S., Toussaint, E.C.: Food access in crisis: food security and COVID-19. Ecolog. Econ. **180** (2021). https://doi.org/10.1016/J.ECOLECON.2020.106859
56. Stern, N., Valero, A.: Research policy, Chris Freeman special issue innovation, growth and the transition to net-zero emissions. Res. Policy **50**(9) (2021). https://doi.org/10.1016/J.RES POL.2021.104293
57. Prosser, L., Thomas Lane, E., Jones, R.: Collaboration for innovative routes to market: COVID-19 and the food system. Agric. Syst. **188** (2021). https://doi.org/10.1016/J.AGSY. 2020.103038
58. Binder, S., et al.: African national public health institutes responses to COVID-19: innovations, systems changes, and challenges. Health Secur. **19**(5), 498–507 (2021). https://doi.org/10. 1089/hs.2021.0094
59. Balázs, K.: Innovation potential embodied in research organizations in Central and Eastern Europe. Soc. Stud. Sci. **25**(4), 655–683 (1995). https://doi.org/10.1177/030631295025004004
60. Furlan, C., Mortarino, C.: Forecasting the impact of renewable energies in competition with non-renewable sources. Renew. Sustain. Energy Rev. **81**, 1879–1886 (2018). https://doi.org/ 10.1016/J.RSER.2017.05.284
61. Palma, I.P., Toral, J.N., Vázquez, M.R. P., Fuentes, N.F., Hernández, F.G.: Historical changes in the process of agricultural development in Cuba. J. Clean Prod. **96**, 77–84 (2015). https:// doi.org/10.1016/J.JCLEPRO.2013.11.078
62. Qureshi, T.M., Ullah, K., Arentsen, M.J.: Factors responsible for solar PV adoption at household level: a case of Lahore, Pakistan. Renew. Sustain. Energy Rev. **78**, 754–763 (2017). https://doi.org/10.1016/J.RSER.2017.04.020
63. Etzkowitz, H.: An innovation strategy to end the second great depression. Eur. Plan. Stud. **20**(9), 1439–1453 (2012). https://doi.org/10.1080/09654313.2012.709060
64. S. Ghazinoory, S. Nasri, F. Ameri, G. A. Montazer, and A. Shayan, "Why do we need 'Problem-oriented Innovation System (PIS)" for solving macro-level societal problems?,'" Technol Forecast Soc Change, vol. 150, Jan. 2020, doi: https://doi.org/10.1016/j.techfore. 2019.119749
65. Knox, D., Castles, S.: Education with production-learning from the third world. Int. J. Educ. Dev. **2**(1), 1–14 (1982). https://doi.org/10.1016/0738-0593(82)90062-1

66. Chung, S.: Innovation in Korea (2003).https://doi.org/10.1016/B978-008044198-6/50062-0
67. Stern, N., Valero, A.: Innovation, growth and the transition to net-zero emissions. Res. Policy 50(9), 104293 (2021). https://doi.org/10.1016/j.respol.2021.104293
68. Taalbi, J.: Innovation in the long run: perspectives on technological transitions in Sweden 1908–2016. Environ. Innov. Soc. Transit 40, 222–248 (2021). https://doi.org/10.1016/J.EIST. 2021.07.003
69. Sharif, N.: Emergence and development of the National Innovation Systems concept. Res. Policy 35(5), 745–766 (2006). https://doi.org/10.1016/j.respol.2006.04.001
70. Trinugroho, I., Law, S.H., Lee, W.C., Wiwoho, J., Sergi, B.S.: Effect of financial development on innovation: roles of market institutions. Econ. Model 103 (2021). https://doi.org/10.1016/ J.ECONMOD.2021.105598
71. Tulum, Ö., Lazonick, W.: Financialized corporations in a national innovation system: the U.S. Pharmaceutical Industry. Int. J. Polit. Econ. 47(3–4), 281–316 (2018). https://doi.org/ 10.1080/08911916.2018.1549842
72. Friz, K., Günther, J.: Innovation and economic crisis in transition economies. Eurasian Bus. Rev. 11(4), 537–563 (2021). https://doi.org/10.1007/s40821-021-00192-y
73. Kapetaniou, C., Samdanis, M., Lee, S.H.S.H.: Innovation policies of Cyprus during the global economic crisis: aligning financial institutions with National Innovation System. Technol. Forecast Soc. Change 133, 29–40 (2018)
74. Sharif, N.: An examination of recent developments in Hong Kong's innovation system: 1990 to the present. Sci. Public Policy 33(7), 505–518 (2006). https://doi.org/10.3152/147154306 781778768
75. Yang, C.: Government policy change and evolution of regional innovation systems in China: evidence from strategic emerging industries in Shenzhen. Environ. Plan. C Gov. Policy 33(3), 661–682 (2015). https://doi.org/10.1068/C12162r
76. Knight, P.T., Routti, J.: E-Development and consensus formation in Finland. J. Knowl. Econ. 2(1), 117–144 (2011). https://doi.org/10.1007/s13132-010-0023-6
77. Květoň, V., Horák, P.: The effect of public R&D subsidies on firms' competitiveness: regional and sectoral specifics in emerging innovation systems. Appl. Geogr. 94, 119–129 (2018). https://doi.org/10.1016/J.APGEOG.2018.03.015
78. Ranga, M.: Stimulating R&D and innovation to address Romania's Economic Crisis: a bridge too far? Eur. Plan. Stud. 20(9), 1497–1523 (2012). https://doi.org/10.1080/09654313.2012. 709145
79. Rho, S., Lee, K., Kim, S.H.S.H.: Limited catch-up in China's semiconductor industry: a sectoral innovation system perspective. Millennial Asia 6(2), 147–175 (2015). https://doi. org/10.1177/0976399615590514
80. Gkypali, A., Kokkinos, V., Bouras, C., Tsekouras, K.: Science parks and regional innovation performance in fiscal austerity era: Less is more? Small Bus. Econ. 47(2), 313–330 (2016)
81. Filippetti, A., Archibugi, D.: Innovation in times of crisis: national systems of innovation, structure, and demand. Res. Policy 40(2), 179–192 (2011). https://doi.org/10.1016/J.RES POL.2010.09.001
82. Khan, H.A.: Technology and economic development: the case of Taiwan. J. Contemp. China 13(40), 507–521 (2004). https://doi.org/10.1080/1067056042000213373
83. Kurniati, A., Rosskam, E., Afzal, M.M., Suryowinoto, T.B., Mukti, A.G.: Strengthening Indonesia's health workforce through partnerships. Public Health 129(9), 1138–1149 (2015). https://doi.org/10.1016/J.PUHE.2015.04.012
84. Etzkowitz, H., Ranga, M.: A trans-Keynesian vision of innovation for the contemporary economic crisis: 'picking winners' revisited. Sci. Public Policy 36(10), 799–808 (2009). https://doi.org/10.3152/030234209X481950

85. Greeson, J.K., Gyourko, J., Ortiz, A.J., Coleman, D., Cancel, S.: 'One hundred and ninety-four got licensed by Monday': application of design thinking for foster care innovation and transformation in Rhode Island. Child Youth Serv. Rev. **128** (2021). https://doi.org/10.1016/J.CHILDYOUTH.2021.106166

86. Sourdin, T., Li, B., McNamara, D.M.: Court innovations and access to justice in times of crisis. Health Policy Technol. **9**(4), 447–453 (2020). https://doi.org/10.1016/J.HLPT.2020.08.020

87. Caniato, M., Carliez, D., Thulstrup, A.: Challenges and opportunities of new energy schemes for food security in humanitarian contexts: a selective review. Sustain. Energy Technol. Assess. **22**, 208–219 (2017). https://doi.org/10.1016/J.SETA.2017.02.006

88. Schlaile et al., M.P.: Innovation systems for transformations towards sustainability? Taking the normative dimension seriously. Sustain. (Switz.) 9(12) (2017). https://doi.org/10.3390/su9122253

89. Urmetzer, S., Schlaile, M.P., Bogner, K.B., Mueller, M., Pyka, A.: Exploring the dedicated knowledge base of a transformation towards a sustainable bioeconomy. Sustain. (Switz.) **10**(6), 16–20 (2018). https://doi.org/10.3390/su10061694

90. Bardin, L.: Análise de Conteúdo (2004)

Tools Facilitating Remote Work in the Greek Business Reality and Their Contribution to the Perceived Proximity of the Remote Workers

Eirini Martimianaki[(✉)] [iD] and Ariana Polyviou[iD]

Department of Management, School of Business, University of Nicosia, 2417 Nicosia, Cyprus
imartimianaki@gmail.com, polyviou.a@unic.ac.cy

Abstract. This paper focuses on COVID-19 pandemic and how this new normal has affected employees in Greece. In particular it explores the benefits and challenges, the engagement and productivity, working conditions and management of remote workers. Most of the existing literature has focused on the employees' well-being and productivity in remote working settings. However, little have researchers investigated remote working in the Greek business context, including the IT tools used to perform remote work, their contribution in the efficiency and effectiveness of the communication and collaboration of the remote workers and the perceived proximity of the workers in remote working settings. This research aims to address this gap by exploring the effectiveness of the communication and collaboration of remote workers as well as the level of contribution of the technological tools used in building relationships with their colleagues. In addition, it analyzes the perceived benefits and challenges by remote workers concerning this remote working culture and the digital teamwork. Inspired by the evolving landscape of remote work and recognizing gaps in existing research, the study aims to tackle two central research inquiries. The first delves into the IT tools used for effective communication and collaboration in Greek remote working settings. The second query delves into how these tools bridge the gap between the geographical distance and the perceived proximity of the remote workers. To respond to these questions, the author undertakes a literature review across diverse fields including information systems, organizational studies, economics, human resources, sociology, and psychology. Then the author draws on empirical research through the use of qualitative methods and more specifically, semi-structured interviews. The findings show that younger individuals and those with more remote work experience hold more positive views on this working style compared to older generations and people with much experience. Participants in larger cities or different locations from their workplace also express positivity due to time-saving benefits and increased job opportunities. Effective communication and diverse technological tools contribute to overall satisfaction and efficiency in remote work. The research suggests implementing a hybrid work model in Greek organizations to offer flexibility and enhance work-life balance. Other recommendations include increasing face-to-face meetings, providing newcomer support, conducting regular morning team meetings.

M. Papadaki et al. (Eds.): EMCIS 2023, LNBIP 502, pp. 266–279, 2024.
https://doi.org/10.1007/978-3-031-56481-9_18

Keywords: remote work · IT tools · perceived proximity · communication and collaboration

1 Introduction

The advent of globalization and digitalization has reshaped how individuals live and work, giving rise to the widespread phenomenon of remote working, telecommuting, or work from home. The COVID-19 pandemic further accelerated this trend [25], necessitating safety-driven remote work globally. The terminology surrounding remote work, such as work from home, telecommuting, or e-working, lacks universal consensus. Despite the variety of terms, they all revolve around two core concepts: working outside the organization's premises and leveraging technology for task fulfilment.

Similarly, to the rest of the world, in Greece, many businesses, have recognized the advantages and continue to embrace remote work as a flexible and effective arrangement. Digital teams, formed by remote workers utilizing collaborative digital tools, present new challenges in terms of perceived proximity and the establishment of shared mental models for efficient digital teamwork.

Motivated by the changing landscape of remote work and recognizing research deficiencies, this paper focuses on exploring the IT-enabled remote working practices in the Greek business context and its impact on remote workers' perceived proximity. In particular, the paper aims to shed light on how IT tools employed bridge the gap between the geographical proximity and the perceived proximity of the employees. It examines the effectiveness of communication and collaboration among remote workers, along with the role of technological tools in establishing relationships with colleagues. Additionally, the research investigates the perceived advantages and challenges faced by remote workers in the context of remote work culture and digital teamwork.

This paper is structured as follows; Sect. 2 summarizes related literature on remote working practices, highlighting the benefits and challenges arising for organizations and workers. Section 3 describes the approach followed in this research, including the methods and data collection and analysis process. Then in Sect. 4, the results of are analyzed and in Sect. 5 they are further discussed so as to highlight the research and practical implications arising by this study. Section 6 concludes the work presented in this paper.

2 Literature Review

The evolution of work in the 21st century, catalysed by globalization and digitalization, has ushered in a new era of remote work, telecommuting, and work-from-home arrangements. This paradigm shift, further accelerated by the COVID-19 pandemic, has been explored extensively in the literature. The multifaceted nature of remote work is evident in its perceived benefits, expected challenges, impact on employees' well-being and productivity, the role of IT tools, social interaction dynamics, and the methods of employee monitoring.

Remote work is hailed for its myriad advantages. It not only translates to significant cost savings for organizations by eliminating travel expenses and reducing physical workspace overheads but also offers financial relief for employees who are freed from commuting costs and the need for office attire [20]. This financial advantage also aligns with positive environmental impacts. Beyond the financial aspects, remote work provides employees with time-saving benefits, offering personal hours for family, self-care, and individual pursuits. Parents, in particular, find relief in the flexibility to manage caretaking responsibilities without compromising work obligations. Flexibility is a recurring theme, empowering employees to control their schedules, whether that involves sleeping in, starting and finishing earlier, or compressing work hours over fewer days.

The benefits extend to mental and emotional well-being. Remote work is associated with reduced stress, burnout, and work fatigue, fostering a positive impact on work engagement, job satisfaction, and overall job performance [5]. However, the literature acknowledges divergent perspectives on the relationship between telecommuting and productivity, with some proponents arguing for its enhancement and others remaining sceptical [29].

The COVID-19 pandemic played a pivotal role in propelling remote work into the mainstream. Governments worldwide mandated remote work for safety, and even post-pandemic, many businesses continue to embrace it as part of their culture and strategy. Digital teams have become the norm, utilizing collaboration and communication tools to transcend geographical boundaries. Despite the diverse terminologies used—work from home, telecommuting, or e-working—the essence revolves around physically working outside the organization's premises, facilitated by technology.

The well-being and productivity of remote workers are focal points in contemporary research. Studies emphasize the positive effects on physical and mental health, citing reductions in stress, fatigue, and negative emotions. However, there are counterpoints, with concerns raised about increased exhaustion due to prolonged screen time [27]. The provision of resources for a proper home office environment and technical assistance emerges as a critical factor in enhancing productivity.

The importance of social interaction in remote work settings is a recurring theme. Traditional workplace relationships and a sense of community are foundational to well-being, job satisfaction, and professional development. However, the digital interactions that have become prevalent in remote work raise concerns about changes in the nature of workplace connections, potentially affecting the sense of belonging and togetherness [3]. Researchers highlight the shift towards increased individualization in work conditions, with employees shouldering more responsibility and autonomy, further amplified by the COVID-19 pandemic.

The necessity for employers to monitor remote employees has given rise to new surveillance methods, with tools labelled as "tattle ware" being implemented. These tools range from webcam snapshots taken at regular intervals to software tracking screenshots, login times, and keystrokes [22]. While legal and deemed ethical by managers for improving organizational productivity, concerns about privacy and reduced trust among employees are evident. Some employees are actively seeking ways to evade such surveillance through the use of anti-surveillance software.

Geographical distance emerges as a significant challenge for virtual teams engaged in remote work [30]. Perceived proximity, defined as one's perception of how far another person is, plays a crucial role in the collaboration of remote teams. Research indicates that as geographical distance increases, teamwork quality tends to decrease due to the lack of physical presence and limited social interaction. Companies are addressing this challenge by fostering a shared mental model among team members and leveraging information and communication technology (ICT) tools to enhance communication quality.

In conclusion, the literature review provides a comprehensive picture of the multi-faceted landscape of remote work. It delves into the advantages, challenges, and nuances of this evolving work paradigm. While remote work offers unprecedented flexibility and numerous benefits for both employees and organizations, it also poses managerial, social, and spatial challenges that demand innovative solutions. The integration of IT tools, monitoring mechanisms, and a nuanced understanding of social dynamics are critical for the continued success and adaptation of remote work in our dynamic and ever-changing work environment.

3 Research Approach

This paper follows a qualitative research approach aiming to the subjective experiences, perceptions, and social contexts surrounding the phenomenon of interest [8]. By utilizing a qualitative approach, we can delve into the richness and complexity of these experiences and gain in-depth insights that quantitative methods may not capture adequately [8]. This is implemented using semi-structured interviews with employees of organizations who have been working remotely for at least six months. The research is intentionally restricted to this target interviewees because it aims to investigate the Greek remote working context and how Greek workers respond to this new reality. While Greek people are considered to be very sociable and communicative by nature and culture, found it hard to adapt to this new reality during COVID-19 years. That's why it is very interesting to investigate their progress in adaptation and how this working style has affected their socialization and mindset. Their answers have been recorded, categorized and analyzed afterwards. Interviews were conducted online. Interviewees provided consent to record and transcript the interview. The data presented in this paper has been anonymized. Table 1 presents the demographics of the interviewees included in this research.

The interview agenda included questions on how they perceive the experience of remote working, the IT tools they use, the difficulties they face when dealing with their colleagues virtually only and the perceived benefits. The full agenda employed for this purpose is included in the Appendix of this paper.

The interview data collected for this study were analyzed using a thematic analysis approach. Thematic analysis is a widely used qualitative data analysis method that allows for the identification, organization, and interpretation of patterns, themes, and meanings within the data [8]. The analysis process involved several iterative steps (Fig. 1).

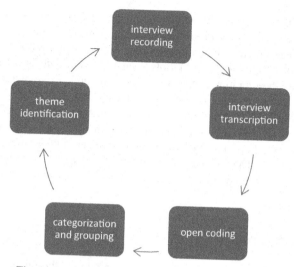

Fig. 1. Steps of the methodological approach followed.

Firstly, all interview recordings were transcribed verbatim to create a textual representation of the data. The transcripts were then carefully reviewed multiple times to gain familiarity with the content and immerse in the data. They are given in Appendix 2.

Next, an initial round of open coding was conducted. This involved systematically reading through the transcripts, line by line, and assigning descriptive codes to meaningful units of data. Codes were generated inductively, allowing patterns and themes to emerge from the data rather than imposing preconceived categories. This process continued until saturation was reached, and no new codes were identified. Following open coding, a process of categorizing and grouping related codes together took place. This involved organizing codes into broader categories or sub-themes based on their similarities and relationships. The researcher constantly compared codes and categories, refining and revising them as needed to ensure accurate representation of the data. Once the coding and categorization were completed, the researcher engaged in a process of theme identification. Themes were overarching patterns or concepts that emerged from the data and captured important aspects of participants' experiences or perspectives. Themes were identified through careful examination of the coded data, comparing similarities and differences, and ensuring they were grounded in the participants' narratives. This is given in Appendix 3 of the paper.

To enhance the rigor and trustworthiness of the analysis, the researcher maintained an audit trail, documenting the decision-making process, and providing justifications for coding choices and theme development. Regular discussions and debriefing sessions were conducted with the research team to ensure the reliability and validity of the identified themes.

4 Analysis

The analysis section delves into multifaceted aspects of remote work, offering a thorough examination based on the experiences and perspectives of the interviewees. It unfolds across three pivotal segments, each shedding light on distinct dimensions of remote work.

Overall, the interview process demonstrated that the interviewees exhibit a diverse range of attitudes towards remote working. While some express neutrality, others display a negative sentiment, citing fatigue after prolonged remote work. However, the majority holds a positive outlook, emphasizing benefits such as time saved on commuting and reduced fatigue. Notably, interviewee F underscores the importance of physical contact when creating something from scratch. Overall, the attitudes vary, reflecting the nuanced perspectives of the participants.

Several factors contribute to the transition to remote work. Company policies, pandemic-related restrictions, job nature, time-saving, and personal circumstances play pivotal roles. Some interviewees attribute remote work to company policies or pandemic-induced mandates. The nature of the job emerges as a critical factor, with certain roles better suited for remote work. Time-saving and convenience, particularly in combining tasks, also drive the shift. An interviewee even highlights the essential role remote work played in their recovery from health issues.

The analysis explores the plethora of technological tools employed during remote work, ranging from communication platforms like Slack and Zoom to project management tools such as Trello and Asana. The COVID-19 pandemic significantly impacted tool adoption, with new tools introduced post-pandemic. The majority, however, used these tools before the pandemic, indicating a pre-existing reliance on technology for remote collaboration. The study underscores the crucial role technology played in maintaining effective communication and collaboration during remote work.

The perceptions on work-life balance vary among interviewees. Some note no significant impact, while others highlight positive effects, such as time savings, better sleep, and increased personal time. However, concerns are raised regarding potential overworking, digital distractions, and difficulties in setting boundaries between work and personal life. The study emphasizes the nuanced impact of remote work on work-life balance, influenced by individual circumstances and preferences.

Participants note several differences between remote work and office work, primarily in the shift to online interactions and virtual meetings. While some express preferences for remote work, citing increased focus and comfort, others acknowledge challenges, such as distractions at home. The absence of direct contact and limited personal interaction with colleagues is highlighted as a significant difference, impacting work-related discussions.

Interviewees report various positive changes and gains from remote work. The elimination of geographical limitations, time and energy savings from commuting, increased flexibility, and enhanced productivity are commonly cited benefits. Remote work also enables individuals to work for companies abroad, fostering a global job market. The positive impact of technology on productivity, sleep quality, and job creation is emphasized.

Negative changes and losses associated with remote work include the lack of inter-action with colleagues for socialization, potential overworking, blurred boundaries between work and personal time, and challenges in building relationships. Concerns about digital distractions, constant accessibility to work notifications, and the impact on work-life balance are raised. The study acknowledges the individualized nature of these drawbacks, dependent on personality, mindset, and preferences.

Interviewees present varied perspectives on relationships with colleagues in the con-text of remote work. While some emphasize the availability of tools for communica-tion, others find building relationships challenging. The lack of direct contact is noted as a significant difference, impacting personal interactions. The study underscores the importance of technology in facilitating communication but acknowledges challenges in building trust and collaborating without physical interactions.

The majority of interviewees express increased productivity during remote work, citing reduced distractions and a more focused environment. Some report no signifi-cant change in productivity, attributing any variations to personal growth and experi-ence. None report feeling more productive in the office due to distractions and fatigue. The study emphasizes the positive impact of remote work on productivity, reflecting individual preferences and job roles.

The analysis reveals that remote work generally increases flexibility and autonomy for interviewees. They express more control over work schedules and decision-making. While some note no change in autonomy, others highlight increased independence, especially in solving problems. Greater flexibility in managing time and work sched-ules is widely appreciated, allowing individuals to align work with personal needs and preferences.

The study identifies communication challenges in remote work, including miscom-munication leading to conflicts, the misinterpretation of humour, and concerns about internet connectivity. Power cuts and the need for alternative internet access locations are noted, emphasizing the reliance on stable connectivity. The study acknowledges the var-ied nature of communication challenges, including difficulties in immediate assistance, collaboration with colleagues from different countries, and the impact of technology on relationships.

In conclusion, the analysis provides a nuanced understanding of remote work, cap-turing the diverse experiences and perspectives of the interviewees. It emphasizes the role of technology in facilitating communication, the impact on work-life balance, and the varied attitudes towards remote work. The study recognizes the individualized nature of both positive gains and negative challenges associated with remote work, shedding light on the complex dynamics of this evolving work paradigm.

5 Discussion

This section reflects on the findings presented in the previous section and discusses them more extensively while it also highlights the contribution of this work with regards to the existing literature.

5.1 Dimensions Shaping Remote Work

Interviewees attitudes towards telecommuting vary and there are numerous factors affecting it. The results show that age variations and the years of experience in working remotely, the size of the city they currently live in may be relevant to attitudes towards remote working. More specifically, based on the research findings, younger people tend to have more positive attitudes towards remote working because they are more accustomed with technology and can handle it in a better manner. Gen Z and Millennials have more positive attitude because technology is a familiar tool for them and they don't have trouble realizing its benefits and work with it. This finding is despite the challenges noted by existing literature on the impact of prolonged screen time to ther remote working experience [27]. On the contrary, older generations and particularly from the age of 30 + tend to view remote working more critically, pointing their need for interaction with colleagues and the isolation that follows.

Moreover, people with more years of working experience in this working style seem more likely to have developed a negative attitude to remote working in contrast to newcomers who experience the benefits of saving time and the comfort of their home. This is aligned with findings of existing literature noting that social interaction is an important for workers [3]. Thus, this conclusion indicates that prolonged remote working and lack of social interaction in the physical workspace is related to negative perceptions towards remote working.

Another key conclusion of this work is the relevance in the city of residence. Candidates who live in big cities or in a different city from the one where their company is located shape a positive attitude towards remote working. The first category saves a lot of time from commuting, which means more time for personal hobbies and family and the second one is offered more job opportunities even away from their hometown and experience the benefit of choice without being limited to a place. While Zamani & Pouloudi [30], demonstrates the relevance of proximity to virtual teams engaged in remote work, the findings of this research add to this body of knowledge by shedding light on the relevance of proximity of remote workers working in big cities or away from their hometown.

The findings of this study indicate that communication with colleagues and building stronger relationships even besides working hours is a subject to personality trait. It depends on the person's intentions and attitudes towards his/her colleagues and their desire to keep in touch with them beyond working hours. This finding emphasizes the subjective aspect of the matter. Consequently, based on our research the remote working style itself, even though it doesn't favor the building of stronger relationships due to its remote nature, does not actually affect the quality of human relationships. Based on the findings, technology has provided a variety of tools to humanity in order to work remotely and at the same time stay in touch more easily with one another and has offered numerous benefits in the communication and collaboration of the colleagues and people in general. The technological tools that facilitate remote work are neutral based on indicators in their everyday usage by humanity, and they are not responsible for their reasonable use, however people tend to develop attitudes towards this working style and towards technology, either positive or negative. So, it is a subjective issue whether the use of technology and remote working style has positively or negatively affected human relations and distance.

Technology has played a significant role in enabling easier communication and offering numerous benefits in terms of collaboration among colleagues and people in general. Technological tools provide various means of staying connected, such as video calls, instant messaging, and social media platforms. However, the usage of these tools ultimately depends on individuals and their choices. The technology itself is neutral and not responsible for how it is utilized by humanity. It is the attitudes and behaviors of individuals that shape their experiences and interactions with technology. Some individuals may develop positive attitudes towards remote work and technology, recognizing the advantages they offer in terms of communication and collaboration. On the other hand, some individuals may develop negative attitudes, possibly blaming technology for any emotional distance that may arise between colleagues and even family members. It is important for individuals to take responsibility for their own actions and choices in utilizing technology and maintaining relationships. In summary, the research findings suggest that the quality of communication and relationships among colleagues is primarily influenced by personal intentions, attitudes, and desires. Remote work, although not inherently conducive to building stronger relationships, does not directly impact the quality of human relationships. Technology provides various tools for communication and collaboration, but it is up to individuals to use them in a reasonable and responsible manner.

It's also important to note that the majority of candidates who use a great number of technological tools to perform their work remotely tend to have positive attitude towards this working style and rate the communication efficiency between colleagues with higher marks in contrast to the ones who use a few tools for communication and collaboration. By having a greater selection of tools, candidates are more likely to find options that suit their specific communication and collaboration needs. This leads to a more positive perception of remote work, as the tools effectively facilitate efficient and effective communication. As a result, these candidates rate the communication efficiency between colleagues with higher marks, reflecting their satisfaction with the remote work setup. In contrast, candidates who use only a few tools may experience limitations in their communication options and collaborative capabilities. They may rely on a single tool or method, which may not fully meet their needs or provide the same level of effectiveness as a more diverse set of tools. This limited access to communication channels and collaborative features may result in lower ratings for communication efficiency and a less positive attitude towards remote work. Overall, the availability and utilization of a greater number of technological tools in remote work can contribute to a more positive perception of the working style and higher ratings for communication efficiency among colleagues. It emphasizes the importance of providing a diverse and comprehensive toolkit for remote work to support effective communication and collaboration.

5.2 Research Implications

Our exploration into attitudes towards remote work and its influencing factors has significant implications for both organizations and individuals navigating the remote work landscape. These findings offer actionable insights, guiding businesses to tailor remote work strategies and foster inclusive environments. Furthermore, our research sheds light on the pivotal role of technology and individual attitudes in shaping communication dynamics within remote teams. These implications serve as a roadmap for effectively navigating the evolving world of remote work.

The research identifies key implications are summarized in Table 1.

Table 1. Research Implications across different dimensions of remote working.

Dimensions	Example
Age and Experience Influence Attitude	Younger individuals, particularly Gen Z and Millennials, exhibit more positive attitudes towards remote work due to their tech familiarity. In contrast, those aged 30 + may view remote work critically, emphasizing the importance of in-person interactions
City of Residence Matters	Candidates in big cities or residing away from their workplace show more positive attitudes towards remote work. This finding, absent in existing literature, emphasizes the influence of geographical location on remote work attitudes
Limited Impact of Gender and Company Variables	Gender, company type, size, and position show no substantial impact on remote work attitudes. This absence of findings in existing literature indicates the novelty and significance of our research
Subjectivity in Communication and Collaboration	Effective communication and relationship-building in remote settings depend on individual intentions and attitudes. The subjective nature of communication efficiency emphasizes the need for tailored strategies
Technological Tools Boost Communication Efficiency	A diverse set of technological tools positively correlates with higher ratings for communication efficiency. This underscores the importance of providing a comprehensive toolkit for remote work to support effective communication and collaboration

5.3   Practical Implications

Drawing from these findings, practical suggestions for organizations in Greece looking to facilitate remote work emerge. These are described in Table 2.

Table 2. Practical Implications across different dimensions of remote working.

Dimensions	Example
Implement a Hybrid Model	Incorporate a hybrid work model to balance flexibility with in-person collaboration, addressing diverse preferences
Increase Face-to-Face Meetings	Foster better understanding and reduce miscommunication by incorporating more frequent face-to-face meetings
Provide Newcomer Support	Ensure proper onboarding and support for newcomers to facilitate effective integration into the team
Conduct Morning Team Meetings	Regular morning meetings can enhance communication and alignment within the team
Hold Frequent Video Call Meetings	Regular video call meetings can reduce feelings of isolation and enhance interaction among team members
Dedicated Personnel for Colleague Connections	Assign personnel dedicated to fostering communication and collaboration among team members
Enhance Communication and Collaboration	Consider changes to prioritize and enhance communication and collaboration, potentially incorporating new tools and strategies
Improve Internet Connection	Invest in reliable internet infrastructure to minimize disruptions and ensure smooth remote work operations

These practical implications address challenges and enhance communication, collaboration, and productivity in the Greek business context. Additionally, promoting the use of various technological tools and allowing employees the freedom to choose their remote work days aligns with personality traits and interests, contributing to a holistic approach to remote work. The responsibility lies with organizations to foster effective communication and collaboration through diverse technological tools and frequent meetings, irrespective of physical distance.

6   Conclusion

The evolving dynamics of modern life and work necessitate innovative approaches to adaptation. Remote work, emerging as a transformative force, is poised to supplant traditional work models. Various definitions have been ascribed to this "new normal," converging on the notion of working from home aided by technology.

This research focuses on assessing the influence of IT tools on remote work within the Greek business landscape. Specifically, it delves into their impact on communication, collaboration, and perceived proximity among remote workers. Despite extensive studies on the repercussions of the COVID-19 pandemic on remote work, a discernible gap exists in scrutinizing the specific IT tools used by Greek remote workers and their role in bridging the gap between geographical and perceived proximity.

The literature review underscores the myriad benefits of remote work in the Greek context, encompassing cost savings, diminished commuting stress, and an improved work-life balance. It emphasizes the positive impacts on mental well-being, job satisfaction, and engagement. Concurrently, the review acknowledges managerial challenges related to communication and performance management in remote settings.

The study employs a dual-method approach, combining a thorough literature review with primary research via qualitative methods. Semi-structured interviews with 17 Greek remote workers provide nuanced insights, analysed through thematic analysis. Findings indicate varying attitudes toward remote work based on factors like age, remote work experience, and city of residence. Younger individuals and those with extensive remote work experience exhibit more positive attitudes. Moreover, participants residing in large cities or distant from their company express optimism due to time-saving benefits and increased job opportunities.

However, limitations include the exclusively Greek context and a modest sample size of 17 interviewees. Future research could broaden the scope to encompass other countries or larger geographical areas, such as Europe, exploring companies with multicultural employees facing greater geographical and cultural differences. This expansion would contribute to a more comprehensive understanding of remote work dynamics and IT tools' impact in diverse contexts.

Appendix

Interviewee Agenda.

Gender:
Age:
Years of working experience:
Name of Company currently working in:
Size of company:
Position/role:
Industry:

Open-ended questions.

1. Can you describe your experience with remote working? How long have you been working remotely, and what led you to pursue this work style?

2. How do you stay connected with your colleagues while working remotely?

3. Can you describe the tools or platforms you use for communication and collaboration? How do they enable you to work remotely?

4. Which of these were introduced before, during and after the pandemic?

5. In what ways, does technologies post challenges for work and family life?

6. Does the use of technology for conducting your work make you feel emancipated (i.e., set free from time, place and device restrictions in completing a task)? Why?

7. You are currently conducting your work remotely through the use of modern technologies/tools. How does that differ, if at all, from how you conduct your work when being physically at the office?

8.Reflecting on your everyday workday, what has changed for the better and for the worse in your work-life after remote working?

9.How much control or flexibility do you currently have on your overall work schedule? What and who affects this?

10.What is the impact of using technology on your everyday productivity? Are you equally productive as you would have been if you were working at your office's physical location?

11.How has remote working impacted your ability to work autonomously?

12.How is the use of these technologies in the context of your work influencing your work-life balance?

13.What do you identify as the biggest gains and the biggest losses? How has technology/tools used contributed to this?

14.How does remote working impact your ability to build relationships with colleagues.

15.How has technology/tools used contributed to this?

16.Can you describe a time when you faced a communication or collaboration challenge while working remotely?

17.How do you overall feel about the change of working style?

18.If you had the chance to change something in order to enhance your communication and collaboration with your colleagues what would that be?

19.On a scale of 1–10, how effective do you think the communication tools you use for remote working are?

References

1. Adams, A., Lugsden, E., Chase, J., Arber, S., Bond, S.: Skill-mix changes and work intensification in nursing. Work Employ Soc. 14(3), 541–555 (2000)
2. Alvarez-Torres, F.J., Schiuma, G.: Measuring the impact of remote working adaptation on employees' well-being during COVID-19: insights for innovation management environments (2022)
3. Anand, A.A., Acharya, S.N.: Employee engagement in a remote working scenario. Int. Res. J. Bus. Stud. 14(2), 119–127 (2021)
4. Archer-Brown, C., Marder, B., Calvard, T., Kowalski, T.: Hybrid social media: employees' use of a boundary-spanning technology. N. Technol. Work. Employ. 33(1), 74–93 (2018)
5. Bailyn, L.: Freeing work from the constraints of location and time. N. Technol. Work. Employ. 3(2), 143–152 (1988)
6. Barnes, A.: The construction of control: the physical environment and the development of resistance and accommodation within call centres. N. Technol. Work. Employ. 22(3), 246–259 (2007)
7. Bicakci, K., Uzunay, Y., Khan, M.: Towards zero trust: the design and implementation of a secure end-point device for remote working. In: 2021 International Conference on Information Security and Cryptology (ISCTURKEY), pp. 28–33 (2021)
8. Braun, V., Clarke, V.: Using thematic analysis in psychology. Qual. Res. Psychol. 3 (2006)
9. De Vincenzi, C. et al.: Consequences of COVID-19 on employees in remote working: challenges, risks and opportunities an evidence-based literature review. Int. J. Environ. Res. Public Health 19(18) (2022)

10. Driscoll, D.: Introduction to primary research: observations, surveys, and interviews. In: Lowe, C., Zemliansky, P. (Eds.), Writing spaces: Readings on writing, vol. 2, pp. 153–174 (2010)
11. Flores, M.F.: Understanding the challenges of remote working and its impact on workers. Int. J. Bus. Market. Manag. (IJBMM) 4(11), 40–44 (2019)
12. Gratton, C., Jones, I.: Research methods for sports studies (2nd ed.). Taylor & Francis (2010)
13. Hubbard, M., Bailey, M.J.: Mastering microsoft teams: end-user guide to practical usage, collaboration, and governance. Apress (2018)
14. Ilag, B.N.: Introduction: Microsoft Teams. In: Introducing Microsoft Teams (pp. 1–42), Apress (2018)
15. Ilag, B.N.: Managing and controlling Microsoft Teams. In: Understanding Microsoft Teams Administration (pp. 37–229), Apress (2020)
16. Ilag, B.N.: Tools and technology for effective remote work. Int. J. Comput. Appl. 174(21), 13–16 (2021)
17. Ilag, B.N.: Employee engagement in a remote working scenario. Int. Res. J. Bus. Stud. 14(2), 119–127 (2021)
18. Kathleen, S., Sven, S., Claudia, N.B., Frank, E.: Fulfilling remote collaboration needs for new work. Procedia Comput. Sci. 191, 168–175 (2021)
19. Kiesewetter, J., et al.: Implementing remote collaboration in a virtual patient platform: usability study. JMIR Med. Educ. 8(3), e24306 (2022)
20. Kłopotek, M.: The advantages and disadvantages of remote working from the perspective of young employees. Organizacja i Zarządzanie: kwartalnik naukowy (2017)
21. Lal, B., Dwivedi, Y.K., Haag, M.: Working from home during Covid-19: doing and managing technology-enabled social interaction with colleagues at a distance. Inf. Syst. Front. (2021)
22. Laker, B., Godley, W., Patel, C., Cobb, D.: How to monitor remote workers—ethically. MIT Sloan Management Review (2020)
23. Manokha, I.: Covid-19: teleworking, surveillance and 24/7 work. some reflexions on the expected growth of remote work after the pandemic. Political Anthropol. Res. Int. Soc. Sci. (PARISS) 1(2), 273–287 (2020)
24. Mokhtar Azizi, Z., Cochrane, J., Thurairajah, N., Mokhtar Azizi, N.S.: Remote working in construction: assessing the affordance of digitization (2022)
25. Papadaki, M., Karamitsos, I., Themistocleous, M.: Covid-19 Digital test certificates and blockchain. J. Enterp. Inf. Manag. 34, 993–1003 (2021). https://www.researchgate.net/public ation/353272635_ViewpointCovid-19_digital_test_certificates_and_blockchain
26. Phillips, S.: Working through the pandemic: accelerating the transition to remote working. Bus. Inf. Rev. 37(3), 129–134 (2020)
27. Rañeses, M.S., et al.: Investigating the impact of remote working on employee productivity and work-life balance: a study on the business consultancy industry in Dubai, UAE. Int. J. Bus. Administr. Stud. 8(2), 63–81 (2022). https://doi.org/10.20469/ijbas.8.10002-2
28. Rodman, A.: The impact of remote workers on crisis, risk, and business continuity management. J. Bus. Contin. Emer. Plan. 15(3), 214–224 (2022)
29. Sullivan, C.: Remote Working and Work-Life Balance, pp. 275–290. In Work and Quality of Life, Springer, Dordrecht (2012)
30. Zamani, E.D., Pouloudi, N.: Shared mental models and perceived proximity: a comparative case study (2022)

Digital Leadership in Cross-Cultural Organizations: Insights from Swiss Healthcare Companies

Mahdieh Darvish$^{(\boxtimes)}$, Luca Laule, Laurine Pottier, and Markus Bick

ESCP Business School Berlin, Heubnerweg 8-10, 14059 Berlin, Germany
mdarvish@escp.eu

Abstract. The pervasive influence of digitalization is undeniable in the contemporary landscape. However, available tools and resources are often underutilized by many leaders. Accordingly, an effective digital leadership is vital for complex organizations in today's interconnected world. Despite considerable attention to digital leadership, a noticeable research gap exists concerning its application in cross-cultural settings. Therefore, this study examines the role of digital leadership in Swiss healthcare companies, which serve as prime examples of global players grappling with cross-cultural complexities. The healthcare sector strives for efficiency through digital technologies and diverse global team management. Moreover, Switzerland, known for its advanced development, regulatory framework and multinational healthcare environment, offers an ideal setting for the research in this context. Conducting semi-structured expert interviews, this study provides insights to better understand the digital leadership strategies and challenges. Our findings highlight that effective digital leadership extends beyond technical skills, encompassing communication, collaboration, and cultural competencies.

Keywords: Digital Leadership · Healthcare Industry · Cross-Cultural Organizations

1 Introduction

In today's increasingly interconnected world, where digital transformation and cross-cultural collaboration are common, the importance of effective digital leadership in complex organizations has never been more vital [1]. *Digital Leadership* is defined as "an ethical and agile mindset that quickly responds to changes and learns from them, fostering a trust-based culture that values people and its diversity, coaching them to collaborate and thrive in a digital scenario" [2]. In the context of the Swiss healthcare sector, digital leadership has emerged as an exponentially vital facet of navigating cross-cultural landscapes, driven by organizations' efforts to harness digital technologies for gaining a competitive edge and proficiently overseeing diverse global teams [3, 4]. Switzerland stands as a preeminent global hub in the healthcare sector. Consequently, the examination of digital leadership in cross-cultural contexts within Swiss healthcare corporations offers invaluable insights that hold significance for both academic research and practical application [5].

The original version of the chapter has been revised. The initial publication had a typing error in the name of the second author. A correction to this chapter can be found at
https://doi.org/10.1007/978-3-031-56481-9_23

M. Papadaki et al. (Eds.): EMCIS 2023, LNBIP 502, pp. 280–291, 2024.
https://doi.org/10.1007/978-3-031-56481-9_19

The burgeoning field of digital leadership has garnered substantial attention from researchers and practitioners, leading to a wealth of global research endeavors. These investigations encompass a wide array of subjects, ranging from diversity, equity, and inclusion to communication strategies, trust establishment, and cultural intelligence, all within the context of digital leadership. Various research efforts have highlighted the pivotal role of understanding the effective digital leadership [2, 6]. Studies have demonstrated that in organizations engaged in cross-cultural operations, adept digital leadership is a requisite for the successful execution of digital transformations and the preservation of a competitive edge in the global marketplace [2, 6]. Furthermore, leaders who are capable of navigating the intricacies of cross-cultural environments while harnessing digital technologies are more likely to accomplish prosperous digital transformations [7].

However, there is a noticeable dearth of research specifically dedicated to digital leadership within healthcare sector. Given Switzerland's advanced development and regulatory environment as well as multinational healthcare landscape, it presents an ideal setting for addressing this research gap. Therefore, this study examines digital leadership practices in Swiss healthcare corporations operating within cross-cultural contexts. Through qualitative interviews with nine employees from two prominent Swiss healthcare firms, we aim to shed light on the unique challenges faced by these organizations and the innovative solutions they have devised. Moreover, this study provides valuable insights for similar organizations operating in cross-cultural digital landscapes and offer guidance to managers navigating these complexities by answering the following research questions (RQs):

RQ1: Which digital leadership strategies promote effective collaboration within healthcare corporations?
RQ2: What are the primary challenges encountered by healthcare corporations when implementing digital leadership strategies?

The main objective of our research is to gain a deeper understanding of the perceptions of digital leadership in healthcare corporations. Therefore we apply a qualitative research approach through an explorative literature review. Semi-structured expert interviews are then conducted to answer the afore mentioned RQs within a real-life context. Nine interviews are conducted with employees from two prominent healthcare corporations in Switzerland, comprising top-level executives, directors, and senior managers. Our study contributes to the body of knowledge on digital leadership in healthcare by exploring the driving reasons to accelerate different strategies as well as investigating the significant limitations and challenges that multinational healthcare organizations face in this context. Moreover, our results indicate that effective digital leadership necessitates not only digital competencies, such as heightened digital awareness, but also cultural competencies, including a greater recognition of cultural distinctions. Furthermore, our research pinpointed notable challenges faced by digital leaders, such as managing the intricacies of establishing trust in a digital environment. Practical insights from our study hold substantial value for training and recruiting leaders to possess crucial competencies for successful digital leadership.

The remainder of this paper is organized as follows: The literature review provides detailed background information about digital leadership, its implications, especially in cross-cultural contexts, as well as the characteristics of the Swiss healthcare environment. Section 3, elaborates on the methodological approach. Finally, the results and discussion section present the major findings and the concluding section addresses the limitations and future research.

2   Literature Review

In the context of ongoing digital transformation, leaders face the need to redefine their roles, with a key challenge being the effective integration of digital tools into leadership practices. In international corporations, effective leadership is essential, particularly for navigating complex intercultural situations and leveraging digital technologies for improved business outcomes. Studies identify three core dimensions of digital leadership: strategic, operational, and people-oriented. Operational leadership involves executing digital transformation plans and ensuring efficient use of digital resources [8]. Strategic leadership focuses on defining the vision, direction, and objectives of digital transformation. People-oriented leadership encompasses talent management, fostering a culture supporting digital transformation, promoting innovation, motivating teams, and facilitating collaboration. To facilitate communication, collaboration, and decision-making in multinational organizations, digital leaders must understand the cultural, political, and legal contexts they operate within [9]. Success in digital transformation relies on bridging cultural gaps and promoting intercultural communication and collaboration [10]. More recently, the COVID-19 pandemic has significantly accelerated digital transformation in multinational enterprises, with remote work and digital collaboration evolving rapidly [11]. Accordingly, digital leaders must adapt to these changes and use digital technology to ensure business continuity [12].

2.1   Digital Leadership in (Swiss) Cross-Cultural Landscape

In Swiss organizations, cross-cultural workplaces are commonplace due to the country's diverse population and its long history of welcoming immigrants [13]. Switzerland is well-regarded for its "neutral" culture, characterized by direct and explicit communication, with emotions seldom displayed in the workplace, as per Erin Meyer (2014)'s Cultural Map [14]. Additionally, according to Hofstede (2001)'s cultural dimensions theory, multinational corporations operating within Swiss culture may be influenced by its tendencies toward individualism, masculinity, and uncertainty avoidance [15]. Understanding and adapting to these cultural nuances is crucial for effective cross-cultural communication [16]. Recognizing and accommodating cultural differences is imperative for facilitating successful cross-cultural communication. Moreover, cultural miscommunication and language barriers can lead to reduced productivity [17]. To align with the direct and explicit communication style prevalent in neutral cultures, a digital leader in a Swiss organization should tailor their communication approach [14, 15]. Furthermore, they must be cognizant of how cultural values such as independence or ambiguity may influence their decision-making and leadership style [15].

The utilization of technology to foster cross-cultural communication and collaboration is a salient topic in the literature. Digital tools can serve as bridges to span physical and cultural divides, facilitating more efficient communication and collaboration transcending cultural boundaries [18].

Digital leaders frequently face numerous challenges when endeavoring to implement digital leadership strategies. These challenges encompass issues such as a deficiency in digital literacy among leaders [19], resistance to change [20], the intricacy and uncertainty associated with the adoption of digital technologies [21], as well as insufficient resources and infrastructure [22].

2.2 Digital Leadership in (Swiss) Healthcare Corporations

The significance of digitalization within the healthcare sector remains unquestionable [23]. Healthcare leaders can leverage digital technologies to gather and analyze vast quantities of patient data, enabling informed decision-making and the tailoring of treatments to individual patient needs and preferences. The customization of therapies based on each patient's medical history and lifestyle can substantially enhance patient experiences and outcomes [22].

Healthcare executives can enhance the coordination of care between specialists and doctors by utilizing digital platforms to share patient information and treatment plans. This can potentially lead to improved customer outcomes and more effective treatment programs [24]. Additionally, the adoption of digitalization can streamline administrative tasks and reduce errors, thereby enhancing operational efficiency and yielding cost savings [25]. Seseli et al. (2023) emphasized that digitalization can mitigate errors by eliminating manual data entry and the potential for transcription mistakes [26]. Furthermore, a notable advantage of digitalization is its capacity to enhance customer engagement [27].

Switzerland presents promising prospects for the advancement of digital leadership within the intercultural contexts of healthcare corporations, given its well-established healthcare system [28]. The country's highly skilled workforce holds great potential for digital leadership within the Swiss healthcare sector. Combining private and public sectors, Switzerland offers an average gross monthly wage of 6,500 Swiss Francs for individuals working in the field of human health [29].

Switzerland's multicultural character presents a valuable opportunity for the application of digital leadership within the healthcare sector [30]. Additionally, Switzerland boasts a cutting-edge healthcare system, globally recognized for its quality and efficiency [31]. As a result, it serves as an optimal environment for contemporary healthcare organizations to cultivate and enhance their digital solutions. Another pivotal advantage for the healthcare industry in Switzerland is its good regulatory framework [28].

3 Methodology

The main objective of this study is to investigate the significance of digital leadership within healthcare corporations in Switzerland, with a particular emphasis on identifying existing strategies and challenges, especially in the context of cross-cultural diversity.

To achieve this, we adopted an explorative approach consisting of a brief review of related works along with semi-structured expert interviews. Given the evolving and multifaceted nature of digital leadership within healthcare organizations, employing an exploratory research approach is deemed appropriate as it allows for an in-depth investigation and understanding of the complex factors and dynamics in this context.

Moreover, semi-structured interviews were chosen as the primary method for data collection due to their flexibility and capacity to delve into specific areas of interest while maintaining a consistent framework across all interviews [32]. The data was collected from leaders in prominent healthcare corporations in Switzerland, with a primary focus on uncovering the strategies and challenges associated with the implementation of digital leadership within a multicultural framework, as well as its impact on communication, collaboration, diversity, equity, and inclusion.

3.1 Data Collection

We applied a purposeful sampling strategy to recruit interview partners and ensuring relevance and diversity within the sample (Table 1), including the top-level executives, directors, and senior managers, all possessing expertise in the domains of digital leadership, cross-cultural communication, and collaboration.

Table 1. Overview: Interview Partners.

Interviewee	Position	Type of Company
IP 1	President, Personal Care and Aroma	Human Nutrition & Health Corporation
IP 2	Chief Group Regulatory, Hazard Communication & Information Office	Multinational healthcare company (Pharmaceuticals and Diagnostics)
IP 3	Global Head of Regulatory Affairs, Health, Nutrition and Care	Human Nutrition & Health Corporation
IP 4	Head of Global Regulatory, Personal Care and Aroma	Human Nutrition & Health Corporation
IP 5	Analytical Project Leader (Senior Principal Scientist)	Multinational healthcare company (Pharmaceuticals and Diagnostics)
IP 6	Product Management Director	Human Nutrition & Health Corporation
IP 7	Program Director Business Service Transformation	Human Nutrition & Health Corporation
IP 8	Global Head of Regulatory Affairs	Human Nutrition & Health Corporation
IP 9	Global Regulatory Affairs	Human Nutrition & Health Corporation

In total, nine semi-structured interviews were conducted online, in English, lasting 20–60 min, during March and April 2023.

3.2 Data Analysis

All interview transcripts were analyzed following the qualitative content analysis approach outlined by Mayring (2000). A two-fold approach was employed. First, a deductive approach was utilized to categorize the data. The coding categories were initially defined in accordance with the method proposed by Mayring (2014) to ensure a thorough and reliable content analysis. We adopted the three fundamental components of the Information Systems (IS) discipline, namely (1) organization, (2) management, and (3) information technology [33], as categories to provide a comprehensive overview of the data collected in this study. In second step, we applied inductive coding for further classification via the frequency count of relevant words. Doing this, the most relevant concepts were identified as two further coding categories: (4) relevant digital leadership strategies and (2) key challenges. An overview of the categories and representative quotations are provided in Table 2.

Table 2. Coding Categories.

Category	Definition
1. Organization	People, structure, business processes, politics, culture, and customers make up the organization in the context of IS [21].
2. Management	Management is in responsibility of overseeing performance and managing expenditures. Providing direction and vision as well as managing stakeholders are additional managerial responsibilities [34].
3. Information Technology	The IS infrastructure is made up of hardware, software, data management, networking, and ICT technology, along with the necessary specialists [21].
4. Relevant Digital Leadership Strategies	*"Digital Leadership is a leadership model that executes digital transformation within the organization can digitize the work environment and learning cultures of organizations."* [6]
5. Key Challenges	Prominent challenges to overcome

4 Findings and Discussion

In alignment with the literature, experts emphasized that the organization's landscape is a key factor while implementing digital leadership strategies. Considering the health industry, COVID-19 was the prominent example in this regard. The experts highlighted how the recent pandemic contributed to the digital transformation in their organizations based on the availability of the resources. IP4 explained that the digital resources were not necessarily used, known or even available to the employees in the pre-pandemic era. IP1 highlighted that COVID-19 accelerated the use of digital tools and, therefore, the digital transformation in healthcare.

However, IP2 commented on his company being more advanced in digital transformation even before the pandemic. But indeed, the whole industry had to adapt faster to new ways of communicating, collaborating, and thus, an accelerated digital transformation that happened due to an imposed situation.

4.1 Comparison of the two Companies

Considering the three initial coding categories, the main similarities and differences of the two companies regarding the implementation of digital leadership strategies are identified as illustrated in Fig. 1.

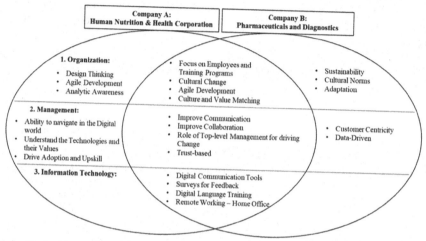

Fig. 1. Similarities and Differences of the Companies regarding the implementation of digital leadership strategies.

Organization. All experts emphasized that the success of digital leadership initiatives within healthcare companies often hinges on a strong focus on employees and training programs, fostering a culture of continuous learning and adaptation. Cultural change in this context becomes imperative in healthcare organizations that embrace digital transformation, as it necessitates a shift towards more agile and innovative practices. Moreover, experts highlighted that the adoption of agile developments in healthcare enhances organizational responsiveness to evolving markets or even black swan events such as COVID-19.

In the context of digital leadership, culture and value matching between leaders and employees can serve as a catalyst for fostering a collaborative and innovative work environment. Effective training programs, aligned with the cultural change agenda, play a pivotal role in nurturing the skills and mindsets required for successful implementation of digital leadership strategies in Swiss healthcare companies.

Management. Experts from both companies emphasized the important role of improved communication and collaboration among employees, but also with the management for fostering a more synergic work environment while going through digital transformation. Trust-based relationships among team members are essential for improving collaboration, as they create an atmosphere of openness and mutual support, facilitating effective problem-solving and decision-making. However, the role of top-level management in driving change is pivotal, as they provide the vision, leadership, and strategic direction necessary for implementing initiatives to improve communication and collaboration throughout the organization. Therefore, experts highlighted that the commitment of the top-level management to building trust-based relationships among employees can serve as a catalyst in this context.

Information Technology. Experts highlighted that digital communication tools in workplaces have revolutionized how organizations connect and collaborate, enabling seamless interactions among remote teams and contributing to the growing trend of remote working or home office arrangements. As remote working and home office setups become prevalent the effective implementation of digital communication tools and corresponding training programs plays a critical role in maintaining productivity, cohesion, and employee engagement within organizations.

4.2 Relevant Digital Leadership Strategies and Key Challenges

Analyzing the term frequencies, two additional categories, including three relevant subcategories each, are identified as follows: Relevant Digital Leadership Strategies: (1) Increasing acknowledgement of cultural differences, (2) Improving cross-cultural communication and collaboration, and (3) Increasing digital awareness, and Key Challenges: (1) Time zones differences around the world, (2) Training employees regarding digital technologies, and (3) Earning Trust Digitally.

Additionally, supportive representative quotations for each subcategory are provided in Table 3.

Table 3. Representative supportive data for each subcategory.

Category	Examples
Relevant Digital Leadership Strategies	
(1) Increasing acknowledgement of cultural differences	"I think it goes both to culture and knowing your people." (IP3)
(2) Improving cross-cultural communication and collaboration	"Communicate with them with certain tools like we started using on workshops." (IP4)
(3) Increasing digital awareness	"One is the ability to understand the technology which is here." (IP7)
Key Challenges	
(1) Time zones differences around the world	"The main problem is the time zone." (IP2)
(2) Training employees regarding digital technologies	"We always have some internal training." (IP8)
(3) Earning Trust Digitally	"It takes more time to build trust digitally." (IP3)

Relevant Digital Leadership Strategies. All experts highlighted the relevant digital leadership strategies while mostly discussing two strategies; increasing acknowledgement of culture differences as well as increasing digital awareness. One expert, president of personal care and aroma division (IP1), elaborated: *"There are the introverted cultures, as well as the extroverted cultures, you have the cultures that respectfully don't ever challenge anyone, and you have the cultures that it's a part of them to always challenge for. You can't approach or lead everyone the same way"*. This is in alignment with the Meyer's culture map (2014) which highlights the cultural differences such as American explicit and French implicit communication as well as American egalitarian and French more hierarchical way of leading [14].

However, another expert, chief group regulatory, hazard communication & information office, emphasized more on the organization's culture: *"The answer is the culture is unified. It's not too complicated anymore. You know, of course, we all have different cultures, but in business, the culture is very unified"*. (IP2)

Moreover, all experts highlighted that increasing digital awareness and digital competencies of the individuals is always relevant regardless of the culture. One expert, director digital change and communication, elaborated that there are three abilities in order to be able to be a digital leader: (1) the ability and willingness to understand the technologies, (2) the ability to understand and picture what do technologies bring to someone's self, and (3) the ability to drive adoption and upskilling. (IP3)

In summary, these findings are in alignment with the literature as digital leaders, in addition to digital competencies, also need to have a deep understanding of the unique cultural, political, and regulatory environments in which they operate to facilitate effective communication, coordination, and decision-making [9]. Hence, it is evident that digital leadership transcends mere technological proficiency. In addition to technology, a digital leader must also have a profound understanding of their team, including their cultural backgrounds. This understanding is crucial for enabling effective communication and collaboration that respects diverse cultures. Furthermore, it is strongly advised for digital leaders to maintain an open-minded attitude and be prepared to adapt when necessary.

Key Challenges. Considering the key challenges, experts highlighted different time zones as the main problem that employees face with seemingly no solution. While employees collaborate and work together from very different locations it is not possible to manage time in a satisfactory manner for everyone within reasonable working hours as in any case, one of the locations will be compromised. One respondent, product management director, underlined:

"Well, that is the time zone differences are a big challenge. And then especially people in Asia suffer on that perspective quite a bit". (IP6)

Moreover, training employees properly by keeping up with the implementation of technologies is another main challenge addressed during the interviews. Experts highlighted the necessity of ongoing trainings as a challenge which has to be managed respectively.

Furthermore, experts addressed the issue of building trust digitally which seems to be a lot harder than if you are physically present in a company. All respondent have at least one team member that they have never met due to their different location. In addition to the complexity of earning trust in digital settings, there is the difficulty of understanding how to gain the trust of a certain individual knowing that culture impact this individual's way of earning trust. Two experts, Analytical Project Leader and Global Regulatory Affairs, elaborated: *"Their manager is moved somewhere else and then they get a new one from Europe. So, you have to make your space, you have to build that trust".* (IP9)

"The challenge was basically to convince people at the outset that this was useful and that just because we didn't see each other around a table, we couldn't communicate or work together. That was the first thing. That's something that's culturally linked on trust as well. And this confidence that people work there, even when they are at home. We had to earn it first because it was not acquired." (IP5)

Indeed, the key challenges identified in the literature can be categorized as: (1) lack of digital literacy among leaders, (2) resistance to change, (3) uncertainty and complexity of digital technologies, (4) inadequate resources and infrastructure, and (5) lack of alignment between digital and organizational strategies. Nonetheless, the findings from this study offer insight into the pragmatic challenges encountered in the routine work environment. For instance, the disparity in time zones presents an ongoing complication, impeding seamless collaboration and daily communication, with no readily apparent solutions. Furthermore, the effective training of employees, while keeping pace with the continuous evolution of technology, presents a formidable challenge necessitating meticulousness, time investment, and the engagement of skilled professionals. Finally, it is worth noting that establishing digital trust is contingent upon an awareness of the specific trust dynamics inherent to different cultures, coupled with proficiency in the mastery of digital tools.

In a nutshell, the relevant digital leadership strategies as well as key challenges in this context can be categorized as technology related and human related. In addition to digital competencies, cross cultural but also personal communicational competencies need to be fostered in order to benefit from digital leadership initiatives, and hence, accelerate the digital transformation in an organization.

5   Conclusion, Limitations and Future Research

The findings of this research extend the research on digital leadership in several ways as it set out to shed light on digital leadership strategies as well as challenges in cross-cultural companies, in the case of Swiss healthcare corporations as prime examples in this context. First, the conducted literature review provided an overview of the research in the context of digital leadership, both in general as well as in cross-cultural healthcare companies. Furthermore, nine expert interviews were conducted, which provided insights into better understanding the implementation strategies for digital leadership as well as challenges in routine work environments from the practitioners view.

In summary, three strategies and three main challenges have been identified as follows: (1) increasing acknowledgement of cultural differences, (2) improving cross-cultural communication and collaboration, and (3) increasing digital awareness, as well as (1) Time zones differences around the world, (2) training employees regarding digital technologies, and (3) earning trust digitally.

Moreover, this paper provides new insights for companies to successfully develop digital leadership strategies and anticipate as well as overcome the challenges in this regard. This contributes new knowledge and important implications for organizations and provides a useful guideline for managers to evaluate whether the digital leadership is suitable from the company's perspective.

In the realm of cross-cultural organizations, maintaining a keen awareness of technological advancements is crucial, especially due to the high demand for operational efficiency. Additionally, an essential attribute of effective digital leadership lies in the ability to empathize with others and understand their perspectives. Furthermore, it is imperative that organizations embrace the consideration of time zone disparities as a standard practice. This approach underscores the importance striving for the highest possible level of inclusion.

This research, like research in general has certain limitations. First, the research was limited to nine interviews with experts in multinational healthcare companies. Future empirical evidence from employees, and organizations implementing digital leadership strategies would allow more generalization. Furthermore, building upon the results provided herein, future research may narrow down each strategy and its impacts on the specific outcomes to develop deeper insights into potentials and challenges. Finally, considering the exploratory nature of this study, in-depth surveys and analyses can be conducted to confirm the findings.

References

1. Westerman, G., Bonnet, D., McAfee, A.: Leading digital: turning technology into business transformation. Harvard Business Press (2014)
2. Tigre, F.B., Curado, C., Henriques, P.L.: Digital leadership: a bibliometric analysis. J. Leadersh. Organ. Stud. **30**, 40–70 (2023)
3. Brynjolfsson, E., McAfee, A.: The second machine age: Work, progress, and prosperity in a time of brilliant technologies. WW Norton & Company (2014)
4. Schein, E.H.: Organizational culture and leadership. John Wiley & Sons (2010)
5. AG, BAK Economics, Grass, M., Fry, S., Vaterlaus, S.: The importance of the pharmaceutical industry for Switzerland (2017)
6. Sağbaş, M., Erdoğan, F.A.: Digital leadership: a systematic conceptual literature review. İstanbul Kent Üniversitesi İnsan ve Toplum Bilimleri Dergisi **3**, 17–35 (2022)
7. Martínez-Caro, E., Cegarra-Navarro, J.G., Alfonso-Ruiz, F.J.: Digital technologies and firm performance: The role of digital organisational culture. Technol. Forecast. Soc. Chang. **154**, 119962 (2020)
8. Abbu, H., Mugge, P., Gudergan, G., Kwiatkowski, A.: Digital leadership-character and competency differentiates digitally mature organizations. In: 2020 IEEE International Conference on Engineering, Technology and Innovation (ICE/ITMC), pp. 1–9. IEEE (2020)
9. Kayworth, T., Leidner, D.: The global virtual manager: a prescription for success. Eur. Manag. J. **18**, 183–194 (2000)
10. Gierlich-Joas, M., Hess, T., Neuburger, R.: More self-organization, more control—or even both? Inverse transparency as a digital leadership concept. Bus. Res. **13**, 921–947 (2020)
11. Soto-Acosta, P.: COVID-19 pandemic: Shifting digital transformation to a high-speed gear. Inf. Syst. Manag. **37**, 260–266 (2020)
12. Sheninger, E.: Digital leadership: changing paradigms for changing times. Corwin Press (2019)

13. Ravasi, C., Salamin, X., Davoine, E.: Cross-cultural adjustment of skilled migrants in a multicultural and multilingual environment: an explorative study of foreign employees and their spouses in the Swiss context. Int. J. Human Resour. Manag. **26**, 1335–1359 (2015)
14. Meyer, E.: The culture map: breaking through the invisible boundaries of global business. Public Affairs (2014)
15. Hofstede, G.: Culture's recent consequences: using dimension scores in theory and research. Int. J. Cross Cult. Manag. **1**, 11–17 (2001)
16. Gudykunst, W.B., Ting-Toomey, S., Chua, E.: Culture and interpersonal communication. Sage Publications, Inc (1988)
17. Taras, V., Rowney, J., Steel, P.: Half a century of measuring culture: review of approaches, challenges, and limitations based on the analysis of 121 instruments for quantifying culture. J. Int. Manag. **15**, 357–373 (2009)
18. Gertler, M.S.: "Being there": proximity, organization, and culture in the development and adoption of advanced manufacturing technologies. Econ. Geogr. **71**, 1–26 (1995)
19. Machin-Mastromatteo, J.D.: Information and digital literacy initiatives. SAGE Publications Sage UK: London, England. Inf. Develop. **37** (2021)
20. Scholkmann, A.B.: Resistance to (digital) change: Individual, systemic and learning-related perspectives. Digital Transf. Learn. Organ. 219–236 (2021)
21. Hess, T., Matt, C., Benlian, A., Wiesböck, F.: Options for formulating a digital transformation strategy. MIS Quart. Executive **15** (2016)
22. Schwertner, K.: Digital transformation of business. Trakia J. Sci. **15**, 388–393 (2017)
23. Carboni, C., Wehrens, R., van der Veen, R., de Bont, A.: Conceptualizing the digitalization of healthcare work: a metaphor-based Critical Interpretive Synthesis. Soc. Sci. Med. **292**, 114572 (2022)
24. Groves, P., Kayyali, B., Knott, D., van Kuiken, S.: The'big data'revolution in healthcare: accelerating value and innovation (2016)
25. Gobble, M.M.: Digital strategy and digital transformation. Res. Technol. Manag. **61**, 66–71 (2018)
26. Seseli, E.M.I., Risakotta, K.A., Bawono, A.: The role of accounting digitization in entrepreneurial success in west java: quantitative study of efficiency, accuracy, cost reduction, customer satisfaction, and data security. ES Account. Finan. **1**, 82–94 (2023)
27. Kushwaha, P.: Digitalization & Customer Engagement: Challenges, Opportunities & Future Prospects. International Bulletin of Management and Economics, VII, 8–15 (2017)
28. Pietro, C.: World Health Organization: Switzerland: health system review, 1817–6127 (2015)
29. Federal Statistical Office: Section Wages and Working Conditions. https://www.bfs.admin.ch/bfs/en/home/statistics/work-income/wages-incomeemployment-labour-costs/wage-levels-switzerland.html
30. Windisch, U.: Beyond multiculturalism: Identity, intercultural communication, and political culture—The case of Switzerland. In: Language, Nation and State: Identity Politics in a Multilingual Age, pp. 161–184. Springer (2004). https://doi.org/10.1057/9781403982452_7
31. Daley, C., Gubb, J., Clarke, E., Bidgood, E.: Healthcare Systems: Switzerland. Civitas Health Unit, London (2012)
32. Cachia, M., Millward, L.: The telephone medium and semi-structured interviews: a complementary fit. Qual. Res. Organ. Manag.: Int. J. **6**, 265–277 (2011)
33. Laudon, K., Laudon, J.: Management Information Systems: International Edition, 11/E. Citeseer (2009)
34. Seufert, S., Meier, C.: Digitale Transformation: Vom Blended Learning zum digitalisierten Leistungsprozess ‚Lehren und Lernen'24. Jahrestagung der Gesellschaft für Medien in der Wissenschaft (GMW) 29, 2016 (2016)

Smart Cities

Designing Services for an ICT Platform to Support City Learning for Developing Smart Cities

Pradipta Banerjee[✉] and Sobah Abbas Petersen

Department of Computer Science, Norwegian University of Science and Technology, 7034 Trondheim, Norway
{pradipta.banerjee,sap}@ntnu.no

Abstract. Cities experiencing rapid urbanisation and technological innovations face multifaceted challenges in demographics, economics, society, and the environment. To address these challenges and ensure sustainable development, the concept of "digital governments" has emerged, leveraging Information and Communication Technologies (ICTs). However, these technocratic approaches lack a citizen-centric focus, potentially widening the digital divide among city residents. Consequently, there has been a shift from "digital government" to "digital governance," emphasising collaboration among key stakeholders and utilisation of ICT to promote public value. The potential of digital governance lies in supporting city learning, wherein a city learns as an innovation ecosystem for its sustainable citizen-centric development and innovation. There is a lack of focus on how ICT can support city learning as an innovation ecosystem. Based on the Design Science Research Method, this study describes our research-in-progress in designing and developing a suitable ICT platform design, which focuses on citizen engagement and participation to support city learning for transforming cities towards sustainable citizen-centric smart cities. Using service blueprints, we illustrate the services that need to be supported by an ICT platform to support city learning.

Keywords: ICT Platform Design · City Learning · Innovation Ecosystem · Digital Governance · Smart Cities

1 Introduction

Cities have been crucial for the economic growth of nations and the world in general. By the year 2050, it is expected that 600 cities in the world will be responsible for 60% of the global GDP [1]. All cities in the world collectively constitute only 2% of the planet's land space, where around 55% of the world's population resides. By 2050, 68% of the world's population is expected to reside in cities [2]. These facts do not only reveal the growing importance of cities, but also highlight the need for cities to be capable of continuously dealing with evolving demographic, economic, social and environmental challenges.

M. Papadaki et al. (Eds.): EMCIS 2023, LNBIP 502, pp. 295–308, 2024.
https://doi.org/10.1007/978-3-031-56481-9_20

Cities comprise diverse entities such as citizens, organisations, the city administrator (representing the local city government), service systems and physical infrastructure [3]. The functioning of cities relies on interconnected social, ecological and technological systems, with various service systems catering to citizens' needs in areas such as water, food, health, education, recreation, electricity, communication and transport [4]. The United Nations' Sustainable Development Goal (SDG) 11 aims to achieve "sustainable cities and communities" that prioritise people's well-being [5]. It advocates for utilising Information and Communication Technology (ICT) and diverse approaches to enhance the overall quality of life, urban functionality and competitiveness. SDG 11 also highlights the importance of considering economic, social, environmental and cultural aspects to meet the needs of both current and future generations. However, until now, the focus of city development efforts has mostly been on leveraging advanced technologies to enhance urban functionality, often overlooking the social and environmental dimensions of urban development [6, 7].

City governments are working on improving their efficiency by providing citizens with better services and making cities smarter using ICT that includes Big data analytics, sensor technologies and IoT, machine learning and cognitive computing. Such efforts have led to the emergence of the concept of "digital government" that aims to make cities "smarter" with the help of new technologies to ensure economic, social and environmental sustainability [8]. However, a top-down implementation of technologies can adversely increase the digital divide among the citizens of a city, and the mere use of digital technologies by city governments may not always be optimal. Such technocratic approaches to city development lack a citizen-centric focus [8].

It is crucial to emphasise collaboration among various stakeholders, including citizens, to ensure the economic, social and environmental dimensions for sustainable city development [6]. These concepts highlight the significance of continuous learning about local contextual challenges and opportunities through interactions and feedback for designing interventions that can yield desired results [9]. These observations have revealed the need for a shift in focus from the "digital government" paradigm to "digital governance" [10]. Digital governance is utilising ICTs to promote public value through collaborations among key stakeholders in the city [11]. It is about creating public value by promoting participatory governance and inclusiveness. The emphasis on collaborations indicates that partnerships and collaborations are crucial for creating public value.

A city can be perceived as a dynamic innovation ecosystem resembling a living organism that continuously evolves and adapts through learning from the interactions and feedback among its diverse components and responding accordingly. An innovation ecosystem can be described as a dynamic framework comprising various elements such as actors, activities, artefacts, institutions and relationships [12]. The potential of digital governance lies in supporting city learning, wherein a city learns as an innovation ecosystem for sustainable citizen-centric development and innovation of a city. City learning has been described as the interactive learning process in which a city as an ecosystem learns from its elements, referred to as "city learning from within itself" and also from other cities, termed "city-to-city learning" or "learning across cities" [3]. However, we find that there is a lack of emphasis on how ICT can support city learning as an innovation ecosystem.

The motivation for this study is to understand the requirements of an ICT platform, hereafter referred to as a platform, to support city learning that can effectively aid in digital governance for developing sustainable citizen-centric smart cities. Even though existing enterprise solutions could have utilised platforms to support similar aspects of learning in the context of those enterprises, a comprehensive platform design specifically tailored for cities as innovation ecosystems remain absent. A comprehensive plat- form design tailored for cities as innovation ecosystems are required to facilitate city learn- ing in alignment with open innovation principles while protecting security and privacy concerns for the relevant stakeholders. We identify the services required by the users (stakeholders of a city ecosystem) from a platform to effectively support city learning for transforming cities towards sustainable citizen-centric smart cities. Based on this motivation, this study's Research Objective **(RO)** is: Designing services for a platform to support city learning.

In this study, we will focus only on the city learning from within itself. The word "support", in this case, refers to addressing required services to support city learning between the relevant stakeholders of a city. Citizens residing in a city are the end users of the services in the city. The citizens need to be empowered and motivated so that they can drive the changes for citizen-centric transformations of the city to address its emerging challenges [13]. This can be realised through participatory digital governance by supporting a city to evolve as a learning innovation ecosystem. In order to focus on citizen-centricity (user-centric), the Design Research Cycle (DRC) [14] has been followed, with inspirations from Service Design [15]. Based on our research-in-progress, we present a service blueprint to illustrate the services that need to be supported by a platform design to support city learning.

The rest of this study is structured as follows: Sect. 2 describes the background and motivation; Sect. 3 describes the research approach; Sect. 4 describes the requirement analysis process and the user scenarios to identify the services; Sect. 5 describes the platform design and Sect. 6, we describe the steps towards validation of the platform design. Finally, we concluded our study in Sect. 7.

2   Background and Motivation

Cities are complex ecosystems [9] where interconnected elements can interact and pos- itively/negatively influence each other. Sustainable city development requires consider- ing stakeholders' contextual needs and developing innovations based on feedback and experiences. Learning is an integral part of the innovation process in cities, wherein knowledge of contextual requirements, challenges and opportunities are the primary steps for innovating a solution for any system. Sustainable innovations in a city re- quire contextual knowledge to address emerging economic, social, and environmental changes because every city has distinct requirements, resources, and administrative set- tings, resulting in heterogeneous knowledge gained from different contexts [16]. In the following subsections, we present a brief overview of digital governance in the context of cities and city learning to underline the motivation behind focusing on supporting city learning for developing sustainable citizen-centric smart cities through effective digital governance.

2.1 Digital Governance

Digital governance has been envisioned as a model for digital technology-supported participatory governance [10], which relies on collaboration with citizens for city transformations. Digital governance goes beyond providing electronic services to engaging citizens in decision-making using ICT and leads towards a citizen-centred governance model [7, 17]. The pursuit of inclusiveness, trustworthy infrastructure, transparency and accountability have been identified as the main challenges of digital governance [18].

An important element of digital governance is social participation, facilitated by new ICT capabilities such as social networking applications, Artificial Intelligence (AI) and Big Data, which have increased the level of connectedness between a city's government, the private sector and the citizens. This has also impacted the level of involvement of the citizens in the decision-making processes of a city [11]. Along with this comes the increase in the need for the citizens to be engaged and to have a say in how their city should evolve. As a result, city governments are left with the challenge of determining the best ways to engage the citizens.

Digital governance also involves a new style of leadership, with new models for funding and partnerships, such as with universities [19]. Cities have improved the services they provide by using new technologies and interacting with citizens in different ways. Up-to-date information can be provided to citizens through social media. Models such as crowdsourcing of ideas and crowdfunding activities for community development have emerged over the years. New services have been developed for citizens to report about situations in the city, such as malfunctioning street lights. S.R. Greenberg has also identified in [19] a number of trends and examples of new city services that create public value for the citizens, mostly through interactions with the citizens.

2.2 City Learning

The conceptualisation of cities as learning territories describes cities or urban territories as spaces where people can learn [20]. The concept of Learning Cities [21] promotes Lifelong Learning for citizens, anytime, anywhere, utilising digital technologies [22]. The growing emphasis on smart cities has led to a greater focus on utilising technology to enable Lifelong Learning, giving rise to the concept of smart city learning. The focus is the learning experiences of individuals in urban areas within the broader context of smart cities. This concept aims to enhance various critical elements essential for regional competitiveness, including mobility, environment, people, quality of life and governance [23]. However, individual knowledge growth alone does not guarantee sustainable innovations for a city. While Learning Cities support long-term learning within the community following the Lifelong Learning paradigm, city learning as an innovation ecosystem goes beyond this concept.

In recent times, the idea of learning in cities has been explored through Living Labs, which involve citizens in open innovation through processes such as co-design, co-creation and feedback [24]. These Living Labs consider governance and draw inspiration from innovation models such as the Triple Helix [25]. However, Living Labs lack a holistic view of cities as ecosystems and lacks a standardised framework for addressing city learning as an innovation ecosystem.

Learning about contextual requirements, challenges and opportunities is necessary for a city's innovation process. Learning can occur from the internal elements within the city and also from the external elements such as other cities. City learning refers to the process through which a city can obtain or gain contextual knowledge for its development and innovation through the interactions between its stakeholders [3, 9]. City learning can be supported by ensuring citizen participation and engagement through collaborative activities within a city involving citizens, public/private institutions, and the city administrator. Thereby, effective digital governance can be achieved by promoting public value through collaborations among key stakeholders in a city through the use of ICT to support city learning as an innovation ecosystem. Digital governance supporting city learning as an innovation ecosystem can contribute to sustainable citizen-centric solutions for emerging challenges in cities.

3 Research Approach

We have used the DRC presented by W. Kuechler and V. Vaishnavi in [14] to guide our research process for this study. The DRC enhances the knowledge capture and production parts described in the Design Science Research (DSR) method [26, 27]. DRC provides a general model describing each process step of DSR. The first step is problem awareness, which is also the starting point of DSR and is similar to identifying an initial problem that has to be solved [26]. The next phase is the suggestion phase, where the possible solutions to the problem are suggested by conceptual translation from the theoretical domains to the design domain through analogical reasoning. The designed IT artefact is then developed and evaluated, and conclusions are drawn from it. The whole process is iterative, and every process step can lead to improvements in prior steps through knowledge flows, which are called circumscription and operation and goal knowledge. The DRC is illustrated in Fig. 1.

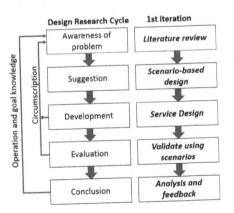

Fig. 1. Research Approach

For the problem awareness step of our study, we have conducted a literature review to identify the research gaps and the high-level requirements for the design. In the suggestions step, we have used scenario-based design [28] to understand how citizens would use a platform. It helped to understand the use of the platform in context and from the users' (citizens') perspectives. These two steps helped to identify a set of services that could be supported by a platform and how citizens would like to use them. In this study from our research-in-progress, we describe the first iteration of the development step, which is inspired by ideas from service design [15]. Service design is focused on value creation, and it takes the perspective of the system's user and aims to support designs that would provide the user with a positive experience. [29] describes the relationship between DSR and service design as complementary, where service design provides a conceptual framework to design for value co-creation within service systems, while DSR provides the methods and tools to understand and envision new forms of value co-creation through the service. Hence, the service design approach emphasises the citizen-centric approach that is desired for supporting city learning through a participatory approach. We started by creating personas to understand the different groups of citizens and to be able to focus on their specific interests and needs.

The first iteration of the DRC has helped to identify a set of services that should be provided on the platform. These have been described as a service blueprint, which describes the user journey and the touch points of the service as the main artefacts that are required to enable the service. This blueprint will be used to define the functional requirements for the platform and the services.

4   Requirement Analysis

The detailed study by P. Banerjee and S.A. Petersen in [9] of existing frameworks for city development, considering the aspects of learning in cities, highlights the need to address the concept of a city as a complex ecosystem that can learn and innovate from within itself for its development and innovation, drawing insights from individual, group and system levels. It identifies three segments of the city ecosystem as drivers of city learning, which are 1) humans, 2) technological systems and 3) the natural environment. However, the human segment, representing the societal aspects of a city, has been identified as the active initiator that is responsible for the knowledge flows through interactions between and across individual citizens, academic/business institutions and government/administrative bodies. It also identifies key aspects for driving city transformations through learning, including leadership-driven planning, citizen motivation and participation, citizen engagement and empowerment, information collection and sharing, co-design, reflection on experience and information analysis, and feedback.

User scenarios, based on intuitive reasoning from the existing literature, have been described as a part of the requirements analysis. Analysis of the user scenarios helped to identify the services that should be supported by a platform that could support city learning through citizen participation and interactions.

4.1   User Scenarios

We describe user scenarios from the citizens' and the implementing authority's perspectives to identify the services of a platform to support city learning. We use the scenarios to visualise how a platform can support the dialogic interactions between the relevant stakeholders in a city. The specific scenarios have been constructed based on the analysis of the results in [3], which presents a detailed literature review of existing frameworks for developing cities that have considered the aspect of learning in cities. The analysis helped us identify some of the processes described as supporting city learning. The user scenarios described here are about sharing personal knowledge and ideas, analysing individual inputs, participating in decision-making processes and providing feedback on the implemented city transformations.

Scenario 1: Sharing Knowledge and Ideas With Others. Kristine is a 35-year-old nurse in a city who loves to spend time outdoors amidst fresh air. She has always felt that there has been a lack of green space close to her apartment complex. She wishes that green spaces had been present in her neighbourhood. Kristine learnt from social media about a public platform managed by the local city government that could be accessed both through web browsers and a mobile App, where she could post about her wishes and opinions about the city. Kristine's input would be visible to other citizens, the local government and service providers. She finds that the platform's purpose is to share and collect information from both the service providers and the users to generate relevant knowledge to transform the city in an inclusive and sustainable manner. She finds that she can check the performance of existing service systems in a city from the platform, view ideas and wishes posted by other citizens, and also share personal ideas and choices with others on the platform. She finds hope that her idea may become a reality. She creates an account on the platform and posts on the open forum that she wishes there were more parks or green spaces in this part of the city. She finds that she can also choose to share her opinion on the open forum on the platform anonymously.

The main user group in this scenario are the citizens. The main requirements for services identified are easy access for the citizens through multiple interfaces (e.g. web and mobile devices), the possibility to create personal profiles, post ideas and wishes, contribute anonymously, protect privacy and view other citizens' ideas and wishes.

Scenario 2: Participating in Decision-Making. Kristine gets a notification from the platform (where she has already registered her profile) that a new survey is being conducted on the platform that she could respond to. The topic is about a community lot that the municipality wants to develop, and they want the opinions of the citizens to ensure citizen-centric decision-making. As this is a topic Kristine is interested in, she decides to take the survey. In the survey, Kristine gets a list of 7 different options for the community lot development. Each option also displays information on the pros and cons associated with the corresponding development based on how it would affect other services/facilities or the natural environment in the city. She sees that her opinion of a park has also been considered in one of the options there. The response for each option could be given on a Likert scale of 1 to 10. Kristine responded to the option with the park idea with a 10 on the Likert scale since this is exactly what she wished for. For the other options, she responded as per her liking for each option idea. After a few days, when the survey was closed, Kristine receives a notification that she could now see the results from the survey and reflect on how the other people have responded and the most popular choices. Kristine feels motivated and also empowered that she has been a part of the decision-making process that will affect the future transformations in the city.

The main user groups in this scenario are the citizens and a survey administrator, who could be a public authority. The main services identified that should be supported by the platform are the option to create and administer surveys, support the Likert scale for survey responses, set a time for opening and closing surveys and send notifications to survey participants, who could be the users registered on the platform. Citizens are able to receive notifications about events on the platform and respond to surveys.

Scenario 3: Processing the Citizens' Inputs. Hans is a 40-year-old urban planner who works for the city government. He has been engaged in planning a community lot based on the opinions of the citizens in the area. Hans has been given access to the platform where citizens from diverse walks of life have already expressed their opinions about the community lot. Hans has received the responses from the survey. However, he wants to rerun the survey again a few times to see if the survey respondents reassess their choices after reflecting on the previous survey session results. He thought the iterative process may lead to more rationalised and convergent results. He, therefore, reopens the survey and makes the results from the first iteration of the survey and the pros and cons of each choice in the survey available to the citizens. He sees that the winning idea has changed from the first survey to the latest one. He decides how many iterations of the survey he wants to conduct. From the final survey session, Hans picks the most popular choice after due consideration of economic feasibility from the implementer's perspective and starts planning the community lot. When the community lot development is complete, Hans observes and analyses for a certain period of time from an expert urban planner's point of view regarding what has been the impact of the development on the neighbourhood areas and if the community lot affected the performance of any other service systems in the city and the city in general. He updates the information from his analyses on the platform so that the citizens can view and assess the impact of the community lot development.

The main user groups in this scenario are the citizens and a survey administrator. The main requirements for services that were identified are that the platform should support the possibility of running a survey several times, display information related to survey options, such as the pros and cons related to each option in a survey, make the results of previous iterations of the survey available to the participants of the survey and make them aware of the iterations of the survey, updating of the performance status or outcome of a developmental activity after its completion by the corresponding authority responsible for the initiation of the development. The need for supporting roles that take the initiative, such as survey administrator, is highlighted.

Scenario 4: Providing Feedback. Ida is a 25-year-old university student who lives near the recently developed community lot. Overall, she is very happy with the development. However, during winter, she notices how dark it gets and thinks that there should be more street lights in the area so that everyone can feel safe walking there during the dark hours. She heard that the community lot was based on the ideas of citizens who used a platform for sharing ideas. Ida decided to join the platform to share her opinions and hopes that she can contribute to the area around her. Since she wishes there were more streetlights, she writes a post about the lights in her area and how she wishes there were more. She hopes that this idea will be taken into consideration for the city's development.

The main user group in this scenario is the citizens. A main service identified to be supported by the platform is an option to make a request and share an idea directed to an explicit user group, e.g. an administrative authority such as the city council, and not just other users on the platform.

5 Service Design

Based on the user scenarios, it is evident that a city can effectively learn as an innovation ecosystem by enabling the dialogic processes between a city's stakeholders and utilising the knowledge generated from the interaction between them. Seamless information exchange is crucial for collaboration. From the first scenario, we see that citizens should be able to share their needs, aspirations, and ideas with relevant authorities and among themselves to generate new ideas collectively. They also need options to view the performance statuses of existing service systems based on which they can contribute their ideas for enhancing existing performance. It also highlights the fact that citizen's privacy concerns must be addressed to ensure their uninhibited participation in the collaborative development of their city. The second scenario illustrates that in order to empower the citizens, they should be able to participate in the decision-making process for transforming their city. This gives them a sense of ownership of the transformations through the collaborative processes that can motivate them to participate and engage while being aware of the existing scenarios of their city.

The third scenario about processing the citizens' inputs highlights that the implementing authorities, represented by the survey administrator, should be able to reflect on citizens' inputs and feedback. Implementing authorities who may initiate and carry out the innovation and development for a city can be individual entrepreneurs, public/private institutions, or the city administrator. To ensure a citizen-centred approach, implementing authorities need to design service systems (for new or improved services) based on citizen concerns, aspirations and ideas or align their plans with citizen inputs. Collaboration between implementing authorities and citizens is essential to finalise relevant innovation and development. The fourth scenario highlights the necessity of feedback from the citizens based on learnings from their experiences and interactions to support the iterative process of innovation and development. This can ensure a sense of accountability for the city transformations among all the participating user groups through transparency of the interactions and their outcomes. Based on this understanding, we can identify three high-level requirements for platform design to support the city learning process: 1) Citizen participation and engagement, 2) Support communication between the stakeholders, and 3) Transparency.

We analysed the user scenarios and requirements to support city learning through the continuous dialogic processes between its relevant stakeholders. Each user group among a city's stakeholders (citizens, relevant service designers, entrepreneurs, companies or institutions running specific services) require a set of services on a platform to support interactions and participation among citizens. We highlight the services different user groups require to be supported on a platform to support city learning.

General services for all user groups, e.g. citizens, private/public bodies (relevant service designers, entrepreneurs, companies or institutions providing services in a city):

- Access to a platform through both a mobile app and web browsers.
- Create a personal/institutional profile on the platform.
- Login to personal/institutional profile passwords.
- View and assess the performance status of existing service systems in a city.
- View the unstructured ideas and requests from citizens to have the overview of contextual requirements/challenges of the city citizens for planning policy/service designs for the city.

Services for Citizens:

- Easy access to public services and information about them.
- Share ideas, wishes, opinions and requirements using the free text.
- Possibility to contribute content on the platform anonymously to address the pri- vacy concerns of the citizens.
- Access to ideas, wishes, opinions and requirements shared by others.
- Possibility to provide an opinion on other people's ideas.
- Receive promotional notifications about the platform and its features to join the platform for the collaborative transformations of the city.
- Notify/inform relevant authorities regarding their requests/ideas.
- Receive update notifications for new surveys posted on the platform.
- Answer Likert scale-based surveys after reflecting on all the options and additional information associated with them.

Services for private/public bodies (relevant service designers, entrepreneurs, companies or institutions running certain services in a city):

- Possibility to obtain citizens' opinions in multiple ways, e.g. through surveys or new ideas.
- Possibility to assess citizens' ideas and their popularity among other citizens.
- Publish a survey with multiple design alternatives for a project aiming to address specific requirements/challenges based on the understanding of the unstructured ideas and requests from citizens and/or the present status of different service systems.
- Assign a Likert scale to all possible design alternatives along with their corresponding pros and cons, reflecting how their implementation will likely impact related service systems in the city.
- View survey results to get the popularity details of the alternatives among the citizens.
- Option of republishing a survey with additional information on the popularity of each alternative based on how the citizens have reacted to the previous survey round.
- Update the performance of the service systems on the platform after implementing the finally selected design alternative for the concerned project.

The city administrator has three additional functions over the other implementing authorities, which are:

- Hosting and maintaining the platform.

- Providing a mechanism for assessing citizens' ideas and opinions posted on the platform as unstructured inputs or through structured online surveys and speeding up the information flows by using Natural Language Processing (NLP) tools such as Open AI GPT-4 [30], Google Cloud Natural Language API (NLAPI) [31], Amazon Comprehend [32] and IBM Watson Natural Language Understanding [33], which can process large sets of unstructured/unstructured data to generate a discrete set of ideas by extracting key concepts, entities, and sentiments, and identifying and grouping similar ideas into discrete categories through their techniques such as automated text summarisation, sentiment analysis, topic modelling.
- Promoting the platform among all the stakeholders in a city.

5.1 Service Blueprint

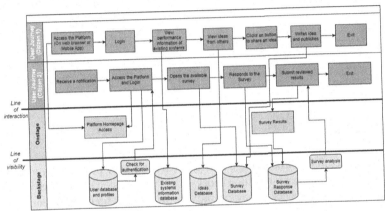

Fig. 2. Service Blueprint from the citizen perspective

We have developed user journeys and a service blueprint for the platform to identify the main components of a platform that could support the services identified in the earlier sections. The primary objective of a service blueprint is to provide service providers and designers a thorough understanding of their services, including resources and processes that are both visible and invisible for the users, which are required to deliver a satisfactory customer/user experience to the user. This understanding is essential for innovating and developing products/services while serving several purposes, such as identifying weaknesses, streamlining processes and coordinating future changes. The service blueprints from the citizen and relevant implementing authority perspectives are shown in Figs. 2 and 3, respectively, to highlight different stages of interactions and also the underlying activities involved that are represented through various layers for the provided service, here by the platform [34]. Multiple databases have been identified in the blueprints to present an objective view of the platform design that would help visualise aspects of data separation and fault isolation between different modules, and support the platform's scalability. Due to limited space, we have only described two services regarding the interactions between the citizens and city administrators: one from the citizens' perspectives and one from administrators of city services. We have not dealt with assessing the ideas and opinions of citizens for planning future innovations. The blueprints are simplified, and we are in the process of elaborating them to specify the functional requirements of the platform.

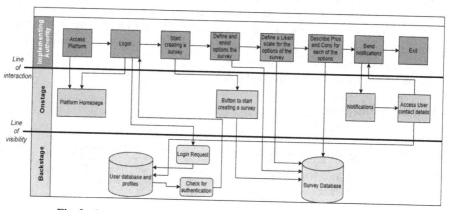

Fig. 3. Service Blueprint from the implementing authority perspective

6 Validation

The services identified through the user scenarios need to be provided by a platform to support city learning as an innovation ecosystem. The next step in our work is to analyse the blueprint designs to check if they can provide the services for the different user groups among a city's stakeholders that have been identified through the user scenarios. The blueprints will be used to develop an interactive prototype to design the user interface, which will be used to validate the user interface and the functionalities of the platform. The interactive prototype will be developed using a rapid prototyping tool such as Figma [35].

7 Conclusion

In this study, based on our research-in-progress, we have discussed the design of a platform to support city learning, where the city is considered a learning and innovation ecosystem. Learning about the needs of the citizens by engaging them is an important part of digital governance. Hence, encouraging the participation of citizens in making decisions about a city's services and its transformation processes are important aspects of digital governance. As such, the research objective of this study is to describe the design of services that should be supported on a platform to support city learning to aid digital governance for developing sustainable citizen-centric smart cities.

User scenarios about how different stakeholders would use the envisaged platform have been used to understand the needs for the services that should be made available on the platform. The activities presented by the user scenarios reflect the complete cycle of the iterative process of innovation through learning from knowledge generation to its implementation and feedback thereof for further innovation. The services are described using service blueprints. We highlight that supporting city learning from interactions, ideas, knowledge, experiences and reflective analyses of its stakeholders with a citizen-centric focus by empowering citizens through their participation, motivation and engagement can lead to effective digital governance.

The main limitation of this work is the lack of validation of the user scenarios and the service designs. This will be the next step in our research. We plan to develop an interactive prototype that supports the services that we have identified through the user scenarios and validate both the services and the user interface for the platform on multiple technologies, such as mobile devices and web browsers.

Acknowledgements. We are grateful to Helene Elling, Master's student in the Department of Computer Science at NTNU, for providing her insightful support in preparing the scenarios used in this study.

References

1. Koop, S.H., van Leeuwen, C.J.: The challenges of water, waste and climate change in cities. Environ. Dev. Sustain. **19**(2), 385–418 (2017)
2. Correia, F., Erfurth, P., Bryhn, J.: The 2030 agenda: the roadmap to globallization (2018)
3. Banerjee, P., Petersen, S.A.: How cities can learn: key concepts, role of ICT and research gaps. In: Dascalu, M., Mealha, O., Virkus, S. (eds.) Smart Learning Ecosystems as Engines of the Green and Digital Transition, pp. 53–73. Springer Nature Singapore (2023). https://doi.org/10.1007/978-981-99-5540-4_4
4. Krueger, E.H., Constantino, S.M., Centeno, M.A., Elmqvist, T., Weber, E.U., Levin, S.A.: Governing sustainable transformations of urban social-ecological-technological systems. NPJ Urban Sustain. **2**(1), 10 (2022)
5. United Nations: Sustainable cities and communities (2015). https://www.globalgoals.org/goals/11-sustainable-cities-and-communities/
6. Ahdekivi, V., Ghanbari, H., Rossi, M.: Building climate resilience in smart cities using open data services (2023)
7. Saxena, K.B.C.: Towards excellence in e-governance. Int. J. Public Sect. Manag. **18**(6), 498–513 (2005)
8. Zheng, L., Kwok, W.M., Aquaro, V., Qi, X.: Digital government, smart cities and sustainable development. In: Proceedings of the 12th International Conference on Theory and Practice of Electronic Governance, pp. 291–301 (2019)
9. Banerjee, P., Petersen, S.A.: Learning in cities from within and across cities: a scoping review. Triple Helix (forthcoming) (2023)
10. Erkut, B.: From digital government to digital governance: are we there yet? Sustainability **12**(3), 860 (2020)
11. Chen, Y.C.: Managing Digital Governance-Issues, Challenges, and Solutions. Routeledge (2017)
12. Granstrand, O., Holgersson, M.: Innovation ecosystems: a conceptual review and a new definition. Technovation **90**, 102098 (2020)
13. Oliveira, A·, Campolargo, M.: From smart cities to human smart cities. In: 2015 48th Hawaii International Conference on System Sciences, pp. 2336–2344. IEEE (2015)

14. Kuechler, W., Vaishnavi, V.: A framework for theory development in design science research: multiple perspectives. J. Assoc. Inf. Syst. **13**(6), 3 (2012)

15. Blomkvist, J., Holmlid, S., Segelström, F.: Service design research: yesterday, today and tomorrow (2010). https://onlinelibrary.wiley.com/doi/pdf/10.1111/jocn.13651

16. Calzada, I.: Replicating smart cities: the city-to-city learning programme in the Replicate EC-H2020-SCC project. Smart Cities **3**(3), 978–1003 (2020)

17. Luciano, E.M., Wiedenhöft, G.C., dos Santos, F.P.: Promoting social participation through digital governance: identifying barriers in the Brazilian public administration (2018). https://doi.org/10.1145/3209281.3209376

18. Barbosa, L.S.: Digital governance for sustainable development. In: Digital Nations–Smart Cities, Innovation, and Sustainability: 16th IFIP WG 6.11 Conference on e-Business, e-Services, and e-Society, I3E 2017, Delhi, India, 21–23, November 2017, Proceedings 16, pp. 85–93. Springer (2017). https://doi.org/10.1007/978-3-319-68557-1_9

19. Greenberg, S.R.: Using innovation and technology to improve city services. Report, IBM Center for the Busines of Government (2015)

20. Komninos, N.: Intelligent cities: innovation, knowledge systems and digital spaces. Routledge (2013)

21. Longworth, N.: Making Lifelong Learning work: Learning cities for a learning century. Routledge (2019)

22. Gianni, F.V., Divitini, M.: Technology-enhanced smart city learning: a systematic mapping of the literature. IxD&A Inter. Design Archit. (s) **27**, 28–43 (2015)

23. Hollands, R.G.: Will the real smart city please stand up?: Intelligent, progressive or entrepreneurial? In: The Routledge companion to smart cities, pp. 179–199. Routledge (2020)

24. Lucchesi, G.P., Rutkowski, E.W.: Living labs: science, society, and co-creation. Ind. Innov. Infrastr. 706–715 (2021)

25. Etzkowitz, H., Leydesdorff, L.: The Triple Helix–University-industry-government relations: a laboratory for knowledge based economic development. EASST Rev. **14**(1), 14–19 (1995)

26. Hevner, A.R., March, S.T., Park, J., Ram, S.: Design science in information systems research. MIS Quart. **28**(1), 75–105 (2004). https://www.jstor.org/stable/pdf/25148625.pdf

27. Beck, R., Weber, S., Gregory, R.: Theory-generating design science research. Inf. Syst. Front. **15**, 637–651 (2013). https://www.researchgate.net/publication/251237390_Theory-Generating_Design_Science_Research

28. Rosson, M.B., Carroll, J.M.: Scenario-based design. In: Human-Computer Interaction. CRC Press, 20 (2009)

29. Teixeira, J.G., Patrício, L., Tuunanen, T.: Advancing service design research with design science research. J. Serv. Manag. **30**(5), 577–592 (2019)

30. OpenAI: Open AI GPT-4 (2023). https://openai.com/gpt-4

31. Google: NLAPI (2023). https://cloud.google.com/natural-language?hl=en

32. Amazon: Amazon Comprehend (2023). https://docs.aws.amazon.com/comprehend/

33. IBM:IBMWatsonNaturalLanguageUnderstanding (2023). https://www.ibm.com/products/natural-language-understanding

34. Bitner, M.J., Ostrom, A.L., Morgan, F.N.: Service blueprinting: a practical technique for service innovation. Calif. Manage. Rev. **50**(3), 66–94 (2008)

35. Staiano, F.: Designing and Prototyping Interfaces with Figma: learn essential UX/UI design principles by creating interactive prototypes for mobile, tablet, and desktop. Packt Publishing Ltd (2022)

Role of ICT in City Learning for Developing Smart Cities: A Review of the Literature

Pradipta Banerjee[✉] and Sobah Abbas Petersen

Department of Computer Science, Norwegian University of Science and Technology,
7034 Trondheim, Norway
{pradipta.banerjee,sap}@ntnu.no

Abstract. Cities are continuously challenged by demographic, socio-economic and environmental changes. Cities need to adapt and evolve to respond to scenarios arising within themselves, such as the needs of their citizens, as well as from external or global influences, such as the COVID-19 pandemic. Cities have been compared to complex organisations comprised of individuals, groups and institutions that evolve organically like complex organisms that adapt to changes by learning from their continuous interactions and feedback. The sharing of knowledge and experiences between city elements through these interactions is fundamental to cities for evolving as a learning and innovation ecosystem. The learning process of a city could essentially enhance citizen-centric transformations for developing resilient and sustainable smart cities. Information and Communications Technology (ICT) has also been one of the major drivers for developing smart cities. In this study, we report the results from a Systematic Literature Review to explore the role of ICT in supporting city learning as innovation ecosystems for developing citizen-centric smart cities. The results show that ICT support in the context of city learning has been considered in Living Labs, Lifelong Learning and Digital Platforms for knowledge sharing and information analytics where collaboration is a key concept.

Keywords: City learning · Systematic Literature Review · Smart cities · ICT · Knowledge sharing · Collaboration

1 Introduction

The demographics, socio-economic dynamics and natural environments rapidly evolve in cities. Cities need to keep pace and cope with the rapid changes occurring in and around them to ensure responsible, resilient and sustainable citizen-centric development. The cities must be able to learn to innovate to respond to emerging challenges and opportunities. A common method for innovating cities has been through policy transfers and by sharing and replicating best practices from developed cities onto developing cities [1–4]. Such policy transfers result in mere replications, often overlooking local contextual requirements, limitations and opportunities. The transformation of cities towards sustainable smart cities with a human focus, referred to as Human Smart Cities in [5], is a step toward developing the necessary resilience and adaptability. Human Smart City

© The Author(s), under exclusive license to Springer Nature Switzerland AG 2024
M. Papadaki et al. (Eds.): EMCIS 2023, LNBIP 502, pp. 309–325, 2024.
https://doi.org/10.1007/978-3-031-56481-9_21

has been described in [6] as a concept for improving the quality of life of the citizens, leading to well-being and happiness through services that can be defined as new and innovative "ad hoc" services developed by the local government in collaboration with the citizens and other stakeholders, to tackle "wicked" societal problems.

Diverse entities such as individual citizens, groups of individuals, administrators, organisations/institutions, service systems and physical logistics such as buildings, roads and ecological environment are constituent elements of cities. These elements interact among themselves and exert positive and/or negative influences on each other through different activities, resulting in a complex interconnected system that is referred to as the organically evolving city ecosystem. Different city elements can learn from their experiences and interactions for necessary adaptations and innovations. A city evolves like a Complex Adaptive System [7–10], where the whole system's properties differ from the properties of its elements. In such a system, with every intervention, the system's characteristics change by assimilating the impacts of the temporal intervention, giving rise to a new system [11]. This emergent behaviour of a city necessitates corresponding interventions in the system to be innovated through continuous learning from the feedback of experiences and interactions to cope with emerging challenges and yield desired results. In this study, we refer to city learning from its elements as city learning from within itself. In contrast, for a city learning from other cities, we refer to it as city-to-city learning or learning across cities.

Learning in cities has been discussed in the context of smart cities, where the emphasis is given to Lifelong Learning [12]), describing it as personal growth from learning anytime and anywhere, mainly with the use of Information and Communication Technologies (ICT) [13–17]. In addition, collaboration among actors within a city [18], partnerships and knowledge exchange for learning between all actors within a city [19] have been identified as popular concepts. Central to these ideas is the need to foster knowledge creation, knowledge sharing and innovation. Living Labs (LLs) [20] approach has been used in the urban context [21] which is also inspired by the perspectives of Lifelong Learning [12] and focuses on the collaboration of citizens through activities for co-design, co-creation and feedback [22]. LLs are described as "socio-digital innovation environments in realistic city life conditions based on multi-stakeholder partnerships that effectively involve citizens in the co-creation and co-production of new or reformed public services and infrastructures" [23]. ICT appears to play a central role in all these approaches.

Taking the systemic view of a city as an innovation ecosystem that evolves, it is observed that the elements within a city and their interactions are fundamental for the development of a city [24]. The interactions among its citizens and between the citizens and other entities can significantly impact achieving resilience and developing a sustainable citizen-centric city [25]. The findings in [26] highlight the need to address the concept of a city as a complex ecosystem that can learn and innovate from within itself for its development and innovation, drawing insights from individual, group and system levels. City learning has been described in [27] as the mechanism through which a city as an ecosystem learns from its elements, referred to as "city learning from within itself" and also from other cities, termed "city-to-city learning" or "learning across cities".

Viewing cities as innovation ecosystems, our research objective (**RO**) is to explore the role of ICT that has been discussed in studies for developing cities towards smart cities through citizen-centric innovations considering the aspects of learning in the context of cities. We conducted a Systematic Literature Review (SLR) to achieve our research objective. In addition to identifying the role of ICT in supporting city learning, we also identify the approaches wherein learning in cities has been considered for developing cities through citizen-centric innovations and what has been the role of collaboration and knowledge sharing in them.

The rest of this study is organised as follows: Sect. 2 describes the method; Sect. 3 presents the results from the SLR; Sect. 4 describes the main concepts in the studies, the role of ICT addressed in the context of learning in cities; Sect. 5 presents the conclusion discussing the implications of this study.

2 Methods

To accomplish our objective in the study, we conducted an SLR, which is a form of a literature review suitable for identifying existing international evidence in topics of interest while informing future areas of research [28]. An SLR examines data and findings of authors relative to a specified research question and relevant search criteria.

Search Strategy: The search strings for this SLR were framed considering three aspects of our research objective: a) consideration of a city as an innovation ecosystem, b) the role of ICT in supporting the city's learning for contextual innovations and development of citizen-centric cities, and c) city learning from within and across cities. The search keywords selection process was driven by the PICOC framework, proposed in [29], which helps to develop a comprehensive set of search keywords for quantitative research according to population, intervention, comparison, outcome and context. The PICOC framework helps to find the different elements in the context of the search concepts derived from breaking down the research objective of the study [30] and ensures contextually relevant searches. Table 1a demonstrates the classification of the basis for our search terms. The search strings are shown in Table 1b. Search restrictions for the date of publication were not applied.

Study Selection: We followed the gold standard guidelines from the Preferred Reporting Items for Systematic Review and Meta-Analyses (PRISMA) [31] model for the SLR to filter and select the studies. A systematic search using online research databases for peer-reviewed studies from SpringerNature, ScienceDirect, IEEE, SAGE, ACM, Taylor & Francis, Emerald, Wiley, MDPI, Inderscience and IGI Global, was conducted to identify relevant studies. The research databases were searched based on the search strings. Screening excluded studies that were not written in English, duplicates and studies that did not qualify the inclusion criteria.

To fulfil our research objective (**RO**), we analysed the full-text studies and selected those studies based on the inclusion criteria. We set our inclusion criteria that the study should be discussing:

- Aspects of learning in a city for developing cities towards smart cities through citizen-centric innovations.

Table 1. Framing search keywords.

Population	-
Intervention	Role of ICT to support learning
Comparison	-
Outcome	Human Smart Cities, City-to-City Learning, City Learning
Context	City learning within and across cities, Innovation ecosystems of cities

Table 1a. PICOC framework for framing search keywords

Context	"city learning" OR "city ecosystem" OR "innovation ecosystem" OR "learning innovation" OR "within cities" OR "across cities" OR "cities" OR "learning"
	AND
Intervention	"role of ICT" OR "ICT support" OR "support learning"
	AND
Outcome	"city-to-city learning" OR "city learning" OR "sources of innovation" OR "smart cities" OR "human smart cities"

Table 1b. Search keywords

– Utilisation of ICT support for developing cities towards smart cities.

Studies that did not consider human interactions and presented methods solely based on Machine Learning (ML) were excluded because ML models are based on mathematical optimisation techniques at their core [32], and they alone cannot incorporate and comprehend interdependent variables, while cities continuously evolve through the interactions of interdependent elements. Book reviews, abstract-only studies, and presentations were removed. Additionally, a backward and forward search was performed on eligible full-text studies. We avoid bias in our literature review by keeping the inclusion and exclusion criteria aligned with the study's research objective, conducting a comprehensive and systematic search of multiple peer-reviewed databases using different combinations of keywords through Boolean operators and conducting the review without any constraints for the time range of publications.

The initial search identified 2218 studies from 11 databases (see Fig. 1). Screening the results yielded 1625 unique studies. On manual examination of the retrieved titles and their abstracts and removal of repeated entries and works not in English, 1366 out of the 1625 studies were excluded. Full-text studies were assessed, and 238 studies were removed that dealt with either ICT or how cities learn but did not focus on ICT support in the context of learning in cities. The results presented in this study are based on the final selection of 21 studies.

3   Results

This section describes the general data extracted from the 21 studies, such as the country of the author's affiliation, the publication year, the interdisciplinarity of the publications and how ICT has been used.

On analysis of the publication dates of the selected studies, we find that the topic of ICT support in the context of learning in cities has gained focus in recent years, as shown in Fig. 2. We find that the earliest year for the publication of relevant studies is 2006; a high number of publications are typically observed in 2016 and 2019. The publications are distributed among journals, conferences and book chapters, where the journal publications are more than 50% of the selection, as shown in Fig. 3.

Apart from the conceptual context of cities, they have a physical context associated with them in the real world. To explore the influence of the geographical contexts in the selected studies, we extracted information on the country-wise distribution of the authors' affiliations. We extracted the name of the country of the affiliated institution for each author of the studies. We assigned one point to the affiliation country of every author. We conducted this procedure for all the authors of the selected studies and summed the points of respective countries to determine the country-wise distribution of the author's affiliation, as shown in Fig. 4. We find that there have been conceptual studies of the qualitative aspects of ICT support for city learning from within and across cities from research institutions of mostly European countries where Italy (n = 9), the UK (n = 6) and Belgium (n = 5) are the countries where the highest number of the studies have been conducted. However, we find the cities that were studied within the selected list of studies were from Belgium [33], Finland [34], Israel [35], Italy [34], Tunisia [36] and UK [37]. We also processed the authors' affiliated disciplines to assess the interdisciplinarity dimension involved in these studies. As illustrated in Fig. 5, we find the study interest in this topic is spread over multiple disciplines, where Engineering and Technology appear to have the highest contribution, followed by Computer Science and Informatics and Business Administration and Management.

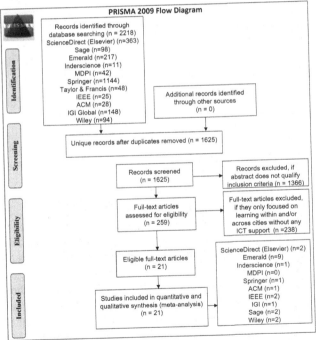

Fig. 1. PRISMA (Preferred Reporting Items for Systemic Reviews and Meta-Analyses [31]) flowchart for study selection.

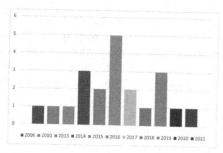

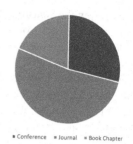

Fig. 2. Year-wise distribution of publications

Fig. 3. Categories of publications

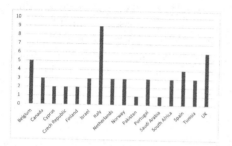

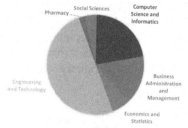

Fig. 4. Country-wise distribution of affiliations of the authors

Fig. 5. Interdisciplinarity of publications

4 Study Analysis

This section presents the analysis of the SLR results regarding ICT's role in developing cities towards smart cities through citizen-centric innovations considering the context of learning in cities.

We analysed the information from the SLR in the context of learning in cities and related utilisation of ICT support and presented the overview in Table 2. We find only one study [37] that discussed the aspect of learning from other cities and two studies [5, 49] that considered both the aspects of learning from within itself and across cities. The remaining eighteen discussed only the aspect of learning from within a city.

We found that terms such as collaboration, networking, knowledge exchange and partnerships have been prevalent in the studies in the SLR. The results show that 20 studies mention collaboration in various contexts such as collaborative platforms [39], collaborative enterprises, services and workgroups [6], collaborative government/governance [39, 44, 46] and collaborative technologies [5, 46]. Eighteen studies mention knowledge exchange, where knowledge transfer [34, 44], knowledge sharing [6, 23, 33, 40, 43] and knowledge networks [39, 45] were some of the concepts that were addressed. We identified that such terms have been integral for describing collaborative networks where the key notions are that of a community and the interplay among stakeholders [50], knowledge creation and sharing [51], partnerships and knowledge exchange [19].

Furthermore, the approaches where ICT support is used in the context of learning in cities highlight the relevance of collaboration and knowledge exchange; e.g. LLs are means for co-creating futures with stakeholders, and digital platforms were reported to support citizen engagement and knowledge exchange. The distribution of terms related to collaborative networks is shown in Fig. 6.

Table 2. Overview of the SLR results analysis

Study	Learning from within a city	Learning across cities	Approaches utilising ICT support for learning in cities			Remarks
			Lifelong Learning	Living Labs (LLs)	Digital platforms and information analytics	
[38]	Y		Y			Proposed a new way to empower technology by incorporating IoTs teaching-learning process of education systems in smart city learning scenarios
[39]	Y		Y		Y	Highlights that the ICT sector is a major carrier of learning spillovers across economies that have opened many opportunities to shape the future of economies and societies and to facilitate the creation of a learning society and competitive economy
[35]	Y		Y			Elaborates on a systematic process to enhance regional innovation in one specific field, namely, formal education through the regional future centre concept
[36]	Y		Y			Proposes design framework for Smart City Learning scenarios based on three main aspects Learner, Contextualized Activity and Space for pedagogical designers and teachers
[40]	Y		Y			Approach is based on Adaptive, Knowledge-based Learning Systems, a Personal Learning and Working Environment and a general Semantic Framework

(continued)

Table 2. (*continued*)

Study	Learning from within a city	Learning across cities	Approaches utilising ICT support for learning in cities			Remarks
			Lifelong Learning	Living Labs (LLs)	Digital platforms and information analytics	
[41]	Y		Y	Y	Y	States that the creation of district labs and networks – socio-technical places aiming at properly requalifying/maintaining/using cities – can build, in such scenery, a strong means of physical requalification and evolution of the social capital in degraded environments, by building a path guiding toward the implementation of "learning cities" as a new way of thinking about modalities, times and places of learning
[6]	Y			Y	Y	Proposes to develop a citizen-driven, smart, all-inclusive and sustainable environment, with a governance framework in which citizens and government engage in active communication with each other
[23]	Y			Y	Y	Provides overviews of the urban governance models emerging in such environments and finally focuses on the challenges posed by these models as a result of integration between the 'technology push' Smart City vision and the 'human pull' Urban LLs concept
[42]	Y			Y	Y	Concludes that given certain criteria are met, LLs with technological support can be a solution for sustainable development and innovation in cities

(*continued*)

Table 2. (*continued*)

Study	Learning from within a city	Learning across cities	Approaches utilising ICT support for learning in cities			Remarks
			Lifelong Learning	Living Labs (LLs)	Digital platforms and information analytics	
[43]	Y			Y		Concluded that co-innovation approaches of LLs aided with technology might represent a relevant way to create "socially acceptable" projects, and more than that, effective "stakeholders-driven" projects. Emphasises the role of intermediation entities such as LLs to the development of complex urban projects and the renewal of project management practices in such contexts through knowledge exploration and open design processes that can benefit urban projects, as well as healthcare projects
[44]	Y				Y	Highlights technology can emphasise citizen centricity for sustainable city developments through collaborative governance models
[37]		Y			Y	Presents the partial results of the EU BlueSCities project that developed the City Blueprint Framework for water and waste and the City Amberprint Framework for energy, transport and ICT that can promote city-to-city learning
[45]	Y				Y	Examines the concepts of learning and diffusion within the context of urban development and sustainable active mobility. A system dynamics (SD) computer simulation model is proposed, which treats learning as a diffusion process in a dynamic way so that key strategies and their effects are investigated over time

(*continued*)

Table 2. (*continued*)

Study	Learning from within a city	Learning across cities	Approaches utilising ICT support for learning in cities			Remarks
			Lifelong Learning	Living Labs (LLs)	Digital platforms and information analytics	
[10]	Y				Y	A qualitative and interpretative approach is adopted to reflect upon the role of technologies in everyday life in the Smart City ecosystem which is defined as a multilevel construct that helps in understanding how technical and technological dimensions of the Smart City can be managed not only as supportive instruments but also as key pillars to support, facilitate and ensure an effective cognitive alignment among all the involved
[5]	Y	Y			Y	Presented a Smart City Ecosystem Architecture for conceiving the adaptive and learning process of transformation, shaped by city leaders and citizens and enabled by digital technologies and information
[46]	Y				Y	Discusses qualitatively how ICT can be incorporated for development through learning and adaptations
[33]	Y				Y	Illustrates that public service delivery, related to the urban space, can be co-designed between the city and its citizens, if different toolkits (technological aids) aligned with the specific capacities and skills of the citizens are provided

(*continued*)

Table 2. (*continued*)

Study	Learning from within a city	Learning across cities	Approaches utilising ICT support for learning in cities			Remarks
			Lifelong Learning	Living Labs (LLs)	Digital platforms and information analytics	
[47]	Y			Y	Y	Through an analysis of 149 SC initiatives from 76 European cities, it provides interesting insights into how participatory models have been introduced in the different areas and dimensions of the cities, how citizen engagement is promoted in SC initiatives, and whether the so-called creative SCs are those with a higher number of projects governed in a participatory way. The findings of this research show that citizen participation in the research literature got mainly promoted through the use of ICTs such as mobile applications, sensors and IoT devices, and open data
[34]	Y				Y	Presents a comparative qualitative study based on interviews for creating an overall picture of learning processes occurring in the organizations involved (universities and local government) within the smart city
[48]	Y				Y	Proposed a framework where communication technologies are perceived as subservient to the Smart City services, providing the means to collect and process the data needed to make the services function. Proposes a new vision in which technology and Smart City services are designed to take advantage of each other in a symbiotic manner According to this new paradigm, named "SymbioCity", Smart City services can indeed be exploited to improve the performance of the same communication systems that provide them with data

(*continued*)

Table 2. (*continued*)

Study	Learning from within a city	Learning across cities	Approaches utilising ICT support for learning in cities			Remarks
			Lifelong Learning	Living Labs (LLs)	Digital platforms and information analytics	
[49]	Y	Y			Y	Presents a systematic review of the literature on smart city big data analytics to highlight how platforms for BigData can support learning in cities

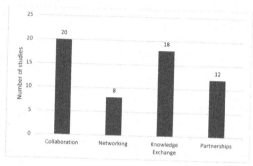

Fig. 6. Distribution of terms related to collaborative networks

4.1 Role of ICT

From Table 2, we find that out of the 21 studies from the SLR, six studies [35, 36, 38–40, 44] discussed how ICT tools can support learning in cities through Lifelong Learning. Among these six studies, two of them [39, 44] cite the use of ICT-enabled platforms for knowledge sharing. Five studies [6, 23, 41–43] mention the capability of ICT in supporting LLs, which are referred to as open innovation systems by enabling knowledge transfers amongst different participating actors in cities for fostering the city's learning process [42]. Among these five studies, [6, 23, 42] have considered a digital platform-oriented ICT support for learning in LLs through information sharing, and [41] encompasses the attributes of both Lifelong Learning and LLs through digital platforms. Eleven studies [5, 10, 33, 34, 37, 44–49] state how ICT-enabled digital platforms can support the learning process in cities through enhancing citizen participation, knowledge exchange and/or Big Data analytics. The distribution of these three approaches, along with their functional overlaps across the studies, is shown in Fig. 7. An overview of the use of smart devices and support for information exchange across these three approaches are provided in Fig. 8.

Lifelong Learning has been discussed mostly in the context of smart cities, which have been acknowledged as a robust ecosystem for learning, e.g. [38]. There is a focus on extending traditional ICT support with sensor data. The study in [39], categorically states that ICT investments would not automatically yield substantial benefits and that they have to be complemented with investments in human capital and organisational and social learning.

The importance of ICT-enabled knowledge exchanges in LLs is shown in [6]. Coinnovation approaches of LLs aided with ICT solutions have been found to support experiential learning, leading to citizen-centric innovation systems in cities [43]. Modern approaches such as LLs for learning in cities provide more attention towards citizen participation while utilising ICT tools. However, challenges remain in ensuring citizen participation and creating harmony between the 'technology push' of the Smart City vision and the 'human pull' from the LLs approach [23].

Digital platforms for citizen engagement to acquire and develop human capital, knowledge sources to promote value creation [44], and information exchange across the different sectors of a city such as energy and water [37] have been identified as areas for ICT support.

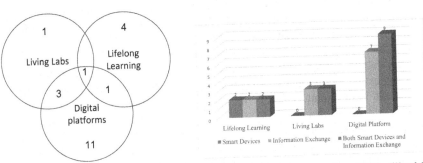

Fig. 7. Utilisation of ICT by approaches that have considered aspects of learning in cities for developing cities

Fig. 8. Distribution of categories of ICT utilised for supporting the approaches that have considered aspects of learning in cities for developing cities

We infer from the analysis that the use of ICT support revolves around communication and engaging citizens to obtain their feedback and incorporating the feedback for sustainable contextual innovation and development of a city. This inference highlights ICT's central role in supporting knowledge transfer and interactions among citizens and the entities within a city. ICT is also identified as a means of creating synergy between the city, the citizens and smart technologies and as a means of enhancing transparency and validating efficiency. Such ICT support is found to be crucial for the formation of collaborative networks and co-creation.

Through this study, we find that information exchange has appeared as the key to learning through knowledge generation across any approach considering the aspects of learning in a city. The observations in [6] re-emphasise the fact that even though technological changes are necessary, they are not enough on their own, and citizen engagement and participation are crucial for sustainable innovations and city transformations [52]. We find that there is a lack of emphasis on ICT support for city learning from within and across cities while considering a city as an innovation ecosystem.

5   Conclusion

In this study, we conducted an SLR to accomplish our research objective of identifying ICT's role in developing cities towards smart cities through citizen-centric innovations considering the aspects of learning in a city. The SLR shows that even though there have been works that have considered learning aspects in a city for its development through citizen-centric innovations while utilising ICT, research is lacking in exploring the utilisation of ICT to support the learning process of a city as an innovation ecosystem. From the SLR, we identify that historically, the perspective for learning in cities has been largely confined within the paradigm of Lifelong Learning. We also find that LLs have presented a modern approach to developing citizen-centric cities following the Lifelong Learning approach. They have incorporated learning through co-design and co-development with the aid of ICT solutions for facilitating interactions and knowledge transfers. Simultaneously, digital platforms for developing cities have also been discussed regarding utilising the potential of ICT solutions through information exchange in open innovation settings. We find from our analysis of the study that collaborative environments and platforms are crucial for city learning. Furthermore, knowledge exchange, networks and partnerships have also been discussed as key concepts in this context.

The practical implications of these insights could benefit the stakeholders, such as public/private institutions, city administrators and service providers, to focus on identifying contextual requirements for supporting city learning as an innovation ecosystem that can aid in designing policies and service systems to mitigate the shortcomings in ensuring economic, social and environmental sustainability for developing responsible and resilient, sustainable citizen-centric smart cities. The theoretical implications would be to support the current limitations in understanding how cities can learn as innovation ecosystems and how ICT can be utilised to support learning from within and across cities for developing sustainable citizen-centric cities.

As a limitation of our work, our study may have some selection bias due to the subjective selection of studies based on manual analysis of abstracts and full-text studies. Furthermore, our study could have been broadened based on the choice of keywords and an expanded set of bibliographic search databases. The studies we obtained from the SLR show that 28% of the studies are from disciplines other than engineering, Computer Science or Information Systems, which were our primary literature sources. Enhancing the search databases to include other disciplines may increase the number of relevant studies.

Our future work would focus on understanding what and how a city can learn as an innovation ecosystem and, based on the understanding, designing and validating a framework for ICT support for city learning as an innovation ecosystem to develop sustainable citizen-centric smart cities.

References

1. European Commission: Creating the Links for Scaling Smart Cities Solutions. H2020, Eindhoven, The Netherlands (2016)
2. Commission, E.: EU Research & Innovation for and with Cities. Belgium, Brussels (2017)

3. Commission, E.: The Making of a Smart City: Policy Recommendations. Belgium, Brussels (2017)
4. Commission, E.: The Making of a Smart City: Replication and Scale-Up of Innovation in Europe. Belgium, Brussels (2017)
5. Hanna, N.K.: Developing Smart Cities, chap. 8, pp. 167–174. Emerald Publishing Limited (2016)
6. de Oliveira, A´., Campolargo, M., Martins, M.: Constructing human smart cities. In: Smart Cities, Green Technologies, and Intelligent Transport Systems, pp. 32–49. Springer (2015)
7. Sanders, T.I.: Complex systems thinking and new urbanism. New urbanism and beyond, pp. 275–279 (2008)
8. Nel, D.H.: Exploring a complex adaptive systems approach to the study of urban change. Ph.D. thesis, University of Pretoria (2015)
9. Ulysses, S.: Complexityscience: the urban is a complex adaptivesystem. In: Defining the Urban, pp. 249–265. Routledge (2017)
10. Caputo, F., Walletzky, L., S̆tepánek, P.: Towards a systems thinking based view for the governance of a smart city's ecosystem: a bridge to link smart technologies and big data. Kybernetes 48(1), 108–123 (2019)
11. van Geert, P.L.: Dynamic systems, process and development. Hum. Dev. 63(3–4), 153–179 (2019)
12. Power, C.N., Maclean, R.: Lifelong Learning: meaning, challenges, and opportunities. In: Skills development for inclusive and sustainable growth in developing Asia-Pacific, pp. 29–42. Springer, Dordrecht (2013)
13. Larsen, K.: Learning cities: the new recipe in regional development. OECD Obs. 217(218), 73 (1999)
14. Commission of the European Union, E.: Information note: The "R3L initiative" European networks to promote the local and regional dimensions of lifelong learning (2003)
15. Longworth, N.: Learning cities, learning regions, learning communities: lifelong learning and local government. Routledge (2006)
16. UNESCO: Key features of learning cities, UIL (UNESCO Institute for Lifelong Learning), in UNESCO, international conference on learning cities. Lifelong Learning for all: Inclusion, prosperity and sustainability in cities (2013)
17. Osborne, M., Kearns, P., Yang, J.: Learning cities: developing inclusive, prosperous and sustainable urban communities. Int. Rev. Educ. 59(4), 409–423 (2013)
18. Gianni, F.V., Divitini, M.: Technology-enhanced smart city learning: a systematic mapping of the literature. IxD&A Inter. Design Archit. (s) 27, 28–43 (2015)
19. Meijer, A., Bolıvar, M.P.R.: Governing the smart city: a review of the literature on smart urban governance, vol. 82, pp. 392–408 (2016)
20. ENOLL: What are living labs (2022). https://enoll.org/about-us/
21. Gebhardt, C.: The impact of participatory governance on regional development pathways: citizen-driven smart, green and inclusive urbanism in the brainport metropolitan region. Triple Helix 6(1), 69–110 (2020)
22. Lucchesi, G.P., Rutkowski, E.W.: Living labs: science, society, and co-creation. Ind. Innov. Infrastr. 706–715 (2021)
23. Concilio, G., Molinari, F.: Living labs and urban smartness: the experimental nature of emerging governance models. In: Handbook of Research on Social, Economic, and Environmental Sustainability in the Development of Smart Cities, pp. 98–111. IGI Global (2015)
24. Granstrand, O., Holgersson, M.: Innovation ecosystems: a conceptual review and a new definition. Technovation 90, 102098 (2020)
25. Campbell, T.: Beyond smart cities: how cities network, learn and innovate. Routledge (2013)
26. Banerjee, P., Petersen, S.A.: Learning in cities from within and across cities: a scoping review. Triple Helix (forthcoming) (2023)

27. Banerjee, P., Petersen, S.A.: How cities can learn: key concepts, role of ICT and research gaps. In: Dascalu, M., Mealha, O., Virkus, S. (eds.) Smart Learning Ecosystems as Engines of the Green and Digital Transition, pp. 53–73. Springer Nature Singapore (2023)

28. Munn, Z., Stern, C., Aromataris, E., Lockwood, C., Jordan, Z.: What kind of systematic review should I conduct? A proposed typology and guidance for systematic reviewers in the medical and health sciences. BioMed. Central 18, 1–9 (2018)

29. Papaioannou, D., Sutton, A., Booth, A.: Systematic approaches to a successful literature review. Systematic approaches to a successful literature review, pp. 1–336 (2016)

30. Mengist, W., Soromessa, T., Legese, G.: Method for conducting systematic literature review and meta-analysis for environmental science research. MethodsX 7, 100777 (2020)

31. Moher, D., Liberati, A., Tetzlaff, J., Altman, D.G., Group*, P.: Preferred reporting items for systematic reviews and meta-analyses: the prisma statement. Ann. Inter. Med. 151(4), 264–269 (2009)

32. Sun, S., Cao, Z., Zhu, H., Zhao, J.: A survey of optimization methods from a machine learning perspective, vol. 50, pp. 3668–3681. IEEE (2019)

33. Van der Graaf, S., Veeckman, C.: Designing for participatory governance: assessing capabilities and toolkits in public service delivery. Information 16(6), 74–88 (2014)

34. Laitinen, I., Piazza, R., Stenvall, J.: Adaptive learning in smart cities–The cases of Catania and Helsinki. J. Adult Contin. Educ. 23(1), 119–137 (2017)

35. Dvir, R., Schwartzberg, Y., Avni, H., Webb, C., Lettice, F.: The future center as an urban innovation engine. J. Knowl. Manag. 10(5), 110–123 (2006)

36. Malek, J., Laroussi, M., Ghezala, H.B.: A design framework for smart city learning scenarios. In: 2013 9th International Conference on Intelligent Environments, pp. 9–15. IEEE (2013)

37. Strzelecka, A., Ulanicki, B., Koop, S., Koetsier, L., Van Leeuwen, K., Elelman, R.: Integrating water, waste, energy, transport and ICT aspects into the smart city concept. Procedia Eng. 186, 609–616 (2017)

38. Gianni, F., Mora, S., Divitini, M.: IoT for smart city learning: towards requirements for an authoring tool. In: SERVE@ AVI, pp. 12–18 (2016)

39. Hanna, N.K.: Why ICT-enabled Transformation?, vol. 2, pp. 15–40. Emerald Publishing Limited (2016)

40. Salerno, S.: Knowledge lifecycle and smart cities learning position paper for DUBAI 2020: smart city learning. In: 2014 International Conference on Web and Open Access to Learning (ICWOAL), pp. 1–4. IEEE (2014)

41. Di Sivo, M., Ladiana, D.: Towards a learning city the neighborhood lab and the lab net. Procedia Soc. Behav. Sci. 2(2), 5349–5356 (2010)

42. Schuurman, D., Baccarne, B., Marez, L.D., Veeckman, C., Ballon, P.: Living labs as open innovation systems for knowledge exchange: solutions for sustainable innovation development. Int. J. Bus. Innov. Res. 10(2–3), 322–340 (2016)

43. Lehmann, V., Frangioni, M., Dube´, P.: Living lab as knowledge system: an actual approach for managing urban service projects? J. Knowl. Manag. 19(5), 1087–1107 (2015)

44. Romanelli, M.: Analysing the role of information technology towards sustainable cities living. Kybernetes 49(7), 2037–2052 (2020)

45. Papageorgiou, G., Demetriou, G.: Investigating learning and diffusion strategies for sustainable mobility. Smart Sustain. Built Environ. 9(1), 1–16 (2020)

46. Hanna, N.K.: Implementing, Learning, and Adapting, vol. 15, pp. 343–371. Emerald Publishing Limited (2016)

47. Corte´s-Cediel, M.E., Cantador, I., Bol´ıvar, M.P.R.: Analyzing citizen participation and engagement in European smart cities. Soc. Sci. Comput. Rev. 39(4), 592–626 (2021)

48. Chiariotti, F., Condoluci, M., Mahmoodi, T., Zanella, A.: Symbiocity: smart cities for smarter networks. Trans. Emerg. Telecommun. Technol. 29(1), e3206 (2018)

49. Soomro, K., Bhutta, M.N.M., Khan, Z., Tahir, M.A.: Smart city big data analytics: an advanced review. Wiley Interdis. Rev.: Data Min. Knowl. Discov. **9**(5), e1319 (2019)
50. Camarinha-Matos, L.M., Afsarmanesh, H., Boucher, X.: The role of collaborative networks in sustainability. In: Collaborative Networks for a Sustainable World: 11th IFIP WG 5.5 Working Conference on Virtual Enterprises, PRO-VE 2010, St. Etienne, France, 11–13 October 2010. Proceedings 11, pp. 1–16. Springer (2010). https://doi.org/10.1007/978-3-642-15961-9_1
51. Camarinha-Matos, L.M., Afsarmanesh, H.: Collaborative networks: value creation in a knowledge society. In: Knowledge Enterprise: Intelligent Strategies in Product Design, Manufacturing, and Management: Proceedings of PROLAMAT 2006, IFIP TC5 International Conference, 15–17 June 2006, Shanghai, China. pp. 26–40. Springer (2006). https://doi.org/10.1007/0-387-34403-9_4
52. Hollands, R.G.: Will the real smart city please stand up?: Intelligent, progressive or entrepreneurial? In: The Routledge Companion to Smart Cities, pp. 179–199. Routledge (2020)

A Systematic Literature Review on Developing Job Profiles and Training Content for Open Data-Driven Smart Cities

Koukounidou Vasiliki[1]([⊠]) [ID], Kokkinaki Angelika[1] [ID], Osta Alain[1,2] [ID], and Tsakiris Theodoros[1] [ID]

[1] University of Nicosia, Nicosia, Cyprus
{koukounidou.v1,osta.a}@live.unic.ac.cy, {kokkinaki.a, tsakiris.t}@unic.ac.cy
[2] Université La Sagesse, Lebanon, Cyprus

Abstract. With the escalating influx of open data daily, the European Commission, as per the European's 2022 Strategic Foresight Report, prioritizes fostering equitable, greener and digital societies. The conversion of conventional services to digital platforms to support citizens in urban areas is now becoming the norm, posing challenges for accessible civic infrastructures. Open data-defined as data that can be freely used, re-used and redistributed by anyoneare expected to bring forth opportunities and societal adjustment challenges. Despite being under investigation for the past two decades, the intersection of Open Data and Smart Cities continues to be a dynamic and developing research area. The progressive implementation of Smart Cities demands the development of new skills and competences. Our systematic literature review uncovers these research gaps highlighting potential areas for development and further study. In addition, this study investigates the legal and policy frameworks regarding Open Data and Open Data for Smart Cities. It explores the impact of Open Data on Smart Cities, as well as the necessary competencies and skills required by data officers to support the implementation of Smart Cities. Furthermore, the research will examine the upskilling and reskilling strategies for data officers in Smart Cities concerning Open Data.

Keywords: Open Data · Smart Cities · Skills · Training · UN SDG 11

1 Introduction

The shift from conventional services to digital platforms aimed at aiding citizens in urban areas has become increasingly prevalent. This transition, however, has created a complex environment for many individuals seeking access to civic infrastructures [35, 47, 54]. As indicated by the Open Data Barometer, the production of open data is increasing rapidly on a daily basis [5, 13, 15, 29]. The European Commission's Strategic Foresight Report for 2022 [17], emphasizes the importance of assisting member states in making their societies more environmentally friendly and digitally advanced in a fair manner. Open data, characterized as data that is freely accessible, reusable and distributable by

M. Papadaki et al. (Eds.): EMCIS 2023, LNBIP 502, pp. 326–337, 2024.
https://doi.org/10.1007/978-3-031-56481-9_22

anyone, is expected to encounter both opportunities and challenges. This is particularly relevant for society, which is compelled to adapt rapidly to numerous changes in their daily lives [27, 28, 50].

There are two dimensions to consider, when referring to open data. The first dimension pertains to data produced and held by public sector bodies at national, regional, and local levels, including content held by ministries, state agencies, municipalities, and organizations funded mostly by or under the control of public authorities such as meteorological institutes. The second dimension relates to research data, which are defined in the Open Data Directive (art2§9) [9] as documents in digital form that are collected or produced in the course of scientific research activities and are used as evidence in the research process or are commonly accepted in the research community as necessary to validate research findings and results [48, 49]. European Commission and many national authorities formulate policies to ensure that research data is openly provided for further use according to FAIR Principles (findable, accessible, interoperable, and reusable) [4, 12].

The Open Data Directive [10], which replaced the Public Sector Information Directive, was implemented by EU Member States in various ways. Cyprus promptly responds to relevant EU directives, exemplified by aligning its legislation to the European Directive [25] to the European Directive 2019/1024/EU of the European Parliament and of the Council of 20 July 2019, on open data and the further use of public sector information [9]. This law replaces the Law on the Further Use of Information of the Public Sector of 2015 (Law 205 (I) / 2015), which is repealed by Law 142 (I)/2021. Numerous actions have been taken at different levels in recent years to support the policies mentioned above.

In Open Science, an ever-evolving legal ecosystem is crucial for comprehensive support. [14, 47]. The most recent development occurred on February 23rd, 2022, when the European Commission proposed the adoption of a Regulation on harmonized rules on fair access to and use of data, known as the Data Act [10]. This legislation is a critical component of the European strategy for data, with the primary objective of making Europe a leader in the data economy by leveraging the potential of the ever-increasing amount of industrial data to benefit the European economy and society [18].

To sustain its global data economy leadership, the European strategy for data aims to enact legislation on data governance, access, and reuse. This includes fostering business-to-government data sharing for the public interest, enhancing availability of high-value publicly held datasets for free reuse, allocating €2 billion in a European High Impact Project to develop data processing infrastructures and governance mechanisms, and ensuring access to secure, fair, and competitive cloud services. These measures aim to bolster business innovation through increased data accessibility. [11, 49].

Since 2017, the EU Commission has required that all research data produced by EU-funded projects be openly accessible and adhere to FAIR principles [14]. As a result, many member states, including Cyprus, have followed the EU Commission's example, and adopted or formulated national policies accordingly [20].

As policies continue to evolve at the levels of both the European Union and its member states, the swift expansion of smart devices and digitalization results in a substantial influx of varied data, commonly known as Big Data [26]. This data presents both challenges and opportunities for enhancing the lives of citizens and cities. A key application of smart cities in contemporary societies is the e-government process [16, 30], which must be tailored to accommodate not only the needs of citizens but also their digital competencies.

However, certain demographic groups are excluded from accessing crucial digital services due to factors such as language barriers, cultural differences, or age. To mitigate this issue, Lister [35] proposes that community-wide digital learning activities should be designed and developed in an inclusive, accessible, and adaptable manner to serve a broad spectrum of citizens.

To meet these needs, trained professionals are urgently required to plan, create, and educate active citizens and build resilience in Smart Cities.

The research objectives include identifying the current practices related to Open Data in both local and EU contexts, exploring the need to establish new job profiles/careers, assessing the required skills, developing training content and material for new job profiles, and providing a training kit for data officers.

This review makes significant contributions to the Open Data and Smart Cities field. It offers a systematic analysis of existing literature, clarifies the relationship between these two aspects and establishes a comprehensive framework for future scholars and practitioners. Furthermore, the study identifies theoretical, methodological, and contextual gaps in knowledge, as well as emerging themes and limitations within existing research. These findings collectively lay the groundwork for a prospective and auspicious research agenda.

The paper's structure is as follows: introduction with research aims and objectives, a detailed account of the systematic review methodology; a descriptive overview of the research, followed by a thematic analysis of the findings. Discussion of theoretical and practical contributions and identification of future research directions.

2   Conceptual Boundaries of the Review

Given the multifaceted nature of the topic, research boundaries are necessary. This research focuses on the relation of Open Data and Smart Cities based on the European Commission's definition that states that "Smart Cities is a place where traditional networks and services are made more efficient with the use of digital solutions for the benefit of its inhabitants and business. A smart city goes beyond the use of digital technologies for better resource use and less emissions. It means smarter urban transport networks, upgraded water supply and waste disposal facilities and more efficient ways to light and heat buildings. It also means a more interactive and responsive city administration, safer public spaces and meeting the needs of an ageing population" [3, 8]. A pivotal factor in Smart Cities establishment, is the development and training of personnel. Developing digital skills for this target population is crucial and essential [46] in order to achieve the goals of implementing digital interactive environments and services. In addition, developing the digital skills of the municipalities' personnel is crucial for all occupational profiles and specifically for those that are actively involved in the development and operation of digital services for a smart city. Hence, this systematic review provides a comprehensive understanding of the topic's interplay and requirements.

We opted for the systematic literature review method to comprehensively analyze existing research, ensuring rigor, transparency, and reproducibility [23, 51]. This approach enables us to draw robust conclusions about both known and unknown aspects in the research domain. As highlighted by several scholars [7, 22] this method offers distinct advantages over traditional narrative reviews: a) it enhances the quality of the review procedure and outcomes [22, 34, 36]; b) allows for reduced bias and errors [21, 45]; c) increases process validity through replicable steps applied during the review process [52]; d) facilitates for data synthesis and literature mapping of a specific research area [19, 32, 55], and; e) often yields frameworks that consolidate existing knowledge [e.g. 19, 21, 39].

2.1 Search Protocol

2.1.1 Question Formulation

A Systematic Literature Review is a way of identifying, evaluating, and synthesizing all available research relevant to a particular research question, topic area, or phenomenon of interest [24, 43].

Accordingly, it begins with a well-formulated research question that contains welldefined terms related to the topic under investigation aiding the selection for review [42–45]. In accordance with this process, two research questions were established: What is the legal/policy framework around open data and open data for smart cities? What are the necessary skills for being a data officer in a smart city?

2.1.2 Inclusion Criteria

The systematic literature review will employ mostly scientific sources of information to find research publications, book chapters, legal documents, and legislation related to the examined topic using key inclusion criteria. The first criterion involves the use of electronic and academic databases, such as EBSCO Business Source Ultimate, ScienceDirect, and Web of Science (WoS), as the primary research sources. Furthermore, due to the novelty of Open Science, grey literature will be incorporated. Over recent decades, systematic literature reviews particularly in the fields of health and medicine have underscored the value of discovering and analysing grey literature to ensure a comprehensive evidence base and counter publication bias [40]. Also, deliverables of EU projects accessible through Open Access e.g. Zenodo and institutional repositories will be examined. The language of sources should be English. The second criterion involves constructing a research list that encompasses abroad range of topics. The third criterion involves confining the search to a period, from 2016 to 2022, to ensure contemporary developments on the research topic. Nonetheless, some earlier articles will be included to provide a historical perspective of the subject under investigation. This approach is anticipated to provide a precise and representative context of relevant academic research. However, journal rankings will not dictate study selection to prevent overlooking important studies and research outputs published in less established journals or across diverse disciplines.

The elimination of papers, guided by assessment criteria, will be based on abstracts' analysis. The review's selection criteria will be aligned with the research questions, objectives, and high-quality evidence within the literature. Establishing clear inclusion and exclusion criteria aims to reduce the potential for reviewer bias. At the next stage, the researcher will review and critically appraise the full text of the selected papers according to predefined quality criteria. This critical appraisal is a crucial part of the review process, as it assesses the validity of the selected studies, provides a justification, and enables readers to assess the systematic review's relevance to their own research to their own research study [7]. The initial article search based on the inclusion criteria yielded 1827 hits in total.

2.1.3 Search Strategy

We conducted searches in the fields of Title, Abstract, and Keyword fields of the afore-mentioned electronic databases, following the approach of many authors [7, 22, 53] as these fields typically hold the relevant search terms. Title, keyword, abstract is a stan-dard practice applied in other Systematic Literature Reviews also. Search strings for each topic were combined using the Boolean AND, OR, NOT operator. Thus, the employed search formula was as follows: ("smart cities" AND "open data") AND (training AND open data AND smart cities). Next, the references of these papers were checked to ensure that any relevant articles that satisfy our criteria were not missed.

2.1.4 Exclusion Criteria

The preliminary selection of potentially pertinent articles underwent further assessment using diverse exclusion criteria. Following the conventions of contemporary system-atic reviews [21, 33, 38, 51], the search prioritized peer-reviewed academic journal publications that included full-texts. Unavoidably, articles or any other material beyond English were excluded and remained in the examination of papers mainly but not strictly

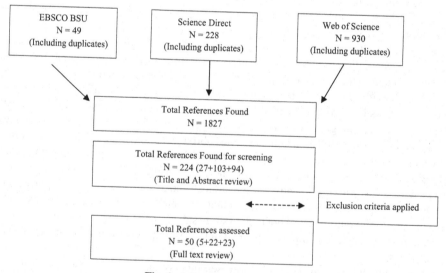

Fig. 1. Literature search results

restricted (in cases with great relevance to the topic) to European geolocations. Duplicates were removed. These exclusion criteria led to a usable sample of 50 articles; this set was further evaluated based on the boundaries of this review (Fig. 1).

3 Descriptive Review of the Literature

The review of the existing literature revealed several trends concerning Open Data and Smart Cities. In the following sections, we present findings on data collection sources, research areas, publication years, article types, author attributes, industrial paper analysis, and employed methodologies. This process facilitates an initial literature synopsis and highlights areas for future investigation.

3.1 Timeframe and Type of Outputs Found

It is noteworthy that certain databases display related outputs beginning in 2017, while others only do so from 2012 onwards. As shown in Fig. 2, there is a consistent upward trend, with a peak in the number of articles occurring in 2020 (n = 13). Despite the significance of open data, the literature is scarce in systematic and structured research that evaluates its impacts on the smart city context [37]. Furthermore, our systematic literature review reveals that the topic of workforce training and skills is predominantly found in grey literature (e.g., in reports like European Centre for the Development of Vocational Training (Cedefop) [6, 7], conference proceedings etc.), and is comparatively less common in academic journals/articles. These findings suggest a scarcity of theory-building studies and purely conceptual papers (n = 43 Journal articles and n = 7 Book, sections) in this research area, which are crucial for establishing a strong theoretical

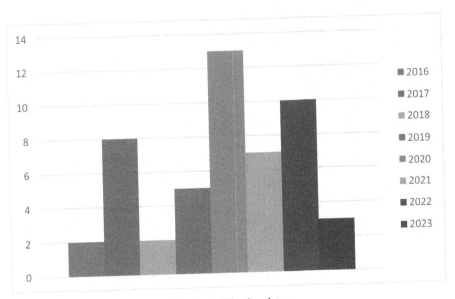

Fig. 2. Publication dates

foundation and extending its boundaries. Consequently, future scholars should prioritize qualitative empirical studies to enrich their comprehension of this evolving realm. This endeavor will foster conceptual contributions, leading to new research directions in the domain.

3.2 Journal Outputs, Fields of Research

As illustrated in Fig. 3, the majority of research output was categorized within the domain of Smart Cities and Smart governance. This is closely followed by the fields of Open Innovation and Open Data. Additionally, research is also incorporated within the areas of Urban development, Urban studies, Policy and Planning. Geographically, the majority of studies, based on the location the first author, originate from Europe and United Kingdom. However, papers from Asia, USA and Canada were also identified. In terms of Journal titles, no = 9 articles were retrieved by the journal "Cities", n = 4 from "Smart Cities" and no = 3 from "Government Information Quarterly". Less than n = 3 articles were retrieved from journals "Data in Brief", "Economics Ecology Socium", "Frontiers in Sustainable Cities", "Future Generation Computer Systems", "Government Information Quarterly"," Information Systems and E-Business Management", "International Entrepreneurship and Management Journal", "International Journal of Information Management", "Internet of Things", "Journal of Intellectual Capital", "Journal of Urban Affairs", and "Journal of Urban Technology "(Fig. 3).

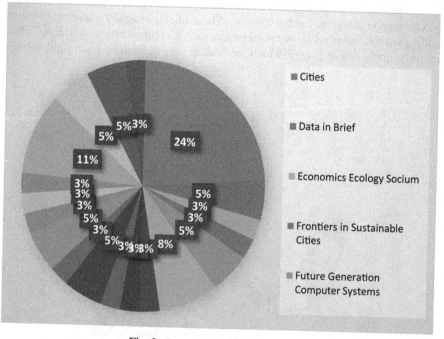

Fig. 3. Journal titles/Articles sources

4 Thematic Analysis of the Literature

In this section, we conduct a thematic analysis of the reviewed outputs to comprehend the theoretical foundation upon which these studies are based or focused. This process aids researchers in identifying and verifying the validity of the outputs in terms of topic association. In addition, identifying the areas of study provides an opportunity to explore beyond the surface meanings of the findings, thereby enhancing our understanding of the topic under discussion. Given that Smart cities and Open data is a relatively new area of studies, the thematic analysis is deemed critical and highly necessary to provide researchers with a clearer perspective on the areas that need further examination.

4.1 Areas of Studies

The analysis revealed that the topic is examined across a broad range of disciplines. This interdisciplinary nature of the research stream is encouraging as it suggests promising trajectory for further evolution and exploration of research gaps within this domain.

More specifically, the majority of the articles identified are mainly theoretical or practical cases dealing with Information and Communication Technologies (ICT) as the fundamental characteristic of smart cities in order to improve the urban functions; Internet of Things (IoT) as "a novel cutting edge technology that proffers to connect a plethora of digital devices endowed with several sensing, actuation, and computing capabilities with the Internet, thus offering manifold new services in the context of a smart city" [2, 56]; Smart City policies, as a necessity for the engagement, establishment and implementation of Urban development and innovation. Some articles are showing use cases, different countries, and cities (e.g., London, Portugal, Spain etc.) of the effective use and impact of Open Data to the implementation of strategies for the development

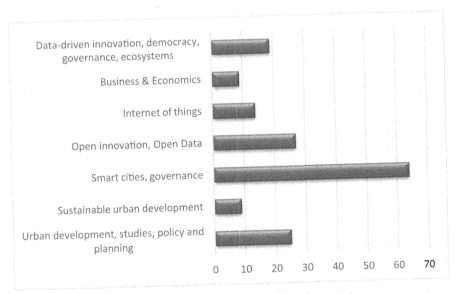

Fig. 4. Field of research based on publication outputs.

of the Smart city framework. Another topic aroused is the citizen science and citizen engagement and their role that evolved to become active agents in decision-making [1]. Additionally, and unavoidably, the topics of e-Government and e-Governance, e-Democracy in combination with the Open Data and Smart Cities, are in the discussion especially into which extent they can make an impact on society (Fig. 4).

5 Conclusion

The 2030 Agenda stands as a significant development milestone for humanity, reflecting our collective commitment in safeguarding the interests of future generations. Smart Cities play an essential role in this development, the development direction of which is based, among other factors, on the fulfilment of the criteria set by the various Sustainable Development Goals (SDGs) [40, 41].

Smart cities place paramount importance on the strategic deployment of Information Communication Technology (ICT) and the Internet of Things (IoT) to enhance provided services and improve the well-being of their citizens. As the relevant stakeholders (e.g., the administrative authorities that manage cities) often have gaps on the knowledge and skills needed to transform their operations, smart city initiatives usually involve a complex set of actors, from local urban authorities and their technical departments to small and large IT firms, academics, and civic organizations, as well as individual citizens [31].

This Systematic Literature Review confirms that the subject matter under investigation is inherently new and continuously evolving. Especially the part of training, knowledge and skills of stakeholders has been an area that is minimally addressed or is indirectly addressed.

In conclusion, this systematic literature review resulted into forming a working definition of Job profiles and competences in relation to Open Data in the context of Smart Cities. This would serve as the foundation for nurturing a cadre of professionals who will be able to provide a diverse skill set to a "360° stakeholders" level. It is important to acknowledge that the selected taxonomy of keywords and databases, though carefully selected, may have missed the inclusion of research outputs due to researchers classifying their research using alternative terms and not the terms specifically targeted with this study.

Acknowledgments. The research leading to the results presented in this chapter has received funding from the ERASMUS+ funded Project Open Data City Officer (OPENDCO) under Grant Agreement No. 2022-1-CY01-KA220-HED-000089196.

References

1. Aguilera, U., Peña, O., Belmonte, O., López-de-Ipiña, D.: Citizen-centric data services forsmarter cities **76**, 234–247 (2017).https://doi.org/10.1016/j.future.2016.10.031
2. Alavi, A.H., Jiao, P., Buttlar, W.G., Lajnef, N.: Internet of Things-enabled smart cities:state-of-the-art and future trends **129**, 589–606 (2018)

3. Albino, V., Berardi, U., Dangelico, R.M.: Smart cities: definitions, dimensions, performance, and initiatives **22**, 3–21 (2015)
4. Caragliu, A., Del Bo, C.F.: Smart innovative cities: the impact of Smart City policies onurban innovation **142**, 373–383 (2019)
5. de Castro Neto, M., Rego, J.S., Neves, F.T., Cartaxo, T.M.: Smart & open cities: Portuguese municipalities open data policies evaluation. In: Presented at the 2017 12th Iberian Conference on Information Systems and Technologies (CISTI) (2017)
6. (Cedefop), E.C. for the D. of V.T.: Skills for green jobs 2018 update. European synthesis report (2018)
7. Christofi, M., Leonidou, E., Vrontis, D.: Marketing research on mergers and acquisitions: a systematic review and future directions (2017)
8. Commission, E.: The Future of the Cities. Opportunities Challenges and the Way Forward. http://publications.jrc.ec.europa.eu/repository/bitstream/JRC116711/the-future-of-cities_online.pdf
9. Commission, E.: Directive (EU) 2019/1024 of the European Parliament and of the Council of 20 June 2019 on open data and the re-use of public sector information. https://eur-lex.europa.eu/eli/dir/2019/1024/oj
10. Commission, E.: Proposal for a Regulation of the European Parliament and of the Council on European data governance (Data Governance Act). https://eur-lex.europa.eu/. legal-content/EN/TXT/?uri=CELEX:52020PC0767
11. Commission, E.: Europe Sustainable Development Report 2021 (2021) (a)
12. Commission, E.: Open Science. Research and innovation. https://research-and-innova-tion.ec.europa.eu/strategy/strategy-2020-2024/our-digital-future/open-science_en
13. Commission, E.: Open Data Maturity Report 2021. https://data.europa.eu/sites/de-fault/files/landscaping_insight_report_n7_2021.pdf
14. Commission, E.: Where is Open Science in Horizon Europe? https://openscience.eu/Open-Science-in-Horizon-Europe?fbclid=IwAR18qn2KAnz0_6SI-%20ozTtlhB95TKR_9cN2mDvWMZCsoHGP7hPHPwksgGyo9w
15. Commission, E.: The European Commission's 2022 Strategic Foresight Report is out! (2022) (b)
16. Commission, E.: Smart cities. https://commission.europa.eu/eu-regional-and-urban-development/topics/cities-and-urban-development/city-initiatives/smart-cities_en
17. Commission, E.: Strategic Foresight Report 2022 (2022) (d)
18. Commission, E.: European legislation on open data. https://digital-strategy.ec.eu-ropa.eu/en/policies/legislation-open-data
19. Crossan, M.M., Apaydin, M.: A multi-dimensional framework of organizational innovation: a systematic review of the literature **47**, 1154–1191 (2010)
20. Cyprus, R. of: National Policy of the Republic of Cyprus for Open Science Practices. https://www.dmrid.gov.cy/dmrid/research.nsf/All/5FEDF16528FC350FC2258974003CA0CE/$file/National%20OS%20policy.pdf?OpenElement
21. Dada,O.: A model of entrepreneurial autonomy in franchised outlets: a systematic review of the empirical evidence **20**, 206–226 (2018)
22. Danese,P., Manfè, V., Romano, P.: A systematic literature review on recent lean research: state-of-the-art and future directions **20**, 579–605 (2018)
23. Deng, T., Zhang, K., Shen, Z.-J. (Max).: A systematic review of a digital twin city: a new pattern of urban governance toward smart cities **6**, 125–134 (2021). https://doi.org/10.1016/j.jmse.2021.03.003
24. van Dinter, R., Tekinerdogan, B., Catal, C.: Automation of systematic literature reviews: a systematicliteraturereview**136**,106589(2021). https://doi.org/10.1016/j.infsof.2021.106589
25. Finance, R. of C.M. of: Open Data Cyprus. https://www.data.gov.cy

26. Fitsilis, P., Kokkinaki, A.: Smart cities body of knowledge (2021)
27. Foundation, O.K.: Open Definition. Defining Open in Open Data, Open Content and Open Knowledge. https://opendefinition.org/
28. Foundation, O.K.: What is Open Data? https://opendatahandbook.org/guide/en/what-is-open-data/
29. foundation, W.W. web: The Open Data Barometer. https://opendatabarome-ter.org/?_year=2017&indicator=ODB
30. Georgiou, I., Nell, J.G., Kokkinaki, A.I.: Blockchain for smart cities: a systematic literature review. Presented at the Information (2020)
31. Karimikia, H., Bradshaw, R., Singh, H., Ojo, A., Donnellan, B., Guerin, M.: An emergenttaxonomy of boundary spanning in the smart city context-The case of smart Dublin 185, 122100 (2022). https://doi.org/10.1016/j.techfore.2022.122100
32. Kauppi, K., Salmi, A., You, W.: Sourcing from Africa: a systematic review and a researchagenda 20, 627–650 (2018)
33. Klang, D., Wallnöfer, M., Hacklin, F.: The business model paradox: a systematic reviewand exploration of antecedents 16, 454–478 (2014)
34. Leonidou, E., Christofi, M., Vrontis, D., Thrassou, A.: An integrative framework of stakeholder engagement for innovation management and entrepreneurship development 119, 245–258 (2020)
35. Lister, P.: Smart learning in the community: supporting citizen digital skills and literacies. In: Presented at the Distributed, Ambient and Pervasive Interactions: 8th International Conference, DAPI 2020, Held as Part of the 22nd HCI International Conference, HCII 2020, Copenhagen, Denmark, 19–24 July 2020, Proceedings 22 (2020)
36. Myers, M.D.: Qualitative research in business and management. 1–364 (2019)
37. Neves,F.T., de Castro Neto, M., Aparicio, M.: The impacts of open data initiatives on smart cities: a framework for evaluation and monitoring 106, 102860 (2020). https://doi.org/10.1016/j.cities.2020.102860
38. Nguyen, D.H., de Leeuw, S., Dullaert, W.E.: Consumer behaviour and order fulfilment inonline retailing: a systematic review 20, 255–276 (2018)
39. Nofal,A.M., Nicolaou, N., Symeonidou, N., Shane, S.: Biology and management: a review, critique, and research agenda 44, 7–31 (2018)
40. Paez,A.: Gray literature: an important resource in systematic reviews 10, 233–240 (2017). https://doi.org/10.1111/jebm.12266
41. Parra-Dominguez, J., Gil-Egido, A., Rodriguez-Gonzalez, S.: SDGs as one of the driversof smart city development: the indicator selection process 5, 1025–1038 (2022).https://doi.org/10.3390/smartcities5030051
42. Rother,E.T.: Systematic literature review X narrative review. 20, v–vi (2007)
43. Themistocleous, M. Cunha, P. Tabakis, E., Papadaki, M.: Towards cross-border CBDC interoperability: insights from a multivocal literature review. J. Enterp. Inf. Manag. Emer. Publ. Limit. 36(5), 1741–0398, 1296–1318 (2023). https://doi.org/10.1108/JEIM-11-2022-0411
44. Tranfield, D., Denyer, D., Smart, P.: Towards a methodology for developing evidence-informedmanagement knowledge by means of systematic review 14, 207–222 (2003)
45. Turnbull, D., Chugh, R., Luck, J.: Systematic-narrative hybrid literature review: a strategyfor integrating a concise methodology into a manuscript 7, 100381 (2023). https://doi.org/10.1016/j.ssaho.2022.100381
46. United Nations Educational, S. and C.O. (UNESCO): Digital skills critical for jobs and social inclusion. https://en.unesco.org/news/digital-skills-critical-jobs-and-social-inclusion
47. United Nations Educational, S. and C.O. (UNESCO): Smart cities: shaping the society of 2030 (2019)

48. United Nations Educational, S. and C.O. (UNESCO): Recommendation on Open Science. https://unesdoc.unesco.org/ark:/48223/pf0000379949?posInSet=10&queryId=63e afee3-babf-42d7-9654-1560daa1ca25
49. United Nations Educational, S. and C.O. (UNESCO): Open data for AI. What now? https://unesdoc.unesco.org/ark:/48223/pf0000385841?posInSet=1&queryId=a7bf30be-b2e9-413d-97b0-beea17dbe7a0
50. Verhulst, S.G., Young, A.: Open Data in Developing Economies Toward Building an Evidence Base on What Works and How. African Minds (2017)
51. Vrontis,D., Christofi, M.: R&D internationalization and innovation: a systematic review, integrative framework and future research directions **128**, 812–823 (2021)
52. Wang,C.L., Chugh, H.: Entrepreneurial learning: past research and future challenges **16**, 24–61 (2014)
53. West,J., Bogers, M.: Leveraging external sources of innovation: a review of research on open innovation **31**, 814–831 (2014)
54. Wildemeersch, D., Jütte, W.: digital the new normal-multiple challenges for the education and learning of adults **8**, 7–20 (2017)
55. Witell, L., Snyder, H., Gustafsson, A., Fombelle, P., Kristensson, P.: Defining service innovation: a review and synthesis **69**, 2863–2872 (2016)
56. Yaqoob, I., Mehmood, Y., Ahmad, F., Adnane, A., Imran, M., Guizani, S.: Internet-of-Things-based smart cities: recent advances and challenges (2017)

Correction to: Digital Leadership in Cross-Cultural Organizations: Insights from Swiss Healthcare Companies

Mahdieh Darvish, Luca Laule, Laurine Pottier, and Markus Bick

Correction to:
Chapter 19 in: M. Papadaki et al. (Eds.):
Information Systems, **LNBIP 502,**
https://doi.org/10.1007/978-3-031-56481-9_19

In the originally published version of chapter 19, the name of the second author had been misspelt. This has been corrected.

The updated version of this chapter can be found at
https://doi.org/10.1007/978-3-031-56481-9_19

Author Index

Printed in the United States
by Baker & Taylor Publisher Services